Spectrum and Dynamics

Volume 52

CRM
PROCEEDINGS &
LECTURE NOTES

Centre de Recherches Mathématiques
Montréal

Spectrum and Dynamics

Proceedings of the Workshop Held in Montréal, QC April 7–11, 2008

Dmitry Jakobson
Stéphane Nonnenmacher
Iosif Polterovich
Editors

The Centre de Recherches Mathématiques (CRM) of the Université de Montréal was created in 1968 to promote research in pure and applied mathematics and related disciplines. Among its activities are special theme years, summer schools, workshops, postdoctoral programs, and publishing. The CRM is supported by the Université de Montréal, the Province of Québec (FQRNT), and the Natural Sciences and Engineering Research Council of Canada. It is affiliated with the Institut des Sciences Mathématiques (ISM) of Montréal. The CRM may be reached on the Web at www.crm.math.ca.

American Mathematical Society
Providence, Rhode Island USA

The production of this volume was supported in part by the Fonds Québécois de la Recherche sur la Nature et les Technologies (FQRNT) and the Natural Sciences and Engineering Research Council of Canada (NSERC).

2010 *Mathematics Subject Classification.* Primary 58Jxx, 37Dxx; Secondary 35Pxx.

Library of Congress Cataloging-in-Publication Data

Spectrum and dynamics : proceedings of the workshop held in Montréal, QC, April 7–11, 2008 / Dmitry Jakobson, Stéphane Nonnenmacher, Iosif Polterovich, editors.
p. cm. — (CRM proceedings & lecture notes / Centre de recherches mathématiques, Montréal ; 52)
Includes bibliographical references.
ISBN 978-0-8218-4778-7 (alk. paper)
1. Differentiable dynamical systems–Congresses 2. Differentiable manifolds–Congresses. 3. Spectral theory (Mathematics)–Congresses. I. Jakobson, Dmitry, 1970– II. Nonnenmacher, Stéphane, 1972– III. Polterovich, Iosif, 1974–

QA614.8.S64 2010
515′.39–dc22 2010020469

∞ The paper used in this book is acid-free and falls within the guidelines
established to ensure permanence and durability.
This volume was submitted to the American Mathematical Society
in camera-ready form by the Centre de Recherches Mathématiques.
Visit the AMS home page at `http://www.ams.org/`

10 9 8 7 6 5 4 3 2 1 15 14 13 12 11 10

Contents

Introduction

The workshop on *Spectrum and Dynamics* was held in April 2008 at the Centre de Recherches Mathématiques in Montréal. The goal of this meeting was to bring together experts working in two different and yet interrelated areas of mathematics: geometric spectral theory and theory of dynamical systems.

The connection between these two fields goes back to the correspondence principle, formulated by Niels Bohr around 1920. It predicts that the behavior of a quantum system in the high energy (semiclassical) limit should reflect the properties of the corresponding classical dynamical system.

One of the most important manifestations of this principle is the quantum ergodicity property satisfied by the Laplace – Beltrami eigenfunctions on negatively curved manifolds: it states that the huge majority of high-energy eigenfunctions are "equidistributed" over the manifold. This property originates from a short paper by A. Shnirelman published in 1974. Since then, the study of "quantum limits" (i.e., the possible asymptotic shapes taken by high-energy eigenfunctions) has been rapidly developing, particularly due to links with number theory and mathematical physics. Some recent achievements in the field of "quantum chaos" are described in the lecture notes of S. Nonnenmacher, that are based on his work with N. Anantharaman and H. Koch. The central result of these notes is a lower bound on the Kolmogorov – Sinai entropy of quantum limits, which implies a "partial delocalization" of all high-energy eigenstates. Such a bound holds for a rather general class of quantum systems, obtained as quantizations of Hamiltonian flows (or diffeomorphisms) of Anosov type, Laplace eigenfunctions on negatively curved manifolds being an important special case. Recent developments related to the celebrated Quantum Unique Ergodicity conjecture of Rudnick – Sarnak are also reviewed in these notes.

The study of quantum limits for differential operators on vector bundles (such as the Hodge Laplacian, the Dirac operator or the Laplacian on forms) is the subject of the survey by A. Strohmaier. In particular, the analogue of Shnirelman's quantum ergodicity theorem for such operators is discussed. The underlying classical system in this setting is the frame flow. Its dynamical properties are quite different from those of the geodesic flow, which makes the topic particularly interesting. The survey is based on the recent joint works of the author with D. Jakobson and S. Zelditch.

An overview of results on classical and quantum dynamical systems associated with transverse Dirac operators on Riemannian foliations is presented in the contribution of Yu. Kordyukov. Well-known results such as Egorov's theorem are extended to the leaf space of a Riemannian foliation, using ideas and notions from noncommutative geometry. The author also considers two examples of Riemannian foliations: a foliation given by the fibers of a fibration, and a linear foliation on the two-dimensional torus.

In his contribution, entitled *Eigenvalue variation and semiclassical concentration*, L. Hillairet considers a semiclassical Schrödinger operator on $\mathbb{R}^d$, and shows that a (semiclassical) limit point of an eigenbranch is necessarily a critical point of the potential. He uses a connection between the behavior of analytic eigenvalue branches, and concentration properties of the corresponding eigenfunctions. Similar ideas are applied to prove an "averaged" remainder estimate in Weyl's law on a smooth compact manifold, where the averaging is taken over an analytic variation of metrics.

The article by H. Lapointe is concerned with spectral asymptotics on tori with Liouville metrics. The geodesic flow on such tori is completely integrable. From the dynamical viewpoint, integrable metrics are "opposite" to the negatively curved metrics, whose geodesic flow is ergodic. Developing the techniques of Colin de Verdière and Bleher – Minasov – Kosygin – Sinai, the author reduces the eigenvalue counting problem for the Laplacian on a Liouville torus to a certain lattice counting problem. This allows him to obtain a van der Corput-type estimate on the error term in Weyl's law for a large class of Liouville tori.

The workshop featured a high level of diversity of topics, which is reflected in the current volume. Some of the talks were "purely spectral," some were "purely dynamical." This gave an opportunity for the participants to learn new mathematics, and, hopefully, to uncover yet unknown links between spectral theory and the theory of dynamical systems.

The article by H. Donnelly is concerned with the spectral properties of complete manifolds of negative curvature. The simplest example of such a manifold is the hyperbolic space of constant negative curvature that has purely continuous spectrum. It has been previously shown by several authors that the absence of embedded eigenvalues also holds for manifolds, whose curvature (as a function of geodesic distance from a given basepoint) converges from below to a negative constant rapidly enough. In the present paper the opposite phenomenon is exhibited: the author constructs examples of complete negatively curved manifolds with eigenvalues embedded in the continuous spectrum.

The paper by B. Helffer and T. Hoffmann-Ostenhof is concerned with the study of spectral minimal partitions. This is a new topic in geometric spectral theory, which is being developed by the authors and their collaborators. The problem is to find a k-partition of a given domain that minimizes the maximum among fundamental tones (i.e., the first Dirichlet eigenvalues) of the subdomains. For $k = 2$, a spectral minimal partition is always given by the nodal domains of a second eigenfunction. The case $k = 3$ is already highly nontrivial, and is considered in detail in the paper. In particular, the authors discuss an intriguing conjecture that the minimal 3-partition of a disk is given by the "Mercedes star."

In his paper, *Asymptotic vertex growth for graphs*, M. Pollicott studies discrete analogues of relatively old results about critical metrics for the entropy functional on Riemannian manifolds, proved by Katok and Besson, Courtois and Gallot. The author considers the entropy on weighted graphs, showing that it is analytic in the edge weights and convex, and determining the extremal weight assignment for the entropy on a given graph. He also studies analogous extremal problems for asymptotic weighted vertex growth (a discrete analogue of the topological pressure) for a real-valued function on the vertices of the graph.

The symbolic coding of the geodesic flow on the Hecke surfaces $G_q\backslash\mathbb{H}$ (where G_q is a Hecke triangle group), is related to nearest λ_q-multiple continued fractions for $\lambda_q = 2\cos(\pi/q)$, discussed in the work by D. Mayer and T. Mühlenbruch. These fractions are generated by interval maps f_q which are conjugate to subshifts over infinite alphabets. They have been introduced for $q = 3$ by Hurwitz and for even q by Nakada. The authors generalize to arbitrary q a result of Hurwitz concerning the G_q- and f_q-equivalence of points on the real line. They prove that the natural extension of f_q can be used as a Poincaré map for the geodesic flow on $G_q\backslash\mathbb{H}$, and allows one to construct the transfer operator for this flow.

In his paper, entitled *Comparing length functions on free groups*, R. Sharp studies pairs of length functions on a free group, associated with pairs of points in the outer space. The author discusses several invariants of these pairs, and describes the connection to an associated "Manhattan curve" analogous to those considered for pairs of hyperbolic surfaces.

The workshop on *Spectrum and Dynamics* was yet another illustration of the unity of mathematics. Analysis, geometry, dynamics, number theory, group theory, mathematical physics — it is hard to find a branch of mathematics that was not present in the talks and discussions!

Many important developments and new ideas were reported at the meeting. For the organizers, it was an interesting and gratifying experience. However, all of this would have been impossible without the daily assistance from the dedicated staff of the Centre de Recherches Mathématiques, in particular, Louis Pelletier and Josée Laferrière. Their indispensable help in the organization of the workshop is greatly appreciated.

Dmitry Jakobson
Stéphane Nonnenmacher
Iosif Polterovich

Centre de Recherches Mathématiques
CRM Proceedings and Lecture Notes
Volume **52**, 2010

Notes on the Minicourse "Entropy of Chaotic Eigenstates"

Stéphane Nonnenmacher

1. Overview and statement of results

1.1. Introduction. These lectures present a recent approach, mainly developed by Nalini Anantharaman and the author, aimed at studying the high-frequency eigenmodes of the Laplace–Beltrami operator $\Delta = \Delta_X$ on compact Riemannian manifolds (X, g) for which the sectional curvature is everywhere negative. It is well-known that the geodesic flow on such a manifold (which takes place on the unit cotangent bundle S^*X) is strongly chaotic, in the sense that it is uniformly hyperbolic (Anosov). This flow leaves invariant the natural smooth measure on S^*X, namely the Liouville measure μ_l (which is also the lift of the Lebesgue measure on X). Studying the eigenstates of Δ is thus a part of "quantum chaos." In these notes we extend the study to more general Schrödinger-like operators, such that the corresponding Hamiltonian flow on some compact energy shell $\mathcal{E}$ has the Anosov property. We also consider the case of quantized Anosov diffeomorphisms on the torus, which are popular toy models in the quantum chaos literature. To set up the problem we first stick to the Laplacian.

For a general Riemannian manifold (X, g) of dimension $d \geq 2$, there exist no explicit, not even approximate expression for the eigenmodes of the Laplacian. One way to "describe" these modes consists in comparing them (as "quantum" invariants) with "classical" invariants, namely probability measures on S^*X, invariant with respect to the geodesic flow. To this aim, starting from the full sequence of eigenmodes $(\psi_n)_{n\geq 0}$ one can construct a family of invariant probability measures on S^*X, called *semiclassical measures*. Each such measure $\mu_{\rm sc}$ can be associated with a subsequence of eigenmodes $(\psi_{n_j})_{j\geq 1}$ which share, in the limit $j \to \infty$, the same macroscopic localization properties, both on the manifold X and in the velocity (or momentum) space: these macroscopic localization properties are "represented" by $\mu_{\rm sc}$. One (far-reaching) aim would be a complete classification of the semiclassical measures associated with a given manifold (X, g).

This goal is much too ambitious, starting from the fact that the set of invariant measures is itself not always well-understood. We will thus restrict ourselves to the

---

2010 *Mathematics Subject Classification.* Primary 81Q50, 37D20, 58J51; Secondary 35P20, 28D20.

This is the final form of the paper.

class of manifolds described above, namely manifolds (X, g) of negative sectional curvature. One advantage is that the classical dynamics is at the same time "irregular" (in the sense of "chaotic"), and "homogeneous." The geodesic flow on such a manifold is (semi)conjugated with a suspended flow over a simple symbolic dynamics (a subshift of finite type over a finite alphabet), which allows one to explicitly construct many different invariant measures. For instance, such a flow admits infinitely many isolated (unstable) periodic orbits γ, each of which carries a natural probability invariant measure μ_γ. The set of periodic orbits is so large that the measures $\{\mu_\gamma\}$ form a dense subset (in the weak-* topology) of the set of invariant probability measures. Hence, it would be interesting to know whether some high-frequency eigenmodes can be asymptotically localized near certain periodic orbits, leading to semiclassical measures of the form

$$\mu = \sum_\gamma p_\gamma \mu_\gamma, \qquad \text{with} \sum_\gamma p_\gamma = 1,\ p_\gamma \in [0,1]. \tag{1.1}$$

This possibility was named "strong scarring" by Rudnick – Sarnak [27], in analogy with a weaker form of "scarring" observed by Heller on some numerically computed eigenmodes [17], namely a "nonrandom enhancement" of the wavefunction in the vicinity of a certain periodic orbit. In the same paper, Rudnick and Sarnak conjectured that such semiclassical measures do not exist for manifolds of negative curvature. More precisely, they formulated the Quantum Unique Ergodicity (QUE) conjecture

Conjecture 1.1 (Quantum Unique Ergodicity [27]**).** *Let (X, g) be a compact Riemannian manifold of negative curvature. Then there exist only one semiclassical measure, namely the Liouville measure μ_l.*

This conjecture rules out any semiclassical measure of the type (1.1). It also rules out linear combination of the form

$$\mu = \alpha\mu_l + (1-\alpha)\sum_\gamma p_\gamma \mu_\gamma, \qquad \alpha \in [0,1). \tag{1.2}$$

The name "quantum unique ergodicity" reminds of a classical notion: a dynamical system (map or flow) is *uniquely ergodic* if and only if it admits a unique invariant measure. In the present case, the classical system is not uniquely ergodic, but the conjecture is that its quantum analogue conspires to be so.

This conjecture was formulated several years after the proof of a general result describing "almost all" the eigenstates (ψ_n).

Theorem 1.2 (Quantum Ergodicity [9, 28, 30]**).** *Let (X, g) be a compact Riemannian manifold such that the geodesic flow is ergodic with respect to the Liouville measure μ_l. Then, there exists a subsequence $S \subset \mathbb{N}$ of density 1, such that the subsequence $(\psi_n)_{n\in S}$ is associated with μ_l.*[1]

The manifolds encompassed by this theorem include the case of negative curvature, but also more general ones (like manifolds where the curvature is negative outside a flat cylindrical part). The proof of this theorem is quite "robust." It has been generalized to many different ergodic systems: Hamiltonian flows ergodic on some compact energy shell [16], broken geodesic flows on some Euclidean domains [14, 32], symplectic diffeomorphism (possibly with discontinuities) on a compact

[1] A subsequence $S \subset \mathbb{N}$ is said to be of density 1 iff $\lim_{N\to\infty} \sharp\{n \in S, n \le N\}/N = 1$.

phase space [6]. This result leaves open the possibility of *exceptional subsequences* $(\psi_{n_j})_{j\geq 1}$ (necessarily of density zero) of eigenmodes with different localization properties.

The QUE conjecture was motivated by partial results concerning a much more restricted class of manifolds, namely compact quotients of the hyperbolic disk $\Gamma\backslash\mathbb{H}$ for which Γ is an arithmetic subgroup of $\mathrm{SL}_2(\mathbb{R})$.[2] Such manifolds admit a commutative algebra of self-adjoint Hecke operators $(T_k)_{k\geq 2}$ which all commute with the Laplacian. It thus makes sense to preferably consider an eigenbasis $(\psi_n)_{n\geq 0}$ made of joint eigenmodes of the Laplacian and the Hecke operators (called Hecke eigenmodes), and the associated semiclassical measures (called Hecke semiclassical measures).[3] Rudnick – Sarnak proved that the only Hecke semiclassical measure of the form (1.2) is the Liouville measure ($\alpha = 1$). Finally, Lindenstrauss [21] showed that for such manifolds, the only Hecke semiclassical measure is the Liouville measure, thus proving an arithmetic form of QUE. He used as an intermediate step a lower bound for the *Kolmogorov – Sinai (KS) entropy* of Hecke semiclassical measures, which he proved in a joint work with Bourgain.

Proposition 1.3 ([5]). *Let X be an arithmetic quotient $\Gamma\backslash\mathbb{H}$. Consider a Hecke semiclassical measure μ_{sc}. Then for any $\rho \in S^*X$ any small $\tau, \epsilon > 0$, the measure of the tube $B(\rho,\epsilon,\tau)$ of diameter ϵ around the stretch of trajectory $[\rho, g^\tau\rho]$ is bounded by*

$$\mu_{\mathrm{sc}}\big(B(\rho,\epsilon,\tau)\big) \leq C_\tau\, \epsilon^{2/9}.$$

As a consequence, for almost any ergodic component μ_{erg} of μ_{sc}, one has

$$H_{\mathrm{KS}}(\mu_{\mathrm{erg}}) \geq 2/9. \tag{1.3}$$

As we will see below, the KS entropy is an affine quantity, therefore $H_{\mathrm{KS}}(\mu_{\mathrm{sc}})$ also satisfies the same lower bound.

1.2. Entropy as a measure of localization. In the previous section we already noticed a relationship between phase space localization and entropy: a uniform lower bound on the measure of thin tubes implies a positive lower bound on the entropy of the measure. For this reason, it is meaningful to consider the KS entropy of a given invariant measure as a "quantitative indicator of localization" of that measure. In Section 3.1 we will give a precise definition of the entropy. For now, let us only provide a few properties valid in the case of Anosov flows [18, Chapter 4].

(1) $H_{\mathrm{KS}}(\bullet)$ is a real function defined on the set of invariant probability measures. It takes values in a finite interval $[0, H_{\max}]$ and is upper semicontinuous. In information-theoretic language, it measures the average complexity of the flow with respect to that measure. The maximum entropy $H_{\max} = H_{\mathrm{top}}(S^*X)$ is also the *topological entropy* of the flow on S^*X, which is a standard measure of the *complexity* of the flow.

(2) a measure μ_γ supported on a single periodic orbit has zero entropy.

(3) Since the flow is Anosov, at each point $\rho \in S^*X$ the tangent space splits into $T_\rho S^*X = E^u_\rho \oplus E^s_\rho \oplus E^0_\rho$, respectively the unstable, stable subspaces and the flow direction. Each of these subspaces is flow-invariant. Let us call $J^u(\rho) = J^u_1(\rho) =$

---

[2]More precisely, Γ is derived from an Eichler order in a quaternion algebra.

[3]It is widely believed that the spectrum of Δ on such a manifold is simple, in which case the restriction to Hecke eigenmodes is not necessary.

$\left|\det g^1_{\restriction E^u(\rho)}\right|$ the unstable Jacobian at the point ρ. Then, any invariant measure satisfies

$$H_{\mathrm{KS}}(\mu) \leq \int \log J^u(\rho)\, \mathrm{d}\mu(\rho). \qquad \text{(Ruelle inequality)} \tag{1.4}$$

The equality is reached iff $\mu = \mu_l$. In constant curvature, one has $H_{\mathrm{KS}}(\mu_l) = H_{\max}$.

(4) the entropy is affine: $H_{\mathrm{KS}}(\alpha\mu_1 + (1-\alpha)\mu_2) = \alpha H_{\mathrm{KS}}(\mu_1) + (1-\alpha)H_{\mathrm{KS}}(\mu_2)$.

Apart from the result of Bourgain – Lindenstrauss (relative to arithmetic surfaces), the first result on the entropy of semiclassical measures was obtained by N. Anantharaman:

Theorem 1.4 ([1]). *Let (X, g) be a manifold of negative sectional curvature. Then, there exists $c > 0$ such that any semiclassical measure μ_{sc} satisfies*

$$H_{\mathrm{KS}}(\mu_{\mathrm{sc}}) \geq c.$$

Furthermore, the flow restricted to the support of μ_{sc} has a nontrivial complexity: its topological entropy satisfies

$$H_{\mathrm{top}}(\operatorname{supp} \mu_{\mathrm{sc}}) \geq \frac{\Lambda^u_{\min}}{2},$$

where $\Lambda^u_{\min} = \lim_{t\to\infty} \inf_\rho \frac{1}{t} \log J^u_t(\rho)$ is the minimal volume expanding rate of the unstable manifold.

The lower bound $c > 0$ is not very explicit and is rather "small." This is to be opposed to the lower bound controlling the complexity of the flow on $\operatorname{supp} \mu_{\mathrm{sc}}$, given in terms of the hyperbolicity of the flow. The lower bound on the KS entropy was improved by Anantharaman, Koch and the author:

Theorem 1.5 ([3,4]). *Let (X, g) be a d-dimensional manifold of negative sectional curvature. Then, any semiclassical measure μ_{sc} satisfies*

$$H_{\mathrm{KS}}(\mu_{\mathrm{sc}}) \geq \int \log J^u(\rho)\, \mathrm{d}\mu_{\mathrm{sc}}(\rho) - \frac{(d-1)\lambda_{\max}}{2}, \tag{1.5}$$

where $\lambda_{\max} = \lim_{t\to\infty} \sup_\rho (1/t) \log|dg^t(\rho)|$ is the maximal expansion rate of the flow.

In the particular case where X has constant curvature -1, this bound reads

$$H_{\mathrm{KS}}(\mu_{\mathrm{sc}}) \geq \frac{d-1}{2} = \frac{H_{\mathrm{top}}(S^*X)}{2}. \tag{1.6}$$

In the constant curvature case, the above bound roughly means that high-frequency eigenmodes of the Laplacian are *at least half-delocalized.* Still, the bound (1.5) is not very satisfactory when the curvature varies much across X; since $\int \log J^u\, \mathrm{d}\mu$ may be as small as $\Lambda_{\min}$, the right-hand side in (1.5) can become negative (therefore trivial) in case $\Lambda_{\min} < (d-1)\lambda_{\max}/2$. The following lower bound seems more natural:

Conjecture 1.6. *Let (X, g) be a manifold of negative sectional curvature. Then, any semiclassical measure μ_{sc} satisfies*

$$H_{\mathrm{KS}}(\mu_{\mathrm{sc}}) \geq \frac{1}{2} \int \log J^u(\rho)\, \mathrm{d}\mu_{\mathrm{sc}}(\rho). \tag{1.7}$$

This bound is identical with (1.6) in the case of curvature -1. Using a nontrivial extension of the methods developed in [4], it has been recently proved by G. Rivière in the case of surfaces ($d = 2$) of *nonpositive curvature* [24, 25], and also by B. Gutkin for a certain class of quantized interval maps [15].

1.3. Generalization to Anosov Hamiltonian flows and symplectic maps. The conjecture 1.6 is weaker than the QUE conjecture 1.1. We expect the bound (1.7) to apply as well to more general classes of quantized chaotic dynamical systems, like Anosov Hamiltonian flows or symplectic diffeomorphisms on a compact phase space. In these notes we will extend the bound (1.5) to these more general Anosov systems (the first instance of this entropic bound actually appeared when studying the Walsh-quantized baker's map [2]). The central result of these notes is the following theorem.

Theorem 1.7. (i) *Let $p(x, \xi; \hbar)$ be a Hamilton function on T^*X (where the smooth manifold X may be either compact, or euclidean near infinity), with principal symbol p_0, such that the energy shell $\mathcal{E} = p_0^{-1}(0)$ is compact, and the Hamiltonian flow $g^t = \mathrm{e}^{tH_{p_0}}$ on $\mathcal{E}$ is Anosov (see Section 2.2). Let $P(\hbar) = \mathrm{Op}_\hbar(p)$ be the $\hbar$-quantization of p. Then, any semiclassical measure μ_{sc} associated with a sequence of null eigenmodes $(\psi_\hbar)_{\hbar\to 0}$ of $P(\hbar)$ satisfies the following entropic bound:*

$$H_{\mathrm{KS}}(\mu_{\mathrm{sc}}) \geq \int_{\mathcal{E}} \log J^u(\rho)\, \mathrm{d}\mu_{\mathrm{sc}}(\rho) - \frac{(d-1)\lambda_{\max}}{2}. \tag{1.8}$$

(ii) *Let $\mathcal{E} = \mathbb{T}^{2d}$ be the $2d$-dimensional torus, equipped with its standard symplectic structure. Let $\kappa\colon \mathcal{E} \to \mathcal{E}$ be an Anosov diffeomorphism, which can be quantized into a family of unitary propagators $\big(U_\hbar(\kappa)\big)_{\hbar\to 0}$ defined on (finite-dimensional) quantum Hilbert spaces $(\mathcal{H}_\hbar)$. Then, any semiclassical measure μ_{sc} associated with a sequence of eigenmodes $(\psi_\hbar \in \mathcal{H}_\hbar)_{\hbar\to 0}$ of $U_\hbar(\kappa)$ satisfies the entropic bound*

$$H_{\mathrm{KS}}(\mu_{\mathrm{sc}}) \geq \int \log J^u(\rho)\, \mathrm{d}\mu_{\mathrm{sc}}(\rho) - \frac{d\lambda_{\max}}{2}. \tag{1.9}$$

In Section 5 we will state more precisely what we meant by a "quantized torus diffeomorphism." Let us mention that the same proof could apply as well to Anosov symplectic maps on more general symplectic manifolds admitting some form of quantization. We restricted the statement to the $2d$-torus because simple Anosov diffeomorphisms on $\mathbb{T}^{2d}$ can be constructed, and their quantization is by now rather standard. As we explain below, their study has revealed interesting features regarding the QUE conjecture.

1.4. Counterexamples to QUE for Anosov diffeomorphisms. For the simplest Anosov diffeomorphisms on $\mathbb{T}^2$, namely the hyperbolic symplectomorphisms of the torus (generalizations of Arnold's "cat map"), the QUE conjecture is known to fail. Indeed, in [12, 13] counterexamples to QUE for hyperbolic symplectomorphisms on the 2-dimensional torus were exhibited, in the form of explicit semiclassical sequences of eigenstates of the quantized map, associated with semiclassical measures of type (1.2) with $\alpha = \frac{1}{2}$. We also showed that, for this particular map, semiclassical measures of the form (1.2) necessarily satisfy $\alpha \geq \frac{1}{2}$.

In the case of toral symplectomorphisms on higher-dimensional tori $\mathbb{T}^{2d}$, Kelmer [19, 20] has exhibited semiclassical measures in the form of the Lebesgue measure on certain co-isotropic affine subspaces $\Lambda \subset \mathbb{T}^{2d}$ of the torus invariant through the map:

$$\mu_{sc} = \mu_{L\restriction\Lambda}.$$

For another example of a chaotic map (the baker's map quantized *à la Walsh*), we were able to construct semiclassical measures of purely fractal (self-similar) nature.

Fact 1.8. *The above counterexamples to QUE all satisfy the entropy bound* (1.7), *and some of them* (*like the measure* (1.2) *with* $\alpha = \frac{1}{2}$) *saturate that bound.*

It is worth mentioning an interesting result obtained by S. Brooks [8] in the case of the "quantum Arnold's cat map". Brooks takes into account the possibility to split any invariant measure μ into ergodic components:

$$\mu = \int_{\mathbb{T}^2} \mu_x \, d\mu(x),$$

where the probability measure μ_x, defined for μ-almost every point x, is ergodic. The affineness of the KS entropy ensures that

$$H_{KS}(\mu) = \int_{\mathbb{T}^2} H_{KS}(\mu_x) \, d\mu(x), \tag{1.10}$$

so to get a lower bound on $H_{KS}(\mu)$ it is sufficient to show that "high-entropy" components have a positive weight in μ. Brooks' result reads as follows:

Theorem 1.9 ([8]). *Let* $\kappa\colon \mathbb{T}^2 \circlearrowleft$ *be a linear hyperbolic symplectomorphism, with positive Lyapunov exponent* λ (λ *is also equal to the topological entropy of* κ *on* $\mathbb{T}^2$). *Fix any* $0 < H_0 < \lambda/2$, *and consider any associated semiclassical measure* μ_{sc}. *Then the following inequality holds:*

$$\mu_{sc}\{x \mid H_{KS}(\mu_x) < H_0\} \leq \mu_{sc}\{x \mid H_{KS}(\mu_x) > \lambda - H_0\}.$$

This result directly implies (through (1.10)) the bound $H_{KS}(\mu_{sc}) \geq \lambda/2$, but it also implies (by sending $H_0 \to 0$) the above-mentioned fact that the weight of atomic components of μ_{sc} is smaller or equal to the weight of its Lebesgue component.

1.5. Plan of the paper. These lectures reproduce most of the proofs of [3, 4] dealing with eigenstates of the Laplacian on manifolds of negative curvature. Yet, we extend the proofs in order to deal with more general Hamiltonian flows of Anosov type (for instance, adding some small potential to the free motion on X). This can be done at the price of using more general, "microlocal" partitions of unity, as opposed to the "local" partition of unity used in [3, 4] (which was given in terms of functions $\pi_k(x)$) only depending on the position variable. This microlocal setting is somehow more natural, since it does not depend on the way unstable manifolds project down to the manifold X. It is also necessary in the case of Anosov maps.

In Section 2 we recall the semiclassical tools we will need, starting with the $\hbar$-pseudodifferential calculus on a compact manifold, and including some exotic classes of symbols. We also define the main object of study, namely the semiclassical measures associated with sequences of null eigenstates $(\psi_\hbar)_{\hbar\to 0}$ of a family of Hamiltonians $\big(P(\hbar)\big)_{\hbar\to 0}$. In the central section 3 we provide the proof of Theorem 1.7(i) that is in the case of an Anosov Hamiltonian flow on a compact manifold X. We first recall the definition of entropies and pressures associated with invariant measures. We then introduce microlocal quantum partitions in Section 3.2, and their refinements used to define quantum entropies and pressures associated with the eigenstates $\psi_\hbar$. We try to provide some geometric intuition on the operators Π_α defining these partitions. We then state the central hyperbolic dispersive estimate on the norms of these operators, deferring the proof to Section 4. We then introduce several versions of "entropic uncertainty principles," from the simplest to the most complex, microlocal form (Proposition 3.12). We then apply this microlocal EUP in order to bound from below quantum and classical pressures associated with our eigenmodes. Section 4 is devoted to the proof of the hyperbolic dispersive estimate. Here we adapt the proof of [23], which is valid in more general situations than the case of geodesic flows. Finally, in Section 5 we briefly recall the framework of quantized maps on the torus, and provide the details necessary to obtain Theorem 1.7(ii).

Acknowledgements. I am grateful to D. Jakobson and I. Polterovich who invited me to give this minicourse in Montréal and to write these notes. Most of the material of these notes were obtained through from collaborations with N. Anantharaman, H. Koch and M. Zworski. I have also been partially supported by the project ANR-05-JCJC-0107091 of the Agence Nationale de la Recherche.

2. Preliminaries and problematics

2.1. Semiclassical calculus on X. The original application of the methods presented below concern the Laplace–Beltrami operator on a smooth compact manifold X of negative sectional curvature. To deal with this problem, one needs to define a certain number of auxiliary operators on $L^2(X)$, which are $\hbar$-pseudodifferential operators on X (ΨDOs), or $\hbar$-Fourier integral operators on X. We will only recall the definition and construction of the former class.

The Hamiltonians mentioned in Theorem 1.7 also belong to some class of $\hbar$-pseudodifferential operators, but the manifold X on which they are defined is not necessarily compact any more. In this setting, the smooth manifold X can be taken as the Euclidean space $X = \mathbb{R}^d$, or be Euclidean near infinity, that is $X = X_0 \sqcup \big(\mathbb{R}^d \setminus B(0,R_0)\big)$, where $B(0,R_0)$ is the ball of radius R_0 in $\mathbb{R}^d$, and X_0 is a compact manifold, the boundary of which is smoothly glued to $\partial B(R_0)$.

2.1.1. *Symbol classes on T^*X and $\hbar$-pseudodifferential calculus.* Let us construct an $\hbar$-quantization procedure on a Riemannian manifold X. To a certain class of well-behaved functions $\big(f(\hbar)\big)_{\hbar\to 0}$ on T^*X (the physical *observables*, referred to as *symbols* in mathematics) one can associate, through a well-defined *quantization procedure* $\mathrm{Op}_\hbar$, a corresponding set of operators $\mathrm{Op}_\hbar\big(f(\hbar)\big)$ acting on $C_c^\infty(X)$. By "well-behaved" one generally refers to certain conditions on the regularity and growth of the function. There are many different types of classes of "well-behaved symbols"; we will be using the class

$$S^{m,k}(T^*X) = \{f(\hbar) \in C^\infty(T^*X), |\partial_x^\alpha \partial_\xi^\beta f(\hbar)| \leq C_{\alpha,\beta}\hbar^{-k}\langle\xi\rangle^{m-|\beta|}\}.$$

Here we use the "Japanese brackets" notation $\langle\xi\rangle \stackrel{\mathrm{def}}{=} \sqrt{1+|\xi|^2}$. The estimates are supposed to hold uniformly for $\hbar \in (0,1]$ and $(x,\xi) \in T^*X$. The seminorms can be defined locally on coordinate charts of X; due to the factor $\langle\xi\rangle^{-|\beta|}$, this class is invariant with respect to changes of coordinate charts on X, and thus makes sense intrinsically on T^*X.

Some (important) symbols in this class are of the form

$$f(x,\xi;\hbar) = \hbar^{-k} f_0(x,\xi) + f_1(x,\xi,\hbar), \qquad f_1 \in S^{m,k'}, \qquad k' < k.$$

In that case, $\hbar^{-k} f_0(x,\xi)$ is called the *principal symbol* of f.

For $X = \mathbb{R}^d$, a symbol $f \in S^{m,k}$ can be quantized using the Weyl quantization: it acts on $\varphi \in \mathcal{S}(\mathbb{R}^d)$ through as the integral operator

$$\mathrm{Op}_\hbar^W(f)\varphi(x) = \int f\Big(\frac{x+y}{2},\xi\Big) e^{i\langle x-y,\xi\rangle/\hbar}\,\varphi(y)\,\frac{dy\,d\xi}{(2\pi\hbar)^d}. \tag{2.1}$$

If f is a real function, this operator is essentially self-adjoint on $L^2(\mathbb{R}^d)$.

If X is a more complicated manifold, one can quantize f by first splitting it into pieces localized on various coordinate charts $V_l \subset X$, through a finite partition of unity $1 = \sum_l \phi_l$, $\operatorname{supp}\phi_l \subset V_l$:

$$f = \sum_l f_l, \qquad f_l = f \times \phi_l.$$

Each component f_l can be considered as a function on $T^*\mathbb{R}^d$, and be quantized through (2.1), producing an operator $\mathrm{Op}_\hbar^{\mathbb{R}}(f_l)$ acting on $C_c^\infty(\mathbb{R}^d)$. A wavefunction $\varphi \in C^\infty(X)$ will be cut into pieces $\tilde\phi_l \times \varphi$, where the cutoffs $\operatorname{supp}\tilde\phi_l \subset V_l$ satisfy $\tilde\phi_l\phi_l = \phi_l$.[4] Our final quantization is then defined as:

$$\mathrm{Op}_\hbar(f)\varphi = \sum_l \tilde\phi_l \times \mathrm{Op}_\hbar^{\mathbb{R}}(f_l)(\tilde\phi_l \times \varphi). \tag{2.2}$$

The image of the class $S^{m,k}(T^*X)$ through quantization is an operator algebra acting on $C_c^\infty(X)$, denoted by $\Psi^{m,k}(T^*X)$. This algebra has nice properties in the semiclassical limit $\hbar \ll 1$. The product of two such operators behaves as a decoration of the usual multiplication:

$$\mathrm{Op}_\hbar(f)\,\mathrm{Op}_\hbar(g) = \mathrm{Op}_\hbar(f \sharp g), \tag{2.3}$$

where $f \sharp g \in S^{m+m',k+k'}$ admits an asymptotic expansion of the form

$$f \sharp g \sim \sum_{j\geq 0} \hbar^j (f \sharp g)_j. \tag{2.4}$$

Here the first component $(f \sharp g)_0 = f \times g$, while each $(f \sharp g)_j \in S^{m+m',k+k'}$ is a linear combination of derivatives $\partial^\gamma f \partial^{\gamma'} g$, with $|\gamma|, |\gamma'| \leq j$. In the case $X = \mathbb{R}^d$ and $\mathrm{Op}_\hbar$ is the Weyl quantization (2.1), $\sharp$ is called the Moyal product. In the case $f \in S(T^*X) = S^{0,0}(T^*X)$, $\mathrm{Op}_\hbar(f)$ can be extended into a continuous operator on $L^2(X)$, and the sharp Gårding inequality ensures that

$$\|\mathrm{Op}_\hbar(f)\|_{L^2} = \|f\|_\infty + \mathcal{O}_f(\hbar).$$

The quantization procedure $f \mapsto \mathrm{Op}_\hbar(f)$ is obviously not unique: it depends on the choice of coordinates on each chart, on the choice of quantization $\mathrm{Op}_\hbar^{\mathbb{R}}$, on the choice of cutoffs $\phi_j, \tilde\phi_j$. Fortunately, this non-uniqueness becomes irrelevant in the semiclassical limit.

[4]Throughout the text we will often encounter such embedded cutoffs. The property $\tilde\phi_l\phi_l = \phi_l$ will be denoted by $\tilde\phi_l \succ \phi_l$.

Proposition 2.1. *In the semiclassical limit $\hbar \to 0$, two $\hbar$-quantizations differ at most at subprincipal order:*

$$\forall f \in S^{m,k}(T^*X), \qquad \mathrm{Op}^1_\hbar(f) - \mathrm{Op}^2_\hbar(f) \in \Psi^{m,k-1}(T^*X).$$

2.2. From the Laplacian to more general quantum Hamiltonians.

2.2.1. *Rescaling the Laplacian.* One of our objectives is to study an eigenbasis $(\psi_n)_{n\geq 0}$ of the Laplace – Beltrami operator on some compact Riemannian manifold (X, g). To deal with the high-frequency limit $n \gg 1$, it turns out convenient to use a quantum mechanical point of view, namely rewrite the eigenmode equation

$$(\Delta + \lambda_n^2)\psi_n = 0, \qquad \lambda_n > 0$$

in the form

$$\Big(-\frac{\hbar_n^2\,\Delta}{2}\psi_n - \frac{1}{2}\Big)\psi_n = 0, \qquad \hbar_n = \lambda_n^{-1}. \tag{2.5}$$

This way, $\hbar_n$ appears as an effective Planck's constant (which is of the order of the *wavelength* of the state ψ_n). The rescaled Laplacian operator

$$-\frac{\hbar^2\Delta}{2} - \frac{1}{2} = P(\hbar)$$

is the $\hbar$-quantization $P(\hbar) = \mathrm{Op}_\hbar(p)$ of a certain classical Hamiltonian

$$p(x,\xi;\hbar) = p_0(x,\xi) + \hbar p_1(x,\xi) + \cdots \in S^{2,0}(T^*X).$$

The principal symbol $p_0(x,\zeta) = |\xi|^2_x/2 - 1/2$ generates (through Hamilton's equations) the motion of a free particle on X. In particular, the Hamilton flow $g^t = \exp tX_{p_0}$ restricted to the energy shell $p_0^{-1}(0) = S^*X$ is the geodesic flow (in the following, we will often denote by $\mathcal{E}$ this energy shell).

Notation 2.2. The Laplacian eigenmodes will often be denoted by $\psi_\hbar$ instead of ψ_n, with the convention that the state $\psi_\hbar$ satisfies the eigenvalue equation

$$P(\hbar)\,\psi_\hbar = \Big(-\frac{\hbar^2\,\Delta}{2} - \frac{1}{2}\Big)\psi_\hbar = 0. \tag{2.6}$$

Definition 2.3. We will call $S \subset (0,1]$ a countable set of scales $\hbar$, with only accumulation point at the origin. A sequence of states indexed by $\hbar \in S$ will be denoted by $(\varphi_\hbar)_{\hbar\in S}$, or sometimes, omitting the reference to a specific S, by $(\varphi_\hbar)_{\hbar\to 0}$.

2.2.2. *Anosov Hamiltonian flows.* In Theorem 1.7 we deal with more general Hamiltonians $p(x,\xi;\hbar)$ on T^*X, where X is compact or could also be the Euclidean space $\mathbb{R}^d$. The (real) symbol p is assumed to belong to a class $S^{m,0}(T^*X)$, and admit the expansion

$$p(x,\xi,\hbar) = p_0(x,\xi) + \hbar^\nu p_1(x,\xi,\hbar), \qquad p_0, p_1 \in S^{m,0}(T^*X),\ \nu > 0,$$

that is p_0 is the principal symbol of p. We could as well consider more general symbol classes, see for instance [11, Section 4.3] in the case $X = \mathbb{R}^d$. We assume that

(1) the energy shell $\mathcal{E} = p_0^{-1}(0)$ is compact, so that $\mathcal{E}_\epsilon \stackrel{\mathrm{def}}{=} p_0^{-1}([-\epsilon,\epsilon])$ is compact as well for $\epsilon > 0$ small enough.
(2) the Hamiltonian flow $g^t = e^{tH_{p_0}}$ restricted to the energy shell $\mathcal{E}$ does not admit fixed points, and is of Anosov type.

The Hamiltonian p is quantized into an operator $P(\hbar) = \mathrm{Op}_\hbar(p) \in \Psi^{m,0}$. The first assumption above implies that, for $\hbar > 0$ small enough, the spectrum of $P(\hbar)$ near zero is purely discrete. We will focus on sequences of normalized null eigenstates $(\psi_\hbar)_{\hbar\to 0}$:

$$P(\hbar)\psi_\hbar = 0. \tag{2.7}$$

Remark 2.4. If $\psi_\hbar$ is a "quasi-null" eigenstate of $P(\hbar)$, that is if $P(\hbar)\psi_\hbar = E(\hbar)\psi_\hbar$ with $E(\hbar) = \mathcal{O}(\hbar^\nu)$, then it is a null eigenstate of $\tilde{P}(\hbar) \stackrel{\text{def}}{=} P(\hbar) - E(\hbar)$, which admits the same principal symbol p_0 as $P(\hbar)$. As a result, Theorem 1.7 is also valid for such sequences of states.

2.3. $\hbar$-dependent singular observables. In the following we will have to use some classes of "singular" $\hbar$-dependent symbols.

2.3.1. "*Isotropically singular*" *observables.* For $\nu \in [0, \frac{1}{2})$, we will consider the class

$$S_\nu^{m,k}(T^*X) = \left\{f(\hbar) \in C^\infty(T^*X), \left|\partial_x^\alpha \partial_\xi^\beta f(\hbar)\right| \leq C_{\alpha,\beta}\hbar^{-k-\nu|\alpha+\beta|}\langle\xi\rangle^{m-|\beta|}\right\}. \tag{2.8}$$

Such functions can strongly oscillate on scales $\geq \hbar^\nu$. The corresponding operators belong to an algebra $\Psi_\nu^{m,k}(T^*X)$ which can still be analyzed using an $\hbar$-expansion of the type (2.4). The main difference is that the higher-order terms $(f \sharp g)_j \in S_\nu^{m+m',k+k'+2j\nu}$. Similarly, the Gårding inequality reads, for $f \in S_\nu^{0,0}$:

$$\|\mathrm{Op}_\hbar(f)\|_{L^2} = \|f\|_\infty + \mathcal{O}_f(\hbar^{1-2\nu}),$$

where the implicit constant depends on a certain seminorm of f.

2.3.2. "*Anisotropically singular*" *observables.* We will also need to quantize observables which are "very singular" along certain directions, away from some specific submanifold (see for instance [29] for a presentation). Consider $\Sigma \subset T^*X$ a compact co-isotropic manifold of dimension $2d - D$ (with $D \leq d$). Near each point $\rho \in \Sigma$, there exist local canonical coordinates (y_i, η_i) such that $\Sigma = \{\eta_1 = \eta_2 = \cdots = \eta_D = 0\}$. For some index $\nu \in [0, 1)$, we define as follows a class of smooth symbols $f \in S_{\Sigma,\nu}^{m,k}(T^*X) \subset C^\infty(T^*X \times (0, 1])$:

- for any family of smooth vector fields $V_1, \ldots, V_{l_1}$ tangent to Σ and of smooth vector fields $W_1, \ldots, W_{l_2}$, we have in any neighbourhood Σ_ϵ of Σ:

$$\sup_{\rho\in\Sigma_\epsilon} |V_1 \cdots V_{l_1} W_1 \cdots W_{l_2} f(\rho)| \leq C\,\hbar^{-k-\nu l_2}.$$

- away from Σ we require $|\partial_x^\alpha \partial_\xi^\beta| = \mathcal{O}(\hbar^{-k}\langle\xi\rangle^{m-|\beta|})$.

Such a symbol f can be split into components f_j localized in neighbourhoods $\mathcal{V}_j \subset \Sigma_\epsilon$, plus an "external" piece $f_\infty \in S^{m,k}(T^*X)$ vanishing near Σ. Each piece f_j is Weyl-quantized in local adapted canonical coordinates (y, η) on $\mathcal{V}_j$ (as in (2.1)), and then brought back to the original coordinates (x, ξ) using Fourier integral operators. On the other hand, f_∞ is quantized as in (2.2). Finally, $\mathrm{Op}_{\Sigma,\hbar}(f)$ is obtained by summing the various contributions. The resulting class of operators is denoted by $\Psi_{\Sigma,\nu}^{m,k}(T^*X)$.

2.3.3. *Sharp energy cutoffs.* We will mostly use this quantization relative to the energy layer $\Sigma \stackrel{\text{def}}{=} \mathcal{E} = p_0^{-1}(0)$, in order to define a family of sharp energy cutoffs. Namely, for some small $\delta > 0$ we will start from a cutoff $\chi_\delta \in C^\infty(\mathbb{R})$ such that $\chi_\delta(s) = 1$ for $|s| \leq e^{-\delta/2}$, $\chi_\delta(s) = 0$ for $|s| \geq 1$. From there, we define, for each $\hbar \in (0,1]$ and each $n \geq 0$, the rescaled function $\chi^{(n)} \in C_c^\infty(\mathbb{R} \times (0,1])$ by

$$\chi^{(n)}(s,\hbar) \stackrel{\text{def}}{=} \chi_\delta(e^{-n\delta}\,\hbar^{-1+\delta}\,s). \tag{2.9}$$

The functions $\chi^{(n)} \circ p_0$ are "sharp" energy cutoffs, they belong to the class $S^{-\infty,0}_{\mathcal{E},1-\delta}$. We will always consider $n \leq n_{\max} = C_\delta|\log \hbar|$, where the constant $C_\delta < \delta^{-1} - 1$, such that $\operatorname{supp}\chi^{(n)}$ is microscopic.

These cutoffs can be quantized in two ways:

(1) we may directly quantize the function $\chi^{(n)} \circ p_0$, into $\mathrm{Op}_{\mathcal{E},\hbar}(\chi^{(n)} \circ p_0) \in \Psi^{-\infty,0}_{\mathcal{E},1-\delta}$.
(2) or we can consider, using functional calculus, the operators $\chi^{(n)}\big(P(\hbar)\big)$. These operators (which generally differ from the previous ones) also belongs to $\Psi^{-\infty,0}_{\mathcal{E},1-\delta}$.

The sequence $(\chi^{(n)})_{0\leq n\leq n_{\max}}$ is an increasing sequence of embedded cutoffs: for each n, we have $\chi^{(n+1)}\chi^{(n)} = \chi^{(n)}$ (equivalently, $\chi^{(n+1)} \succ \chi^{(n)}$). More precisely, we have here

$$\operatorname{dist}\big(\operatorname{supp}\chi^{(n)}, \operatorname{supp}(1-\chi^{(n+1)})\big) \geq \hbar^{1-\delta}e^{\delta n}(e^{\delta/2}-1). \tag{2.10}$$

This distance between the supports implies the following

Lemma 2.5. *For any symbol $f \in S^{m,0}_{\mathcal{E},1-\delta}$ and any $0 \leq n \leq n_{\max}$, one has*

$$\big(\mathrm{Id} - \mathrm{Op}_{\mathcal{E},\hbar}(\chi^{(n+1)} \circ p_0)\big)\,\mathrm{Op}_{\mathcal{E},\hbar}(f)\,\mathrm{Op}_{\mathcal{E},\hbar}(\chi^{(n)} \circ p_0) = \mathcal{O}(\hbar^\infty). \tag{2.11}$$

The same property holds if we replace $\mathrm{Op}_{\mathcal{E},\hbar}(\chi^{(n)} \circ p_0)$ by $\chi^{(n)}\big(P(\hbar)\big)$.

Using the calculus of the class $S^{m,0}_{\mathcal{E},1-\delta}$, one can use the ellipticity of $P(\hbar)$ away from $\mathcal{E}$ to show that, if $(\psi_\hbar)$ is a sequence of null eigenstates of $P(\hbar)$, then

$$\Big(\mathrm{Id} - \chi^{(0)}\big(P(\hbar)\big)\Big)\psi_\hbar = \mathcal{O}(\hbar^\infty), \qquad \hbar \to 0. \tag{2.12}$$

That is, in the semiclassical limit the eigenstate $\psi_\hbar$ is microlocalized inside the energy layer of width $\hbar^{1-\delta}$ around $\mathcal{E}$.

2.4. Semiclassical measures. The $\hbar$-semiclassical calculus allows us to define what we mean by "phase space distribution of the eigenstate $\psi_\hbar$," through the notion of *semiclassical measure*. A Borel measure μ on the phase space T^*X can be fully characterized by the set of its values

$$\mu(f) = \int_{T^*X} f\,d\mu,$$

over *smooth test functions* $f \in C_c(T^*X)$. For each semiclassical scale $\hbar$, one can quantize a test function into a *test operator* $\mathrm{Op}_\hbar(f)$ (which is, as mentioned above, a continuous operator on $L^2(X)$). To any normalized state $\varphi \in L^2(X)$ we can then associate the linear functional

$$f \in C_c^\infty(T^*X) \mapsto \mu_{\hbar,\varphi}(f) \stackrel{\text{def}}{=} \langle \varphi, \mathrm{Op}_\hbar(f)\varphi\rangle.$$

$\mu_{\hbar,\varphi}$ is a distribution on T^*X, which encodes the localization properties of the state φ in the phase space, at the scale $\hbar$. Let us give an example. Using some local coordinate chart near $x_0 \in X$ and a function $\hbar^2 \ll c(\hbar) \ll 1$, we can define a Gaussian wavepacket by

$$\varphi_\hbar(x) \stackrel{\text{def}}{=} C_\hbar\, \chi(x) \exp\Big\{ -\frac{|x-x_0|^2}{c(\hbar)} + \mathrm{i}\frac{x\cdot\xi_0}{\hbar} \Big\}.$$

Here χ is a smooth cutoff equal to unity near x_0, which vanishes outside the coordinate chart, $C_\hbar$ is a normalization factor. When $\hbar \ll 1$, the distribution $\mu_{\hbar,\varphi_\hbar}$ associated with this wavefunction gets very peaked around the point $(x_0, \xi_0) \in T^*X$. If we had used the quantization at the scale $2\hbar$, the measure $\mu_{2\hbar,\varphi_\hbar}$ would have been peaked around $(x_0, \xi_0/2)$ instead.

Since the distribution $\mu_{\hbar,\varphi}$ is defined by duality with respect to the quantization $f \mapsto \mathrm{Op}_\hbar(f)$, it depends on the precise quantization scheme $\mathrm{Op}_\hbar$. In the case $X = \mathbb{R}^d$ and $\mathrm{Op}_\hbar$ is the Weyl quantization, the distribution $\mu_{\varphi,\hbar}$ is called the *Wigner distribution* associated with the state φ and the scale $\hbar$. Fortunately, as shown by Proposition 2.1, this scheme-dependence is irrelevant in the semiclassical limit.

Corollary 2.6. *For any $\varphi \in L^2$, consider the distributions $\mu^1_{\hbar,\varphi}, \mu^2_{\hbar,\varphi}$ defined by duality with two $\hbar$-quantizations $\mathrm{Op}^1_\hbar, \mathrm{Op}^2_\hbar$. Then, the following estimate holds in the semiclassical limit, uniformly with respect to $\varphi \in L^2$:*

$$\forall f \in C^\infty_{\mathrm{c}}(T^*X), \qquad \mu^1_{\hbar,\varphi}(f) - \mu^2_{\hbar,\varphi}(f) = \mathcal{O}_f(\hbar\|\varphi\|).$$

Let $S \subset (0,1]$ be a set of scales. For a given family of L^2-normalized states $(\varphi_\hbar)_{\hbar\in S}$, we consider the sequence of distributions $(\mu_{\hbar,\varphi_\hbar})_{\hbar\in S}$ on T^*X. It is always possible to extract a subset of scales $S' \subset S$, such that

$$\forall f \in C^\infty_{\mathrm{c}}(T^*X), \qquad \mu_{\hbar,\varphi_\hbar}(f) \xrightarrow{S'\ni\hbar\to 0} \mu_{\mathrm{sc}}(f),$$

with μ_{sc} a certain distribution on T^*X. One can show that μ_{sc} is a Radon measure on T^*X [11, Theorem 5.2]. From the above remarks, μ_{sc} does not depend on the precise scheme of quantization.

Definition 2.7. The measure μ_{sc} is called *the* semiclassical measure associated with the subsequence $(\varphi_\hbar)_{\hbar\in S'}$. It is also *a* semiclassical measure associated with the sequence $(\varphi_\hbar)_{\hbar\in S}$.

From now on, we will assume that $\varphi_\hbar = \psi_\hbar$ is a null eigenstate of the quantum Hamiltonian $P(\hbar)$ in Section 2.2: we will then call μ_{sc} a semiclassical measure of the Hamiltonian $P(\hbar)$.

Proposition 2.8. *Any semiclassical measure associated with a sequence $(\psi_\hbar)_{\hbar\to 0}$ of eigenstates of the Hamiltonian $P(\hbar)$ is a probability measure supported on the energy layer $\mathcal{E}$, which is invariant with respect to the geodesic flow g^t on $\mathcal{E}$.*

PROOF. Possibly after extracting a subsequence, we assume that μ_{sc} is the semiclassical measure associated with a sequence of eigenstates $(\psi_\hbar)_{\hbar\in S}$. The support property of μ_{sc} comes from the fact that the operator $P(\hbar)$ is elliptic outside $\mathcal{E}$. As a result, for any $f \in C^\infty_{\mathrm{c}}(T^*X)$ vanishing near $\mathcal{E}$, one can construct a symbol $g \in S^{-\infty,0}(T^*X)$ such that

$$\mathrm{Op}_\hbar(f) = \mathrm{Op}_\hbar(g)P(\hbar) + \mathcal{O}_{L^2\to L^2}(\hbar^\infty).$$

Applying this equality to the eigenstates $\psi_\hbar$, we get $\|\mathrm{Op}_\hbar(f)\psi_\hbar\| = \mathcal{O}(\hbar^\infty)$, proving the support property of μ_{sc}.

To prove the flow invariance, we need to compare the quantum time evolution with the classical one. Denote by $U_\hbar^t = \exp\{-itP(\hbar)/\hbar\}$ the propagator generated by the Hamiltonian $P(\hbar)$: it solves the time-dependent Schrödinger equation, and thus provides the quantum evolution. Let us state **Egorov's theorem**, which is a rigorous form of quantum-classical correspondence in terms of observables:

$$(2.13)\quad \forall f \in C_c^\infty(T^*X), \forall t \in \mathbb{R},$$
$$U_\hbar^{-t}\,\mathrm{Op}_\hbar(f)U_\hbar^t = \mathrm{Op}_\hbar(f \circ g^t) + \mathcal{O}_{f,t}(\hbar), \qquad \hbar \to 0.$$

Since $\psi_\hbar$ is an eigenstate of $U_\hbar^t$, we directly get

$$\mu_{\hbar,\varphi_\hbar}(f) = \mu_{\hbar,\varphi_\hbar}(f \circ g^t) + \mathcal{O}_{f,t}(\hbar) \implies \mu_{\mathrm{sc}}(f) = \mu_{\mathrm{sc}}(f \circ g^t). \qquad \square$$

These properties of semiclassical measures naturally lead to the following question:

Among all flow-invariant probability measures supported on $\mathcal{E}$, which ones appear as semiclassical measures associated with eigenstates of $P(\hbar)$?

To start answering this question, we will investigate the Kolmogorov – Sinai entropy of semiclassical measures. We will show that, in the case of an Anosov flow, the requirement of being a semiclassical measure implies a nontrivial lower bound on the entropy.

3. From classical to quantum entropies

3.1. Entropies and pressures [18].

3.1.1. *Kolmogorov – Sinai entropy of an invariant measure.* In this paper we will deal with several types of entropies. All of them are defined in terms of certain discrete probability distributions, that is *finite* sets of real numbers $\{p_i, i \in I\}$ satisfying

$$p_i \in [0,1], \qquad \sum_{i \in I} p_i = 1.$$

The entropy associated with such a set is the real number

$$(3.1)\qquad H(\{p_i\}) = \sum_{i\in I} \eta(p_i), \qquad \text{where } \eta(s) \stackrel{\text{def}}{=} -s\log s,\ s \in [0,1].$$

Our first example is the entropy $H(\mu, \mathcal{P})$ associated with a g^t-invariant probability measure μ on the energy shell $\mathcal{E}$ and a finite measurable partition $\mathcal{P} = (E_1, \ldots, E_K)$ of $\mathcal{E}$. That entropy is given by

$$(3.2)\qquad H(\mu, \mathcal{P}) = H(\{\mu(E_k)\}) = \sum_{k=1}^{K} \eta\big(\mu(E_k)\big).$$

One can then use the flow g^t in order to *refine* the partition $\mathcal{P}$. For each integer $n \geq 1$ we define the n-th refinement $\mathcal{P}^{\vee n} = [\mathcal{P}]_0^{n-1}$ as the partition composed of the sets

$$E_{\boldsymbol{\alpha}} \stackrel{\text{def}}{=} g^{-n+1}E_{\alpha_{n-1}} \cap \cdots \cap g^{-1}E_{\alpha_1} \cap E_{\alpha_0},$$

where $\boldsymbol{\alpha} = \alpha_0 \cdots \alpha_{n-1}$ can be any sequence of length n with symbols $\alpha_i \in \{1, \ldots, K\}$. In general many of the sets $E_{\boldsymbol{\alpha}}$ may be empty, but we will nonetheless sum over all sequences of a given length n. More generally, for any $m \in \mathbb{Z}, n \geq 1$, we consider the partition $[\mathcal{P}]_m^{m+n-1}$ made of the sets

$$g^{-m} E_{\boldsymbol{\alpha}} = g^{-m-n+1} E_{\alpha_{n-1}} \cap \cdots \cap g^{-m} E_{\alpha_0}, \qquad |\boldsymbol{\alpha}| = n.$$

From this refined partition we obtain the entropy $H(\mu, [\mathcal{P}]_m^{m+n-1}) = H_m^{m+n-1}(\mu)$.

From the concavity of the logarithm, one easily gets

$$\forall m, n \geq 1, \qquad H(\mu, [\mathcal{P}]_0^{n+m-1}) \leq H(\mu, [\mathcal{P}]_0^{n-1}) + H(\mu, [\mathcal{P}]_n^{n+m-1}). \tag{3.3}$$

If the measure μ is g^t-invariant, this has for consequence the *subadditivity property*:

$$H(\mu, \mathcal{P}^{\vee(n+m)}) \leq H(\mu, \mathcal{P}^{\vee n}) + H(\mu, \mathcal{P}^{\vee m}) \tag{3.4}$$

It thus makes sense to consider the limit

$$H_{\mathrm{KS}}(\mu, \mathcal{P}) \stackrel{\text{def}}{=} \lim_{n\to\infty} \frac{1}{n} H(\mu, \mathcal{P}^{\vee n}) = \lim_{n\to\infty} \frac{1}{2n} H(\mu, [\mathcal{P}]_{-n+1}^{n}),$$

the *Kolmogorov – Sinai entropy* of the invariant measure μ, associated with the partition $\mathcal{P}$. The KS entropy *per se* is defined by maximizing over the initial (finite) partition $\mathcal{P}$:

$$H_{\mathrm{KS}}(\mu) \stackrel{\text{def}}{=} \sup_{\mathcal{P}} H_{\mathrm{KS}}(\mu, \mathcal{P}).$$

For an Anosov flow, this supremum is actually reached as soon as the partition $\mathcal{P}$ has a sufficiently small diameter (that is, its elements E_k have uniformly small diameters).

3.1.2. *Pressures associated with invariant measures.* Let us come back to our probability distribution $\{p_i, i \in I\}$. We may associate to it a set of weights, that is of positive real numbers $\{w_i > 0, i \in I\}$, making up a weighted probability distribution. The pressure $p(\{p_i\}, \{w_i\})$ associated with this weighted distribution is the real number[5]

$$p(\{p_i\}, \{w_i\}) \stackrel{\text{def}}{=} -\sum_i p_i \log(w_i^2 p_i) = H(\{p_i\}) - 2 \sum_{i\in I} p_i \log w_i.$$

For instance, in the case of a flow-invariant measure on $\mathcal{E}$ and a partition $\mathcal{P}$, we can select weights w_k on each component E_k, and define the pressure

$$p(\mu, \mathcal{P}, w) \stackrel{\text{def}}{=} H(\mu, \mathcal{P}) - 2 \sum_k \mu(E_k) \log w_k.$$

We want to refine this pressure using the flow. The weights corresponding to the nth refinement can be simply defined as

$$w_{\boldsymbol{\alpha}} = \prod_{j=0}^{n-1} w_{\alpha_j}, \qquad |\boldsymbol{\alpha}| = n.$$

The refined pressure is denoted by $p_0^{n-1}(\mu, \mathcal{P}, w)$. From the subadditivity of the entropies (3.5) one easily draws the subadditivity of the pressures:

$$p_0^{n+m-1}(\mu, \mathcal{P}, w) \leq p_0^{n-1}(\mu, \mathcal{P}, w) + p_0^{m-1}(\mu, \mathcal{P}, w). \tag{3.5}$$

[5]The factor -2 appearing in front of the second term is convenient for our future aims.

3.1.3. *Smoothed partitions near $\mathcal{E}$.* The definition of $H(\mu,\mathcal{P})$ can be expressed in terms of characteristic functions over the partition $\mathcal{P}$. Indeed, if $\mathbb{1}_k$ is the characteristic function on E_k, then the function

$$\mathbb{1}_{\boldsymbol{\alpha}} = (\mathbb{1}_{\alpha_{n-1}} \circ g^{n-1}) \times \cdots \times (\mathbb{1}_{\alpha_1} \circ g) \times \mathbb{1}_{\alpha_0} \tag{3.6}$$

is the characteristic function of $E_{\boldsymbol{\alpha}}$. In formula (3.2) we can then replace $\mu(E_k)$ by

$$\mu(\mathbb{1}_k) \stackrel{\text{def}}{=} \int_{\mathcal{E}} \mathbb{1}_k \, d\mu.$$

Let us assume that the invariant measure μ does not charge the boundary of the E_k (this is always possible by slightly shifting the boundaries of the E_k). Then, for any $\epsilon > 0$, we can approximate the characteristic function $\mathbb{1}_k$ by a smooth function $\pi_k \in C^\infty_{\mathrm{c}}(\mathcal{E}_\epsilon, [0,1])$ supported in a small neighbourhood $\widetilde{E}_k$ of E_k, and such that these K functions form a *smooth partition of unity near* $\mathcal{E}$:

$$\sum_{k=1}^{K} \pi_k(\rho) = \chi_{\epsilon/2}(\rho), \qquad \operatorname{supp} \chi_{\epsilon/2} \subset \mathcal{E}_\epsilon, \qquad \chi_{\epsilon/2} = 1 \text{ near } \mathcal{E}_{\epsilon/2}. \tag{3.7}$$

One can extend the definition of the entropy to the smooth partition $\mathcal{P}_{\mathrm{sm}} = \{\pi_k\}_{k=1,\dots,K}$ and its refinements through the flow. From the assumption $\mu(\partial\mathcal{P}) = 0$, for any $\epsilon' > 0$ and any $n \geq 1$ we can choose $\mathcal{P}_{\mathrm{sm}}$ such that

$$|H(\mu, \mathcal{P}_{\mathrm{sm}}^{\vee n}) - H(\mu, \mathcal{P}^{\vee n})| \leq \epsilon' .$$

To prove a lower bound on the entropy $H_{\mathrm{KS}}(\mu)$ therefore amounts to proving a lower bound on $(1/n)H(\mu, \mathcal{P}_{\mathrm{sm}}^{\vee n})$, uniform with respect to $n \geq 1$ and the smoothing $\mathcal{P}_{\mathrm{sm}}$ of $\mathcal{P}$. The advantage of using a smoothed partition $\mathcal{P}_{\mathrm{sm}}$ is that it is fit for quantization.

3.2. Quantum partitions of unity.

3.2.1. *Definition.* From the smoothed partition $\mathcal{P}_{\mathrm{sm}} = \{\pi_k\}_{k=1,\dots,K}$ we form a *quantum partition of unity* $\mathcal{P}_{\mathrm{sm,q}} = \{\Pi_k = \mathrm{Op}_\hbar(\widetilde{\pi}_k)\}_{k=1,\dots,K}$, where $\mathrm{Op}_\hbar$ is the $\hbar$-quantization (2.2), and the symbols $\widetilde{\pi}_k \in S^{-\infty,0}(T^*X)$ satisfy the following properties:

(1) for each k, the symbol $\widetilde{\pi}_k$ is real, supported on $\widetilde{E}_k$, and admits $\sqrt{\pi_k}$ as principal symbol. The operator Π_k is thus self-adjoint.

(2) the family $\mathcal{P}_{\mathrm{sm,q}} = \{\Pi_k\}_{k=1,\dots,K}$ is a quantum partition of unity microlocally near $\mathcal{E}$:

$$\sum_{k=1}^{K} \Pi_k^2 = \mathrm{Op}_\hbar(\widetilde{\chi}_{\epsilon/2}) + \mathcal{O}(\hbar^\infty), \tag{3.8}$$

where $\widetilde{\chi}_{\epsilon/2} \in S^{-\infty,0}(T^*X)$ satisfies

$$\widetilde{\chi}_{\epsilon/2}(\rho) \equiv 1 \text{ near } \mathcal{E}_{\epsilon/2}, \qquad \operatorname{supp} \widetilde{\chi}_{\epsilon/2} \subset \mathcal{E}_\epsilon, \qquad \|\mathrm{Op}_\hbar(\widetilde{\chi}_{\epsilon/2})\| = 1 + \mathcal{O}(\hbar^\infty) .$$

Notice that $\widetilde{\chi}_{\epsilon/2}$ has for principal symbol $\chi_{\epsilon/2}$ of (3.7).

Remark. Had we simply taken $\Pi_k = \mathrm{Op}_\hbar(\sqrt{\pi_k})$, the above properties would hold only up to remainders $\mathcal{O}(\hbar)$. By iteratively adjusting the higher-order symbols in $\widetilde{\pi}_k$ (and $\widetilde{\chi}_{\epsilon/2}$), we can enforce these properties to any order in $\hbar$.

3.2.2. *Refined quantum partitions.* In the classical framework, the n-refinement of the partition $\mathcal{P}_{\rm sm} = \{\pi_k\}_{k=1,\dots,K}$ was obtained by considering the products $\pi_{\alpha_{n-1}} \circ g^{n-1} \times \cdots \times \pi_{\alpha_0}$, for all sequences $\boldsymbol{\alpha}$ of length n. Egorov's theorem shows that the quantum observable $\mathrm{Op}_\hbar(\pi_{\alpha_j} \circ g^j)$ resembles the quantum evolution $U^{-j} \mathrm{Op}_\hbar(\pi_{\alpha_j}) U^j$, where $U = U_\hbar = \mathrm{e}^{-\mathrm{i}P(\hbar)/\hbar}$ is the Schrödinger propagator (at time unity). For this reason, we define as follows the elements of the n-refined quantum projection:

$$\Pi_{\boldsymbol{\alpha}} \stackrel{\rm def}{=} U^{-n+1}\Pi_{\alpha_{n-1}} U \cdots U \Pi_{\alpha_2} U \Pi_{\alpha_1} U \Pi_{\alpha_0}, \qquad \boldsymbol{\alpha} = \alpha_0 \cdots \alpha_{n-1}. \tag{3.9}$$

We first need to check that these operators still make up a quantum partition of unity near $\mathcal{E}$.

Proposition 3.1. *Take $n_{\max} = [C_\delta |\log \hbar|]$ as in Section 2.3.3. Then, for each $1 \le n \le n_{\max}$, the family of operators $\mathcal{P}^{\vee n}_{\rm sm,q} = \{\Pi_{\boldsymbol{\alpha}}, |\boldsymbol{\alpha}| = n\}$ forms a quantum partition of unity microlocally near $\mathcal{E}$, in the following sense. For any symbol $\chi \in S^{-\infty,0}(T^*X)$ supported inside $\mathcal{E}_{\epsilon/2}$, we have*

$$\begin{aligned} \forall n \le n_{\max}, \qquad \sum_{\alpha_0,\dots,\alpha_{n-1}} \Pi^*_{\boldsymbol{\alpha}} \Pi_{\boldsymbol{\alpha}} = S_n, \qquad \|S_n\| = 1 + \mathcal{O}(\hbar^\infty), \\ (\mathrm{Id} - S_n)\,\mathrm{Op}_\hbar(\chi) = \mathcal{O}(\hbar^\infty). \end{aligned} \tag{3.10}$$

PROOF. The statement is obvious in the case $\{\Pi_k\}_{k=1,\dots,K}$ forms a full resolution of identity (that is, if the left-hand side in (3.8) is equal to the identity modulo $\mathcal{O}(\hbar^\infty)$), as was the case in [1, 3, 4] and will be the case in Section 5. One can then sum over $\sum_{\alpha_{n-1}} \Pi^2_{\alpha_{n-1}} = \mathrm{Id} + \mathcal{O}(\hbar^\infty)$, then over α_{n-2}, etc., to finally obtain $S_n = \mathrm{Id} + \mathcal{O}(\hbar^\infty)$.

In the case of a microlocal partition near $\mathcal{E}$, the sum over the index α_{n-1} leads to a product $\Pi_{\alpha_{n-2}} \mathrm{Op}_\hbar(\widetilde{\chi}_{\epsilon/2}) \Pi_{\alpha_{n-2}}$, where $\widetilde{\chi}_{\epsilon/2}$ is the symbol appearing in (3.8). To "absorb" the factor $\mathrm{Op}_\hbar(\widetilde{\chi}_{\epsilon/2})$, we will insert intermediate cutoffs at each time. We recall that $\chi \prec \mathbb{1}_{\mathcal{E}_{\epsilon/2}} \prec \widetilde{\chi}_{\epsilon/2}$. We consider a sequence of cutoffs $(\chi_j \circ p_0)_{1 \le j \le n_{\max}}$ such that $\mathbb{1}_{\mathcal{E}_{\epsilon/2}} \prec \chi_1 \circ p_0 \prec \chi_2 \circ p_0 \cdots \prec \chi_{n_{max}} \circ p_0 \prec \widetilde{\chi}_{\epsilon/2}$. Since $n_{\max} \sim |\log \hbar|$, the χ_j will necessarily depend on $\hbar$, their derivatives growing like $\hbar^{-\nu}$ for some small $\nu > 0$, so that $\chi^j \circ p_0 \in S^{-\infty,0}_\nu$. The calculus in $S^{-\infty,0}_\nu$ and $\chi_j \prec \chi_{j+1}$ show that, for any $1 \le k \le K$:

$$\forall 1 \le j \le n_{\max} - 1, \qquad \big(\mathrm{Id} - \chi_{j+1}(P(\hbar))\big)\, \Pi_k \chi_j(P(\hbar)) = \mathcal{O}(\hbar^\infty). \tag{3.11}$$

This implies that we can indeed insert intermediate cutoffs in $\Pi_{\boldsymbol{\alpha}}$ with no harm:

$$\begin{aligned} \Pi_{\boldsymbol{\alpha}} \mathrm{Op}_\hbar(\chi) = U^{-n+1} \Pi_{\alpha_{n-1}} U \Pi_{\alpha_{n-2}} U \chi_{n-2}(P(\hbar)) \cdots U \chi_2(P(\hbar)) \Pi_{\alpha_1} \\ \times U \chi_1(P(\hbar)) \Pi_{\alpha_0} \mathrm{Op}_\hbar(\chi) + \mathcal{O}(\hbar^\infty). \end{aligned}$$

We also have

$$\forall j, \qquad (\mathrm{Id} - \mathrm{Op}_\hbar(\widetilde{\chi}_{\epsilon/2})) \Pi_k \chi_j(P(\hbar)) = \mathcal{O}(\hbar^\infty). \tag{3.12}$$

This equation, and the fact that $\chi_j(P(\hbar))$ commutes with the propagator U, results in

$$\begin{aligned}\sum_{\alpha_{n-2}} U^*\Pi_{\alpha_{n-2}} \operatorname{Op}_\hbar(\widetilde{\chi}_{\epsilon/2})\Pi_{\alpha_{n-2}} U\chi_{n-2}\big(P(\hbar)\big) \\ &= \sum_{\alpha_{n-2}} U^*\Pi^2_{\alpha_{n-2}}\chi_{n-2}\big(P(\hbar)\big)U + \mathcal{O}(\hbar^\infty) \\ &= U^* \operatorname{Op}_\hbar(\widetilde{\chi}_{\epsilon/2})\chi_{n-2}\big(P(\hbar)\big)U + \mathcal{O}(\hbar^\infty) \\ &= U^*\chi_{n-2}\big(P(\hbar)\big)U + \mathcal{O}(\hbar^\infty) \\ &= \chi_{n-2}\big(P(\hbar)\big) + \mathcal{O}(\hbar^\infty).\end{aligned}$$

Using (3.11) at each step, the summation over $\alpha_{n-3}, \alpha_{n-4}, \ldots$ finally brings us to

$$\sum_{\boldsymbol{\alpha}} \Pi^*_{\boldsymbol{\alpha}}\Pi_{\boldsymbol{\alpha}} \operatorname{Op}_\hbar(\chi) = \sum_{\alpha_0} \Pi_{\alpha_0}\chi_1\big(P(\hbar)\big)\Pi_{\alpha_0} \operatorname{Op}_\hbar(\chi) + \mathcal{O}(\hbar^\infty).$$

The equation $\big(\mathrm{Id} - \chi_1\big(P(\hbar)\big)\big)\,\Pi_{\alpha_0} \operatorname{Op}_\hbar(\chi) = \mathcal{O}(\hbar^\infty)$ ends the proof. □

3.3. From the refined operators $\Pi_{\boldsymbol{\alpha}}$ to a quantum symbolic measure. Let us turn back to the sequence of eigenstates $(\psi_\hbar)_{\hbar\to 0}$ associated with the semiclassical measure μ_{sc}. The proof of Proposition 2.8 shows that for any cutoff $\chi \in S^{-\infty,0}$, $\chi \equiv 1$ near $\mathcal{E}$, we have

$$\operatorname{Op}_\hbar(\chi)\psi_\hbar = \psi_\hbar + \mathcal{O}(\hbar^\infty).$$

As a result, for any $1 \le n \le n_{\max}$ we have $\sum_{|\boldsymbol{\alpha}|=n} \|\Pi_{\boldsymbol{\alpha}}\psi_\hbar\|^2 = 1 + \mathcal{O}(\hbar^\infty)$. Therefore, modulo an $\mathcal{O}(\hbar^\infty)$ error, the set $\{\|\Pi_{\boldsymbol{\alpha}}\psi_\hbar\|^2, |\boldsymbol{\alpha}| = n\}$ forms a discrete probability distribution. The proof shows that these weights also satisfy a compatibility condition (up to a negligible error):

(3.13) $\quad \forall n \le n_{\max}, \forall \boldsymbol{\alpha} = \alpha_0 \cdots \alpha_{n-1},$

$$\|\Pi_{\alpha_0\cdots\alpha_{n-2}}\psi_\hbar\|^2 = \sum_{\alpha_{n-1}} \|\Pi_{\alpha_0\cdots\alpha_{n-1}}\psi_\hbar\|^2 + \mathcal{O}(\hbar^\infty).$$

If we forget the errors $\mathcal{O}(\hbar^\infty)$, we can interpret the weights $\|\Pi_{\boldsymbol{\alpha}}\psi_\hbar\|^2$ in terms of a certain probability measure $\mu_\hbar$ on the symbolic space $\Sigma = \{1, \ldots, K\}^{\mathbb{Z}}$. Namely, each $\|\Pi_{\boldsymbol{\alpha}}\psi_\hbar\|^2$ corresponds to the weight of that measure on the *cylinder* $[\cdot\alpha_0\alpha_1\cdots\alpha_{n-1}]$:

$$\mu_\hbar([\cdot\boldsymbol{\alpha}]) \stackrel{\mathrm{def}}{=} \|\Pi_{\boldsymbol{\alpha}}\psi_\hbar\|^2. \tag{3.14}$$

Formally, $\mu_\hbar$ can be defined as an equivalence class of $\hbar$-dependent probability measures taking values on cylinders of lengths $n \le n_{\max} = C_\delta|\log\hbar|$, the equivalence relation consisting in equality up to errors $\mathcal{O}(\hbar^\infty)$. We will call such a measure a **symbolic measure**.

The defining property $\Pi_k = \operatorname{Op}_\hbar(\widetilde{\pi}_k)$ shows that each element $\mu_\hbar([\cdot\alpha_0]) = \|\Pi_{\alpha_0}\psi_\hbar\|^2$ approximately represents the microlocal weight of the state $\psi_\hbar$ inside the element E_k of the partition. Further on, for any fixed $n \ge 1$, Egorov's theorem (2.13) and the composition rule (2.3) show that the refined operators $\Pi_{\boldsymbol{\alpha}}$ are still "good" pseudodifferential operators:

$$\Pi_{\boldsymbol{\alpha}} = \operatorname{Op}_\hbar(\pi_{\boldsymbol{\alpha}}) + \mathcal{O}_n(\hbar),$$

where $\mathcal{P}_{\mathrm{sm}}^{\vee n} = \{\pi_{\boldsymbol{\alpha}}\}$ is the n-refinement of the smooth partition $\mathcal{P}_{\mathrm{sm}}$, as in (3.6). As a result, $\mu_\hbar([\cdot\boldsymbol{\alpha}])$ approximately represents the weight of $\psi_\hbar$ inside the refined partition element $E_{\boldsymbol{\alpha}}$.

From the assumption on the sequence $(\psi_\hbar)$, the symbolic measure $\mu_\hbar$ is obviously related with the semiclassical measure μ_{sc}: for any fixed $n \geq 1$ we have

$$\forall \boldsymbol{\alpha}, |\boldsymbol{\alpha}| = n, \qquad \mu_\hbar([\cdot\boldsymbol{\alpha}]) \xrightarrow{\hbar\to 0} \mu_{sc}(\pi_{\boldsymbol{\alpha}}). \tag{3.15}$$

This limit assumes that the "time" n is fixed when taking $\hbar \to 0$. For our purposes, it will be crucial to extend the analysis to times n of logarithmic order in $\hbar$. Before doing so, let us give a crude description of the sets $E_{\boldsymbol{\alpha}}$ when $n = |\boldsymbol{\alpha}|$ grows. Let us assume that the elements E_k are approximately "isotropic" with respect to the stable and unstable directions. The inverse flow g^{-t} has the effect to compress along the unstable directions, and expand along the stable ones. As a result, the set $E_{\alpha_0\alpha_1} = g^{-1}E_{\alpha_1} \cap E_{\alpha_0}$ will be narrower than E_{α_0} along the unstable direction, but keep approximately the same size in the stable one. Iterating this procedure, the set $E_{\alpha_0\cdots\alpha_{n-1}}$ will become very anisotropic for large n: its size along the stable directions will remain comparable with that of E_{α_0}, while its "unstable volume" will be diminished by a factor $J_n^u(\alpha_0\cdots\alpha_{n-1})^{-1}$, where we use the **coarse-grained unstable Jacobian**

$$J_n^u(\alpha_0\cdots\alpha_{n-1}) \stackrel{\text{def}}{=} \prod_{i=0}^{n-1} J^u(\alpha_i), \qquad J^u(k) \stackrel{\text{def}}{=} \min_{\rho\in E_k} J^u(\rho). \tag{3.16}$$

The factor $J_n^u(\boldsymbol{\alpha})^{-1}$ decreases exponentially with n according to the minimal $(d-1)$-dimensional unstable expansion rate $\Lambda^u_{\min}$:

$$\forall n, \forall \boldsymbol{\alpha}, |\boldsymbol{\alpha}| = n, \qquad J_n^u(\boldsymbol{\alpha})^{-1} \leq C e^{-n(\Lambda^u_{\min}-\epsilon)}, \tag{3.17}$$

so the sets $E_{\boldsymbol{\alpha}}$ become very thin along the unstable direction. This anisotropy is as well visible on the refined smooth functions $\pi_{\boldsymbol{\alpha}}$ or the refined symbols $\widetilde{\pi}_{\boldsymbol{\alpha}}$.

3.4. Egorov theorem up to the Ehrenfest time. The Egorov theorem (2.13) can be extended up to times $t \sim C|\log\hbar|$, provided the constant C is not too large. The breakdown occurs when the classically evolved function $f\circ g^t$ shows fluctuations of size unity across a distance $\sim h^{1/2}$: such a function is no more a "nice quantizable observable" (see Section 2.3).

Let us start from a symbol $f \in S^{-\infty,0}(T^*X)$ supported on the energy layer $\mathcal{E}_\epsilon$. We have called $\lambda_{\max}$ the maximal expansion rate of the flow on $\mathcal{E}$. Assume that $\lambda_\epsilon = \lambda_{\max} + \mathcal{O}(\epsilon)$ is larger than the maximal expansion rate on $\mathcal{E}_\epsilon$. It implies that the derivatives of the flow are controlled as follows:

$$\forall t \in \mathbb{R}, \forall \rho \in \mathcal{E}_\epsilon, \qquad \|\partial^\alpha g^t(\rho)\| \leq C_\alpha e^{\lambda_\epsilon|\alpha t|}.$$

As a result, for any symbol $f \in S^{-\infty,0}$ supported inside $\mathcal{E}_\epsilon$, its classical evolution $f_t = f\circ g^t$ satisfy

$$\forall t \in \mathbb{R}, \forall \rho \in \mathcal{E}_\epsilon, \qquad \|\partial^\alpha f_t(\rho)\| \leq C_{f,\alpha} e^{\lambda_\epsilon|\alpha t|}.$$

For $t \sim C|\log\hbar|$ the right hand sides become of order $\hbar^{-C\lambda_\epsilon|\alpha|}$. Therefore, if we want f_t to belong to a reasonable symbol class (see Section 2.3), we must restrict the values of C. Let us define the time

$$T_{\epsilon,\hbar} \stackrel{\text{def}}{=} \frac{(1-\epsilon)|\log\hbar|}{2\lambda_\epsilon},$$

which is about half of what is generally called the *Ehrenfest time* $T_{\mathrm{E}} = |\log\hbar|/\lambda_{\max}$. Take any $\nu \in [(1-\epsilon)/2, \frac{1}{2})$. The above estimates show that for any choice of sequence $\big(t(\hbar)\big)_{\hbar\to 0}$ satisfying $|t(\hbar)| \leq T_{\epsilon,\hbar}$, the family of functions $\big(f_{t(\hbar)}\big)_{\hbar\to 0}$ belongs to the class $S_\nu^{-\infty,0}(T^*X)$ defined in (2.8). In other words, any seminorm of that class is uniformly bounded over the set $\{f_t,\ |t|\le T_{\epsilon,\hbar},\ \hbar\in(0,1]\}$. It is then not surprising that Egorov's theorem holds true up to the time $T_{\epsilon,\hbar}$.

Proposition 3.2 ([3, Proposition 5.1]). *Fix $\epsilon>0$ and $\nu\in\big((1-\epsilon)/2,\frac12\big)$. Take $f\in S^{-\infty,0}$ supported inside $\mathcal{E}_\epsilon$. Then, for any $\hbar\in(0,1]$ and any time $t=t(\hbar)$ in the range $|t|\le T_{\epsilon,\hbar}$, we have*

$$(3.18)\quad U^{-t}\operatorname{Op}_\hbar(f)U^t=\operatorname{Op}_\hbar(\tilde f_t)+\mathcal{O}(h^\infty),\qquad \textit{with } \tilde f_t-f_t\in S_\nu^{-\infty,-(1+\epsilon)/2},\ f_t=f\circ g^t\in S_\nu^{-\infty,0}.$$

Proof. This proposition was essentially proved in [7] in the case of symbols on $T^*\mathbb{R}^d$ driven by some (appropriate) Hamiltonian flow. In that paper, the $\hbar$-expansion of the symbol $\tilde f_t$ was explicitly computed up to any fixed order $\hbar^L$, and the L^2 norm of the remainder was estimated. In [3, Section 5.2] we used the fact that

$$U^{-t}\operatorname{Op}_\hbar(f)U^t-\operatorname{Op}_\hbar(f_t)=\int_0^t \mathrm{d}s\, U^{-s}\operatorname{Diff} f_{t-s}\, U^s,$$

where

$$\operatorname{Diff} f_s=\frac{\mathrm{i}}{\hbar}[P(\hbar),\operatorname{Op}_\hbar(f_s)]-\operatorname{Op}_\hbar(\{p,f_s\})\in\Psi_\nu^{-\infty,-(1+\epsilon)/2},$$

uniformly for $|t|\le T_{\epsilon,\hbar}$. The Calderon – Vaillancourt theorem on $\Psi_\nu^{-\infty,-(1+\epsilon)/2}$ then implies that

$$(3.19)\qquad \|U^{-t}\operatorname{Op}_\hbar(f)U^t-\operatorname{Op}_\hbar(f_t)\|\le C\,|t|\,h^{(1+\epsilon)/2}.$$

In order to prove that $\tilde f_t\in S_\nu^{-\infty,0}$ one can proceed as in [7], that is compute the $\hbar$ expansion of $\tilde f_t$ order by order, taking into account that the quantization is performed on the manifold X, so that higher-order terms also depend on the various choices of local charts and cutoffs. We will not do so in any detail here, since we will mostly use the inequality (3.19). □

We will apply this proposition to the operators $U^{-j}\Pi_k U^j$: in the range $|j|\le T_{\epsilon,\hbar}$ they are still pseudodifferential operators in some class $S_\nu^{-\infty,0}$. The products of these operators can also be analyzed:

Proposition 3.3. *Take any $1>\epsilon>0$ and $\nu\in[(1-\epsilon)/2,\frac12)$. Then the family of symbols $\{\widetilde\pi_{\boldsymbol\alpha},|\boldsymbol\alpha|\le T_{\epsilon,\hbar}\}$ belongs to a bounded set in the class $S_\nu^{-\infty,0}$. Furthermore, the product operators $\Pi_{\boldsymbol\alpha}$ satisfy $\Pi_{\boldsymbol\alpha}-\operatorname{Op}_\hbar(\widetilde\pi_{\boldsymbol\alpha})\in\Psi_\nu^{-\infty,-(1+\epsilon)/2}$.*

Proof. A similar result was proved in [24, Theorem 7.1]. We already know that the symbols $\widetilde\pi_{\alpha_j}\circ g^j$ composing $\widetilde\pi_{\boldsymbol\alpha}$ belong to the class $\Psi_\nu^{-\infty,0}$. Any finite product of those symbols also remains in that class. We need to check that the symbol $\widetilde\pi_{\alpha_0\cdots\alpha_j}$ remains uniformly bounded in the class when increasing j until $T_{\epsilon,\hbar}$.

We start by applying Egorov's theorem to the operator $\Pi_{\alpha_{n-1}}$, then multiply by $\Pi_{\alpha_{n-2}}$:

$$U^{-1}\Pi_{\alpha_{n-1}}U\Pi_{\alpha_{n-2}}=\operatorname{Op}_\hbar(\widetilde\pi_{\alpha_{n-1}}\circ g\times\widetilde\pi_{\alpha_{n-2}})+R_2,\qquad R_2\in\Psi^{-\infty,-1}.$$

The function $\widetilde\pi_{\alpha_{n-2}\alpha_{n-1}}\stackrel{\text{def}}{=}\widetilde\pi_{\alpha_{n-1}}\circ g\times\widetilde\pi_{\alpha_{n-2}}$ is supported in a "rectangle" and satisfies $\|\partial^\beta\widetilde\pi_{\alpha_{n-2}\alpha_{n-1}}\|\le C_\beta e^{\lambda_\epsilon|\beta|}$. Applying the same procedure (evolution and multiplication), we construct a sequence of symbols

$$\widetilde\pi_{\alpha_{n-j}\cdots\alpha_{n-1}}\stackrel{\text{def}}{=}\widetilde\pi_{\alpha_{n-j+1}\cdots\alpha_{n-1}}\circ g\times\widetilde\pi_{\alpha_{n-j}}$$

and operators

$$U^{-j}\Pi_{\alpha_{n-1}}U\Pi_{\alpha_{n-2}}U\cdots\Pi_{\alpha_{n-j}}.$$

The symbols are supported in small rectangles, similar with the elements $E_{\boldsymbol\alpha}$ of the refined partition $\mathcal{P}^{\vee j}$. One iteratively shows that

$$\|\partial^\beta\widetilde{\pi}_{\alpha_{n-j}\cdots\alpha_{n-1}}\| \leq C_\beta e^{\lambda_\epsilon|\beta|j},$$

with constants C_β uniform with respect to j. Therefore, as long as $j \leq T_{\epsilon,\hbar}$, the symbol $\widetilde{\pi}_{\alpha_{n-j+1}\cdots\alpha_{n-1}} \in S^{-\infty,0}_{\nu_j}$ (with uniform constants), where $\nu_j = \lambda_\epsilon j/\log(1/\hbar)$. At the same time, $\widetilde{\pi}_{\alpha_{n-j}} \in S^{-\infty,0}_0$. As a result,

$$U^{-1}\operatorname{Op}_\hbar(\widetilde{\pi}_{\alpha_{n-j+1}\cdots\alpha_{n-1}})U\operatorname{Op}_\hbar(\widetilde{\pi}_{\alpha_{n-j}}) = \operatorname{Op}_\hbar(\widetilde{\pi}_{\alpha_{n-j}\cdots\alpha_{n-1}}) + R_j,$$

and the remainder $R_j \in \Psi^{-\infty,-1+\nu_j}_{\nu_j}$ satisfies

$$\|R_j\|_{L^2\circlearrowleft} \leq C\hbar^{1-\nu_j}, \qquad j = 2,\ldots,n,$$

with a uniform constant C. The sum of all remainders thus satisfies

$$\sum_{j=2}^n R_j \in \Psi^{-\infty,-1+\nu_n}_{\nu_n}, \qquad \Big\|\sum_{j=2}^n R_j\Big\| \leq \sum_{j=2}^n C\hbar^{1-\nu_j} \leq \widetilde{C}\hbar^{1-\nu_n}. \qquad \square$$

Corollary 3.4. *Take any $1 > \epsilon > 0$ and $\nu \in [(1-\epsilon)/2, \frac12)$. Let $\boldsymbol\alpha, \boldsymbol\beta$ be two sequences of length $n \leq T_{\epsilon,\hbar}$. Then the symbols $\widetilde{\pi}_{\boldsymbol\alpha}, \widetilde{\pi}_{\boldsymbol\beta}\circ g^{-n}$ belong to $S^{-\infty,0}_\nu$, and so does their product. The operator*

$$\Pi_{\boldsymbol\beta\cdot\boldsymbol\alpha} \stackrel{\text{def}}{=} \Pi_{\boldsymbol\alpha}U^n\Pi_{\boldsymbol\beta}U^{-n} = U^{-n+1}\Pi_{\alpha_{n-1}}U\Pi_{\alpha_{n-2}}U\cdots\Pi_{\alpha_0}U\Pi_{\beta_{n-1}}U\cdots U\Pi_{\beta_0}U^{-n}$$

belongs to $\Psi^{-\infty,0}_\nu$, and satisfies

$$\Pi_{\boldsymbol\beta\cdot\boldsymbol\alpha} = \operatorname{Op}_\hbar(\widetilde{\pi}_{\boldsymbol\alpha}\times\widetilde{\pi}_{\boldsymbol\beta}\circ g^{-n}) + \Psi^{-\infty,-1+2\nu}_\nu.$$

These results show that, for times $n \leq T_{\epsilon,\hbar}$, the operators $\Pi_{\boldsymbol\alpha}$ (resp. $\Pi_{\boldsymbol\beta\cdot\boldsymbol\alpha}$) are "quasiprojectors" on refined rectangles $E_{\boldsymbol\alpha} \in \mathcal{P}^{\vee n}$ (resp. in the rectangles $E_{\boldsymbol\alpha} \cap g^n(E_{\boldsymbol\beta})$ of the "isotropic" refined partition $\mathcal{P}^{n-1}_{-n}$). Using the fact that $\Pi_{\boldsymbol\beta\cdot\boldsymbol\alpha} = U^n\Pi_{\boldsymbol\beta\boldsymbol\alpha}U^{-n}$, we also draw the

Corollary 3.5. *Take ϵ, ν as above. Then, for any sequence $\boldsymbol\alpha$ of length $|\boldsymbol\alpha| \leq 2T_{\epsilon,\hbar}$, the operator norm $\|\Pi_{\boldsymbol\alpha}\| = \|\widetilde{\pi}_{\boldsymbol\alpha}\|_\infty + \mathcal{O}(\hbar^{1-2\nu})$, which can be close to unity.*

3.5. Hyperbolic dispersive estimates. We will now consider operators $\Pi_{\boldsymbol\alpha}$ for sequences $\boldsymbol\alpha$ *longer than* $2T_{\epsilon,\hbar}$. We recall that $J^u_n(\boldsymbol\alpha)$ is the coarse-grained unstable Jacobian along orbits following the path $\boldsymbol\alpha$ (see (3.16)). Given some small $\delta > 0$, we have constructed in Section 2.3.3 cutoffs $\chi^{(m)}$ supported on intervals of lengths $2e^{m\delta}\hbar^{1-\delta}$, from which we built up sharp energy cutoffs. Our major dynamical result is the following dispersive estimate [3]. We provide its proof in Section 4.

Proposition 3.6. *Choose $\delta > 0$ small, leading to the constant C_δ of Section 2.3.3. Then, there exists $\hbar_\delta > 0$ and $C > 0$ such that, for any $0 < \hbar \leq \hbar_\delta$, any integers $n, m \in [0, C_\delta\log(1/\hbar)]$ and any sequence $\boldsymbol\alpha$ of length n, the following estimate holds:*

$$\big\|\Pi_{\boldsymbol\alpha}\,\chi^{(m)}(P(\hbar))\big\| \leq Ce^{m\delta/2}\,\hbar^{-(d-1+\delta)/2}J^u_n(\boldsymbol\alpha)^{-1/2}. \tag{3.20}$$

From the bound (3.17) on the coarse-grained Jacobians, we see that (3.20) becomes sharper than the obvious bound $\|\Pi_{\boldsymbol{\alpha}}\chi^{(n)}(P(\hbar))\| \leq 1 + \mathcal{O}(\hbar^\infty)$ for times

$$(3.21) \qquad n \geq T_1 \stackrel{\text{def}}{=} \frac{(d-1)\log(1/\hbar)}{\Lambda^u_{\min}} > 2T_{\epsilon,\hbar}.$$

If we specialize Proposition 3.6 to the case $n \approx 4T_{\epsilon,\hbar}$ and insert $U^{-n/2}$ on the right, we obtain the following

Corollary 3.7. *Take $\delta > 0$ small. For $0 < \hbar < h_\delta$, take $\boldsymbol{\alpha}, \boldsymbol{\beta}$ two arbitrary sequences of length $n = \lfloor 2T_{\epsilon,\hbar} \rfloor$. Then,*

$$(3.22) \qquad \|\Pi_{\boldsymbol{\beta}\cdot\boldsymbol{\alpha}}\,\chi^{(n)}(P(\hbar))\| \leq C\,\hbar^{-(d-1+c\delta)/2}\,J^u_n(\boldsymbol{\alpha})^{-1/2}\,J^u_n(\boldsymbol{\beta})^{-1/2},$$

with uniform constants $C, c > 0$.

It is this estimate which we will use in Section 3.8.

3.5.1. *A remark on the sharpness of* (3.22). Let us give a handwaving argument to show that, in the case of a surface ($d = 2$) of constant curvature, the upper bound (3.22) is close to being sharp. This argument was made rigorous in the case of the toy model studied in [2]. Since our argument is sketchy, we set all "small constants" (ϵ, δ) to zero.

The operator $\Pi_{\boldsymbol{\beta}\cdot\boldsymbol{\alpha}}$ is the product of two quasiprojectors, $\Pi_{\boldsymbol{\alpha}}$ associated with the "thin stable" rectangle $E_{\boldsymbol{\alpha}}$, which has length $\lesssim \hbar$ along the unstable direction, and $\Pi_{\boldsymbol{\beta}\cdot}$ associated with the "thin unstable" rectangle $E_{\boldsymbol{\beta}\cdot}$ which has length $\lesssim \hbar$ along the stable direction. The intersection $E_{\boldsymbol{\beta}\cdot\boldsymbol{\alpha}}$ has length $\lesssim \hbar$ along the two directions transverse to the flow, which are symplectically conjugate to each other. As a result, the refined smoothed characteristic function $\pi_{\boldsymbol{\beta}\cdot\boldsymbol{\alpha}}$ does not belong to any "nice" symbol class, and the norm of the operator $\Pi_{\boldsymbol{\beta}\cdot\boldsymbol{\alpha}}$ is not connected with the sup-norm of $\pi_{\boldsymbol{\beta}\cdot\boldsymbol{\alpha}}$.

Since $E_{\boldsymbol{\alpha}}$ has symplectic volume $\lesssim \hbar$, the "essential rank" of $\Pi_{\boldsymbol{\alpha}}$ is of order $\mathcal{O}(1)$: $\Pi_{\boldsymbol{\alpha}}$ resembles a projector on a subspace spanned by *finitely many* normalized "stable states" $s^i_{\boldsymbol{\alpha}}$ localized in $E_{\boldsymbol{\alpha}}$,

$$\Pi_{\boldsymbol{\alpha}} \approx \sum_i s^i_{\boldsymbol{\alpha}} \otimes s^{i*}_{\boldsymbol{\alpha}}.$$

Similarly, $\Pi_{\boldsymbol{\beta}\cdot}$ effectively projects on $\mathcal{O}(1)$ normalized "unstable states" $u^j_{\boldsymbol{\beta}}$ localized in $E_{\boldsymbol{\beta}\cdot}$. The stable and unstable directions are symplectically conjugate to each other, so that stable and unstable states behave like position vs momentum states in the phase space $\mathbb{R}^2$. The product operator

$$\Pi_{\boldsymbol{\beta}\cdot\boldsymbol{\alpha}} = \Pi_{\boldsymbol{\alpha}}\Pi_{\boldsymbol{\beta}\cdot} \approx \sum_{i,j} \langle s^i_{\boldsymbol{\alpha}}, u^j_{\boldsymbol{\beta}}\rangle s^i_{\boldsymbol{\alpha}} \otimes u^j_{\boldsymbol{\beta}}$$

involves the overlaps between the two families of states, which are all of order $\hbar^{1/2}$. It is thus natural to expect $\|\Pi_{\boldsymbol{\beta}\cdot\boldsymbol{\alpha}}\| \sim \hbar^{1/2}$, which is the order of the estimate (3.22).

3.6. Quantum entropy and pressure.

3.6.1. *Back to the symbolic measure $\mu_\hbar$.* We now turn back to the the symbolic measure $\mu_\hbar$ defined in Section 3.3. We recall that for a fixed sequence $\boldsymbol{\alpha}$, $\mu_\hbar([\cdot\boldsymbol{\alpha}])$ approximately measures the weight of the state $\psi_\hbar$ inside the rectangle $E_{\boldsymbol{\alpha}}$. This interpretation is actually possible as long as $\Pi_{\boldsymbol{\alpha}}$ can be interpreted as a quasiprojector on this rectangle, that is for $n \leq 2T_{\epsilon,\hbar}$. Under this condition, we have seen that the only upper bounds at our disposal are trivial:

$$\mu_\hbar([\cdot\boldsymbol{\alpha}]) \leq 1, \qquad n = |\boldsymbol{\alpha}| \leq 2T_{\epsilon,\hbar}.$$

On the other hand, Proposition 3.6 implies that the weights of longer cylinders satisfy nontrivial bounds:

$$\mu_\hbar([\cdot\boldsymbol{\alpha}]) \leq C\hbar^{-(d-1+c\delta)} J_n^u(\boldsymbol{\alpha})^{-1}, \qquad |\boldsymbol{\alpha}| = n \leq C_\delta \log \hbar^{-1}. \tag{3.23}$$

Corollary 3.8. *For times $n > T_1$ (see (3.21)), the measure $\mu_\hbar$ is necessarily distributed over many cylinders of length n. This corresponds to a dispersion phenomenon: the state $\psi_\hbar$ cannot be concentrated in $\mathcal{O}(1)$ boxes $E_{\boldsymbol{\alpha}}$, since each such box has a volume $\ll \hbar^{d-1}$.*

Following Section 3.1, the distribution of the weights $\{\mu_\hbar([\cdot\boldsymbol{\alpha}]), |\boldsymbol{\alpha}| = n\}$ can be characterized by an entropy. Since $\mu_\hbar$ was built from the quantum state $\psi_\hbar$, it is natural to call this entropy a **quantum entropy**:

$$H_0^{n-1}(\psi_\hbar, \mathcal{P}_{\mathrm{sm,q}}) \overset{\mathrm{def}}{=} H_0^{n-1}(\mu_\hbar) = \sum_{|\boldsymbol{\alpha}|=n} \eta\big(\mu_\hbar([\cdot\boldsymbol{\alpha}])\big). \tag{3.24}$$

One can associate a **quantum pressure** to the state $\psi_\hbar$, the quantum partition $\mathcal{P}_{\mathrm{sm,q}}$ and a set of weights $\{w_k, k = 1, \ldots, K\}$. Below we will be dealing with weights of the form $w_{\boldsymbol{\alpha}} = J_n^u(\boldsymbol{\alpha})^{1/2}$. The quantum pressures will also be denoted by $p_0^{n-1}(\psi_\hbar, \mathcal{P}_{\mathrm{sm,q}}, w) = p_0^{n-1}(\mu_\hbar, w)$.

Upper bounds (3.23) on the weights of "long" cylinders have direct consequences on the values of the quantum entropies:

$$H_0^{n-1}(\psi_\hbar, \mathcal{P}_{\mathrm{sm,q}}) \geq n\Lambda_{\min}^u - (d-1+c\delta)\log\hbar^{-1} - \log C, \qquad n \leq C_\delta \log\hbar^{-1}.$$

The right-hand side is positive (and thus makes up a nontrivial lower bound) only for $n > T_1$, that is for "long" times. A similar lower bound on the entropy of "long times" was used in [1] to deduce nontrivial information on the values of the entropies at "short" times, and finally a lower bound on the KS entropy.

3.7. Entropic uncertainty principles. In [3,4] we used a different strategy, which we describe below. Instead of using the upper bounds (3.23) at "long" times $C_\delta \log\hbar^{-1}$, we rather use the bound (3.22) corresponding to "moderately long" times $4T_{\epsilon,\hbar}$. The strategy consists in interpreting the operator on the left-hand side as a "block matrix element" associated with two different quantum partitions. Through a certain *Entropic Uncertainty Principle* (EUP), the bound (3.22) then leads to a lower bound on the pressures $p_0^{n-1}(\mu_\hbar, w)$ at "moderately long times" $n \approx 2T_{\epsilon,\hbar}$ (that is, right below the Ehrenfest time). Using an approximate subadditivity of those pressures, we then get a lower bound for finite time pressures.

The central piece of this method resides in a certain entropic uncertainty principle. Before giving the precise version used for our aims, we first give the simplest example of such a "principle," first proven by Maassen and Uffink [22].

Proposition 3.9 (EUP, level 1 (finite-dimensional projectors)). *Consider two orthonormal bases in the Hilbert space $\mathbb{C}^N$, $e = \{e_i\}_{i=1,\ldots,N}$ and $f = \{f_j\}_{j=1,\ldots,N}$. For any $\psi \in \mathbb{C}^N$ of unit norm, consider the two probability distributions $\{|\langle e_i \mid \psi\rangle|^2, i = 1, \ldots, N\}, \{|\langle f_j \mid \psi\rangle|^2, j = 1, \ldots, N\}$. Then the entropies* *associated with these two distributions satisfy the inequality*

$$H(\psi, e) + H(\psi, f) \geq -2 \log\Big(\max_{i,j} |\langle e_i, f_j \rangle|\Big).$$

For instance, take $e = \{e_i\}$ the standard basis on $\mathbb{C}^N$, and for $f = \{f_j\}$ the "discrete momentum states," related with $\{e_i\}$ through the discrete Fourier transform. All matrix elements satisfy $|\langle e_i, f_j \rangle| = N^{-1/2}$, so the inequality reads

$$H(\psi, e) + H(\psi, f) \geq \log N.$$

The inequality shows that the distributions of "position amplitudes"$\langle e_i, \psi \rangle$ on the one hand, of "momentum amplitudes"$\langle f_j, \psi \rangle$, cannot be both arbitrarily localized. It is hence a form of "uncertainty principle."

If we call ρ_i (resp. τ_j) the orthogonal projector on the state e_i (resp. f_j), then each overlap $|\langle e_i, f_j \rangle|$ can be interpreted as the operator norm $|\langle e_i, f_j \rangle| = \|\tau_j \rho_i^*\|$. This interpretation allows to generalize this "uncertainty principle" as follows:

Proposition 3.10 (EUP, level 2 (quantum partition)). *On a Hilbert space $\mathcal{H}$, consider two quantum partitions of unity, that is two finite sets of bounded operators $\rho = \{\rho_i, i \in I\}, \pi = \{\pi_j, j \in J\}$ satisfying the identities*

$$\sum_{i \in I} \rho_i^* \rho_i = \mathrm{Id}, \qquad \sum_{j \in J} \tau_j^* \tau_j = \mathrm{Id}.$$

To any normalized state $\psi \in \mathcal{H}$ we associate the probability distributions $\{\|\rho_i \psi\|^2, i \in I\}, \{\|\tau_j \psi\|^2, j \in J\}$. Then, the entropies associated with these two distributions satisfy

$$H(\psi, \rho) + H(\psi, \tau) \geq -2 \log\Big(\max_{i \in I, j \in J} \|\tau_j \rho_i^*\|\Big).$$

Because the instability of the flow may not be uniform, the coarse-grained Jacobians $J_n^u(\boldsymbol{\alpha})$ may vary a lot among all n-sequences $\boldsymbol{\alpha}$. For this reason, the estimates (3.22) may also strongly depend on the sequences $\boldsymbol{\alpha}, \boldsymbol{\beta}$. To counterbalance these variations, it is convenient to use *pressures* instead of entropies (see Section 3.1.2).

Proposition 3.11 (EUP, level 2 (quantum weighted partition)). *On a Hilbert space $\mathcal{H}$, consider two quantum partitions of unity $\rho = \{\rho_i, i \in I\}, \tau = \{\tau_j, j \in J\}$ as in Proposition 3.10 and two families of weights $v = \{v_i > 0, i \in I\}$, $w = \{w_j > 0, j \in J\}$. To any normalized state $\psi \in \mathcal{H}$ correspond the probability distributions $\{\|\rho_i \psi\|^2, i \in I\}, \{\|\tau_j \psi\|^2, j \in J\}$.*

Then, the pressures associated with these distributions and weights satisfy

$$p(\psi, \rho, v) + p(\psi, \tau, w) \geq -2 \log\Big(\max_{i \in I, j \in J} v_i w_j \|\tau_j \rho_i^*\|\Big).$$

This version is almost sufficient for our aims. Yet, the quantum partitions of unity we are using are localized near the energy shell $\mathcal{E}$ (see (3.8), (3.10)), and the estimate (3.20) starts with a sharp energy cutoff. For these reasons, the version we will need is of the following form.

Proposition 3.12 (EUP, level 3 (microlocal weighted partition)). *On a Hilbert space $\mathcal{H}$, consider two approximate quantum partitions of unity, that is two finite sets of bounded operators $\rho = \{\rho_i, i \in I\}, \tau = \{\tau_j, j \in J\}$ satisfying the identities*

$$\sum_{i \in I} \rho_i^* \rho_i = S_\rho, \qquad \sum_{j \in J} \tau_j^* \tau_j = S_\tau,$$

and two families of weights $v = \{v_i, i \in I\}, w = \{w_j, j \in J\}$ *satisfying* $V^{-1} \leq v_i, w_j \leq V$ *for some* $V \geq 1$.

We assume that for some $0 < \varepsilon \leq \min(|I|^{-2}V^{-2}, |J|^{-2}V^{-2})$ *the above sum operators satisfy* $0 \leq S_{\rho/\tau} \leq 1+\varepsilon$. *Besides, let* S_{c_1}, S_{c_2} *be two Hermitian operators on* $\mathcal{H}$ *satisfying* $0 \leq S_{c_*} \leq 1+\varepsilon$, *and related with the above partitions as follows:*

$$\|(S_{c_2} - 1)\rho_i S_{c_1}\| \leq \varepsilon, \qquad \forall i \in I, \tag{3.25}$$

$$\|(S_{\rho/\tau} - 1)S_{c_1}\| \leq \varepsilon. \tag{3.26}$$

Let us define the "cone norm"

$$c_{\mathrm{cone}} \stackrel{\mathrm{def}}{=} \max_{i\in I, j\in J} v_i w_j \|\tau_j \rho_i^* S_{c_2}\|. \tag{3.27}$$

Then, for any $\psi \in \mathcal{H}$ *satisfying*

$$\|\psi\| = 1, \qquad \|(\mathrm{Id} - S_{c_1})\psi\| \leq \varepsilon, \tag{3.28}$$

the pressures of ψ *with respect to the weighted partitions* (ρ, v) *and* (τ, w) *satisfy the bound*

$$p(\psi, \rho, v) + p(\psi, \tau, w) \geq -2\log(c_{\mathrm{cone}} + 3|I|V^2\varepsilon) + \mathcal{O}(\epsilon^{1/5}).$$

The implied constant is independent of the weighted partitions or the cutoff operators S_{c_*}.

PROOF. The proof is a slight adaptation of the one given in [3, Section 6] (in the case $\mathcal{U} = \mathrm{Id}$). One considers a bounded operator $T\colon \mathcal{H}^{|I|} \to \mathcal{H}^{|J|}$, and studies the norm of T as an operator $l_p^{(v)}(\mathcal{H}^{|I|}) \mapsto l_q^{(w)}(\mathcal{H}^{|J|})$, with the weighted norms $\|\Psi\|_p^{(v)} = (\sum_i v_i^{p-2}\|\Psi_i\|^p)^{1/p}$ and similarly for $\|T\Psi\|_q^{(w)}$. Notice that the weights are irrelevant when $p = q = 2$. An auxiliary bounded operator $O\colon \mathcal{H} \to \mathcal{H}$ is used to define a cone of states:

$$\mathcal{C}(O, \vartheta) = \{\Psi \in \mathcal{H}^{|I|}, \|O\Psi_i - \Psi_i\| \leq \vartheta\|\Psi\|_2, i \in I\}.$$

The proof of [3, Theorem 6.3], which uses a Riesz – Thorin interpolation argument, shows that for any Ψ in the cone $\mathcal{C}(O, \vartheta)$, one has:

$$\forall t \in [0,1], \qquad \|T\Psi\|_{2/(1-t)}^{(w)} \leq (c_O(T) + |I|V^2\vartheta\|T\|_{2,2})^t \|T\|_{2,2}^{1-t} \|\Psi\|_{2/(1+t)}^{(v)}, \tag{3.29}$$
$$\text{where } c_O(T) = \max_{i,j} v_i w_j \|T_{ji} O\|.$$

We now apply this result to the specific choice

$$\Psi_i \stackrel{\mathrm{def}}{=} \rho_i \psi, \qquad T_{ji} \stackrel{\mathrm{def}}{=} \tau_j \rho_i^*, \qquad O \stackrel{\mathrm{def}}{=} S_{c_2},$$

where the state ψ satisfies (3.28), that is, the cone $\mathcal{C}(S_{c_1}, \varepsilon)$ is not empty (in the opposite case, the statement of the theorem is empty). The relations (3.26) then imply that

$$\|(S_{\rho/\tau} - 1)\psi\| \leq 3\varepsilon. \tag{3.30}$$

As a consequence, the state components $(T\Psi)_i = \tau_i S_\rho \psi = \tau_i \psi + \mathcal{O}(\varepsilon)$.

The same duality argument as in [3, Lemma 6.5] shows that the $l^2 \to l^2$ norm of the operator T takes the value $\|T\|_{2,2} = \big\|\sqrt{S_\rho}\sqrt{S_\tau}\big\|$. Using the spectral theorem, one easily deduces that $\|\sqrt{S_{\rho/\tau}}\psi - \psi\| \leq \sqrt{3\varepsilon}$, so that $\|\sqrt{S_\rho}\sqrt{S_\tau}\psi - \psi\| \leq 4\sqrt{\varepsilon}$, and hence $\|T\|_{2,2} \in [1 - 4\sqrt{\varepsilon}, 1 + 2\varepsilon]$.

The l^2 norm of Ψ is $\|\Psi\|_2^2 = \langle\psi, S_\rho\psi\rangle \in [1-3\varepsilon, 1+\varepsilon]$. From (3.25) and the fact that ψ is in the cone $\mathcal{C}(S_{c_1}, \varepsilon)$, we easily get $\|(O-1)\rho_i\psi\| \leq 2\varepsilon$, so that $\Psi \in \mathcal{C}(O, \vartheta)$ for $O = S_{c_2}$, $\vartheta = 2\varepsilon/(1-3\varepsilon)$. We are now in a position to apply (3.29) to the above data. The constant $c_O(T)$ is equal to the $c_{\rm cone}$ in the statement of the proposition, so we get

$$\forall t \in [0,1], \qquad \|T\Psi\|^{(w)}_{2/(1-t)} \leq (c_{\rm cone} + 3|I|V^2\varepsilon)^t \|T\|_{2,2}^{1-t}\, \|\Psi\|^{(v)}_{2/(1+t)}.$$

Let us now expand this expression when $t \to 0$. We first split the sum $\sum_i \|\Psi_i\|^{2/(1+t)}$ between the terms $\|\Psi_i\| \geq \epsilon$ and the remaining ones:

$$\begin{aligned}\sum_i v_i^{2t/(1+t)} \|\Psi_i\|^{2/(1+t)} &= \sum_{i,>} \|\Psi_i\|^2 - t\sum_{i,>} \|\Psi_i\|^2 (\log v_i^2 \|\Psi_i\|^2) + \mathcal{O}\big((t\log\varepsilon)^2\big) \\ &\qquad\qquad + \sum_{i,<} v_i^{-2t/(1+t)} \|\Psi_i\|^{2/(1+t)} \\ &= \|\Psi\|_2^2 + tp(\Psi, v) + \mathcal{O}\big((t\log\varepsilon)^2\big) + \mathcal{O}(|I|\varepsilon).\end{aligned}$$

We now take the logarithm of this expression, and use $\|\Psi\|_2^2 = 1 + \mathcal{O}(\varepsilon)$ to get

$$\log\|\Psi\|^{(v)}_{2/(1+t)} = \frac{t}{2} p(\Psi, v) + \mathcal{O}\big((t\log\varepsilon)^2 + |I|\varepsilon + t^2 p(\Psi, v)^2\big).$$

We can perform the same manipulations on the left-hand side of (3.29), noticing that $\|T\Psi\|_2^2 = \langle S_\tau S_\pi \psi, S_\pi\psi\rangle \in [1-10\varepsilon, 1+10\varepsilon]$:

$$\log\|T\Psi\|^{(w)}_{2/(1-t)} = -\frac{t}{2} p(T\Psi, w) + \mathcal{O}\big((t\log\varepsilon)^2 + |J|\varepsilon + t^2 p(T\Psi, w)^2\big).$$

We notice that both pressures satisfy simple bound $|p(\bullet,\bullet)| \leq \log(V^2|I|) \leq 3|\log\varepsilon|$, and (from (3.30)) the estimate $(T\Psi)_i = \tau_i\psi_\hbar + \mathcal{O}(\varepsilon)$. Inserting these estimates in the logarithm of (3.29) and using $\|T\|_{2,2} \in [1-4\sqrt{\varepsilon}, 1]$, we get

$$tp(\psi_\hbar, \tau, w) + tp(\psi_\hbar, \rho, v) \geq -2t\log(c_{\rm cone} + 3|I|V^2\varepsilon) + \mathcal{O}(\sqrt{\varepsilon} + (t\log\varepsilon)^2 + (|J|+|I|)\varepsilon).$$

We now need to make some assumptions on the values of t to make the remainder small. If we take $t = \varepsilon^{1/4}$, a simple power counting shows that

$$p(\psi_\hbar, \tau, w) + p(\psi_\hbar, \rho, v) \geq -2\log(c_{\rm cone} + 3|I|V^2\varepsilon) + \mathcal{O}\big(\varepsilon^{1/4}(\log\varepsilon)^2\big). \qquad \square$$

3.8. Our application of the entropy uncertainty principle. We now apply the EUP to our semiclassical framework. Our choice of quantum partitions is determined by the requirement that the operators $\tau_j \rho_i^* S_{c_2}$ are of the form $\Pi_{\boldsymbol{\beta}\cdot\boldsymbol{\alpha}} \chi^{(n)}\big(P(\hbar)\big)$, with $\boldsymbol{\alpha}, \boldsymbol{\beta}$ two sequences of length n close to the Ehrenfest time. We will thus take

$$\begin{aligned}\text{the time} \qquad n &= \lfloor 2T_{\epsilon,\hbar} \rfloor, \\ \tau &= \{\Pi_{\boldsymbol{\alpha}} = \Pi_{\alpha_{n-1}}(n-1)\cdots\Pi_{\alpha_0}, |\boldsymbol{\alpha}| = n\}, \\ \rho &= \{\Pi^*_{\boldsymbol{\beta}\cdot} = \Pi_{\beta_{-n}}(-n)\cdots\Pi_{\beta_{-1}}(-1), |\boldsymbol{\beta}| = n\}, \\ v &= w = \{J_n^u(\boldsymbol{\beta})^{1/2}, |\boldsymbol{\beta}| = n\}, \\ S_{c_1} &= \chi^{(0)}\big(P(\hbar)\big), \qquad S_{c_2} = \chi^{(n)}\big(P(\hbar)\big).\end{aligned} \tag{3.31}$$

The weights v, w have been selected in order to balance the variations of the upper bounds in (3.22). Both the cardinals $|I| = |J| = K^n$ and the upper bound $v_i, w_j \leq e^{n\lambda_{\max}(d-1)/2}$ are $\mathcal{O}(\hbar^{-M})$ for some $M > 0$. Hence, we may take the small parameter $\varepsilon = \hbar^L$ for some fixed exponent $L \gg M$.

With this choice, the assumption (3.26) is satisfied for $\hbar$ small enough according to Proposition 3.1. The assumption (3.25) can be checked by inserting the increasing sequence of cutoffs $\{\chi^{(j)}(P(\hbar)), 1 \leq j \leq n-1\}$ along the sequence $\Pi^*_{\boldsymbol{\beta}}$, similarly as in the proof of Proposition 3.1. The assumption (3.28) holds if one takes $\psi = \psi_\hbar$ a null eigenstate of $P(\hbar)$, see (2.12).

The coefficient $c_{\rm cone}$ is then estimated by the hyperbolic dispersive estimate of Corollary 3.7:

$$c_{\rm cone} \leq C_{\rm cone}(\hbar) \stackrel{\rm def}{=} C\,\hbar^{-(d-1+c\delta)/2}, \qquad \hbar \leq \hbar_\delta. \tag{3.32}$$

The application of Proposition 3.12 to these data gives the following result.

Proposition 3.13 (Applied entropic uncertainty principle). *Take the weighted quantum partitions* $(\rho, v), (\tau, w)$ *defined in* (3.31), *and* L *a large positive number. Then, there exists* $\hbar_L > 0$, $C > 0$ *such that the pressures of the eigenstates* $(\psi_\hbar)$ *associated with these weighted partitions satisfy the inequality*:

$$p(\psi_\hbar, \rho, v) + p(\psi_\hbar, \tau, w) \geq -2\log C_{\rm cone}(\hbar) + C\hbar^{L/3}, \qquad \hbar < \hbar_L. \tag{3.33}$$

Our next task will be to relate the pressures associated to the "long time" partitions ($n = \lfloor 2T_{\epsilon,\hbar}\rfloor$), to pressures associated with "finite time" partitions (n_0 independent of $\hbar$).

3.9. Approximate subadditivity of the quantum pressure. In (3.33) appear two pressures, associated with two types of refined quantum partitions. The partition $\tau = \{\Pi_{\alpha_{n-1}}(n-1)\cdots\Pi_{\alpha_0}, |\boldsymbol{\alpha}| = n\}$ corresponds to the definition of the symbolic measure $\mu_\hbar$ in (3.14), so that the pressure $p(\psi_\hbar, \tau, w)$ can be expressed as the refined pressure $p_0^{n-1}(\mu_\hbar, w)$. On the other hand, the probability weights involved in $p(\psi_\hbar, \rho, v)$ can be rewritten (after relabelling the sequences and using the fact that $\psi_\hbar$ is an eigenmode) $\rho_{\boldsymbol{\beta}} = \|\Pi_{\beta_{n-1}}(-n)\cdots\Pi_{\beta_0}(-1)\psi_\hbar\|^2$, so they correspond to a backwards evolution.[6] We express these weights in terms of a "backwards symbolic measure" $\tilde\mu_\hbar$ similar with $\mu_\hbar$:

$$\tilde\mu_\hbar([\cdot\beta_0\cdots\beta_{n-1}]) \stackrel{\rm def}{=} \|\Pi_{\beta_{n-1}}(-n)\cdots\Pi_{\beta_0}(-1)\psi_\hbar\|^2, \tag{3.34}$$

so that $p(\psi_\hbar, \rho, v) = p_0^{n-1}(\tilde\mu_\hbar, v)$. For fixed $n > 0$, and any $|\boldsymbol{\beta}| = n$, we have $\tilde\mu_\hbar([\cdot\boldsymbol{\beta}]) \xrightarrow{\hbar\to 0} \mu_{sc}(\pi_{\beta_{n-1}\cdots\beta_0})$.

The bound (3.33) concerns the quantum pressures of $\mu_\hbar$ and $\tilde\mu_\hbar$ at the time $n = \lfloor 2T_{\epsilon,\hbar}\rfloor$ close to the Ehrenfest time. These pressures cannot be directly connected with the pressures of the semiclassical measure μ_{sc}. For this aim, we need to deduce from (3.33) a lower bound for the pressures of $\mu_\hbar$ and $\tilde\mu_\hbar$ at *finite* times n_0. This connection will be done by using an approximate version of the subadditivity property (3.5). We present the computations in the case of $\mu_\hbar$, the case of $\tilde\mu_\hbar$ being identical up to exchanging forward and backward evolution.

The symbolic measure $\mu_\hbar$ was defined on cylinders $[\cdot\boldsymbol{\alpha}]$, up to $\mathcal{O}(\hbar^\infty)$ errors which are uniformly controlled as long as $|\boldsymbol{\alpha}| \leq C|\log\hbar|$. This definition was possible thanks to the (approximate) compatibility condition (3.13) . The property (3.3) can be extended to the measure $\mu_\hbar$:

$$\forall n, n_0 \leq C|\log\hbar|, \qquad H_0^{n+n_0-1}(\mu_\hbar) \leq H_0^{n-1}(\mu_\hbar) + H_n^{n+n_0-1}(\mu_\hbar) + \mathcal{O}(\hbar^\infty). \tag{3.35}$$

[6] Notice that, using the fact that $\psi_\hbar$ is an eigenmode of U, these weights can also be written as $\|\Pi_{\beta_{n-1}}(0)\cdots\Pi_{\beta_0}(n-1)\psi_\hbar\|^2$, which is of the same form as $\|\tau_{\beta_{n-1}\cdots\beta_0}\psi_\hbar\|^2$, except for the ordering of the operators.

The second term on the right-hand side can be written as $H_n^{n+n_0-1}(\mu_\hbar) - H_0^{n_0-1}(\sigma_*^{-n}\mu_\hbar)$, where σ refers to a shift of indices. Shift-invariance would mean that $\sigma^{-n}\mu_\hbar = \mu_\hbar$, which would allow us to replace this entropy with $H_0^{n_0-1}(\mu_\hbar)$. How far are we from this invariance? To answer this question we need to compare the weights

$$\mu_\hbar(\sigma^{-n}[\cdot\beta_0\cdots\beta_{n_0-1}]) \stackrel{\text{def}}{=} \sum_{\alpha_0,\dots,\alpha_{n-1}} \mu_\hbar([\cdot\alpha_0\cdots\alpha_{n-1}\beta_0\cdots\beta_{n_0-1}]), \tag{3.36}$$

with the weights $\mu_\hbar([\cdot\beta_0\cdots\beta_{n_0-1}])$. This is achieved in the following

Lemma 3.14. *Let $(\mu_\hbar)_{\hbar\to 0}$, $(\tilde\mu_\hbar)_{\hbar\to 0}$ be the symbolic measures associated with the eigenstates $(\psi_\hbar)_{\hbar\to 0}$. Fix some $n_0\geq 0$, and take n in the range $[0, 2T_{\epsilon,\hbar}-n_0]$. Then, for any $\boldsymbol\beta$ of length n_0, we have*

$$\mu_\hbar(\sigma^{-n}[\cdot\boldsymbol\beta]) = \mu_\hbar([\cdot\boldsymbol\beta]) + \mathcal{O}(\hbar^{\epsilon/2}), \qquad \tilde\mu_\hbar(\sigma^{-n}[\cdot\boldsymbol\beta]) = \tilde\mu_\hbar([\cdot\boldsymbol\beta]) + \mathcal{O}(\hbar^{\epsilon/2}).$$

This lemma means that the measures $\mu_\hbar$, $\tilde\mu_\hbar$ are approximately shift-invariant in the semiclassical limit.

PROOF. We only give the proof for the measure $\mu_\hbar$. Each term on the right-hand side of (3.36) reads

$$\langle \Pi_{\boldsymbol\beta}(n)\Pi_{\boldsymbol\alpha}\psi_\hbar, \Pi_{\boldsymbol\beta}(n)\Pi_{\boldsymbol\alpha}\psi_\hbar\rangle = \langle \Pi_{\boldsymbol\alpha}^*|\Pi_{\boldsymbol\beta}(n)|^2\Pi_{\boldsymbol\alpha}\psi_\hbar, \psi_\hbar\rangle.$$

We first present a short (but false) proof. In order to use (3.10), we try to bring together the operator $\Pi_{\boldsymbol\alpha}$ and its Hermitian conjugate, such as to let appear the sum $\sum_{|\boldsymbol\alpha|=n}\Pi_{\boldsymbol\alpha}^*\Pi_{\boldsymbol\alpha}$. An error appears while commuting $\Pi_{\boldsymbol\alpha}$ with $|\Pi_{\boldsymbol\beta}(n)|^2$. Still, from Proposition 3.3 we know that for any $|\boldsymbol\alpha| = n$, $|\boldsymbol\beta| = n_0$ with $n+n_0\leq 2T_{\epsilon,\hbar}$, the operators $\Pi_{\boldsymbol\alpha}(-n/2)$ and $|\Pi_{\boldsymbol\beta}(n/2)|^2$ belong to the class $\Psi_\nu^{-\infty,0}$, with $\nu\in[(1-\epsilon)/2, \frac12)$. As a consequence, their commutator satisfies $[|\Pi_{\boldsymbol\beta}(n/2)|^2, \Pi_{\boldsymbol\alpha}(-n/2)]\in\Psi_\nu^{-\infty,-1+2\nu}$, an operator of norm $\mathcal{O}(\hbar^{1-2\nu})$. By unitarity of the evolution, the commutator $[|\Pi_{\boldsymbol\beta}(n)|^2, \Pi_{\boldsymbol\alpha}]$ has the same norm. We thus get

$$\sum_{\boldsymbol\alpha}\mu_\hbar([\cdot\boldsymbol\alpha\boldsymbol\beta]) = \langle|\Pi_{\boldsymbol\beta}(n)|^2 S_n\psi_\hbar, \psi_\hbar\rangle + \sum_{\boldsymbol\alpha}\mathcal{O}(\hbar^{1-2\nu}) = \mu_\hbar([\cdot\boldsymbol\beta]) + \mathcal{O}(K^n\hbar^{1-2\nu}).$$

The remainder in the right-hand side is small if n is uniformly bounded, but it becomes very large if $n\approx 2T_{\epsilon,\hbar}$!

To remedy this problem one actually needs to *successively* group together the pairs of operators $\Pi_{\alpha_j}(j)$ and perform the sum over α_j. One starts by grouping together the $\Pi_{\alpha_{n-1}}$:

$$\begin{aligned}
&\langle \Pi_{\boldsymbol\beta}(n)\Pi_{\boldsymbol\alpha}\psi_\hbar, \Pi_{\boldsymbol\beta}(n)\Pi_{\boldsymbol\alpha}\psi_\hbar\rangle \\
&\qquad = \langle \Pi_{\alpha_{n-1}}^*(n-1)|\Pi_{\boldsymbol\beta}(n)|^2\Pi_{\alpha_{n-1}}(n-1)\Pi_{\alpha_0\cdots\alpha_{n-2}}\psi_\hbar, \Pi_{\alpha_0\cdots\alpha_{n-2}}\psi_\hbar\rangle \\
&\qquad = \langle |\Pi_{\alpha_{n-1}}(n-1)|^2|\Pi_{\boldsymbol\beta}(n)|^2\Pi_{\alpha_0\cdots\alpha_{n-2}}\psi_\hbar, \Pi_{\alpha_0\cdots\alpha_{n-2}}\psi_\hbar\rangle \\
&\qquad\quad + \langle \Pi_{\alpha_{n-1}}^*(n-1)[|\Pi_{\boldsymbol\beta}(n)|^2, \Pi_{\alpha_{n-1}}(n-1)]\Pi_{\alpha_0\cdots\alpha_{n-2}}\psi_\hbar, \Pi_{\alpha_0\cdots\alpha_{n-2}}\psi_\hbar\rangle.
\end{aligned}$$

The commutator in the right-hand side is an operator of norm $\mathcal{O}(\hbar^{1-2\nu})$ for the reasons indicated above. The second overlap is then bounded from above by $\|\Pi_{\alpha_0\cdots\alpha_{n-2}}\psi_\hbar\|^2\mathcal{O}(\hbar^{1-2\nu})$, where the implied constant does not depend on $\boldsymbol\alpha$. Using the quantum partition of order $n-1$, we can sum this second term over $\boldsymbol\alpha$, to obtain an error $\mathcal{O}(\hbar^{1-2\nu})$. The first term on the right-hand side can be summed over α_{n-1}, to produce

$$\begin{aligned}\langle \mathrm{Op}_\hbar(\widetilde{\chi}_{\epsilon/2})(n-1)|\Pi_{\boldsymbol\beta}(n)|^2\Pi_{\alpha_0\cdots\alpha_{n-2}}\psi_\hbar, \Pi_{\alpha_0\cdots\alpha_{n-2}}\psi_\hbar\rangle \\ = \langle |\Pi_{\boldsymbol\beta}(n)|^2\Pi_{\alpha_0\cdots\alpha_{n-2}}\psi_\hbar, \Pi_{\alpha_0\cdots\alpha_{n-2}}\psi_\hbar\rangle + \mathcal{O}(\hbar^\infty),\end{aligned}$$

where we used the fact that $\widetilde{\chi}_{\epsilon/2} \equiv 1$ on the microsupport of $\Pi_{\alpha_0\cdots\alpha_{n-2}}\psi_\hbar$ (see the proof of Lemma 3.1). Apart from the errors, we are now left with the sum

$$\sum_{\alpha_0,\ldots,\alpha_{n-2}} \langle |\Pi_{\boldsymbol\beta}(n)|^2\Pi_{\alpha_0\cdots\alpha_{n-2}}\psi_\hbar, \Pi_{\alpha_0\cdots\alpha_{n-2}}\psi_\hbar\rangle.$$

This sum can be treated as above, by bringing $\Pi_{\alpha_{n-2}}(n-2)$ to the left, commuting it with $|\Pi_{\boldsymbol\beta}(n)|^2$, and summing over α_{n-2}. It produces another error $\mathcal{O}(\hbar^{1-2\nu})$. Iterating this procedure down to α_0, we get

$$\sum_{\boldsymbol\alpha} \mu_\hbar([\cdot\boldsymbol{\alpha\beta}]) = \mu_\hbar([\cdot\boldsymbol\beta]) + \mathcal{O}(n\hbar^{1-2\nu}).$$

By selecting ν appropriately, we get the statement of the lemma. □

Coming back to the subadditivity equation (3.35) in the case n_0 fixed, $n = \lfloor 2T_{\epsilon,\hbar}\rfloor - n_0$, we get

$$\begin{aligned}H_0^{n-1}(\mu_\hbar) &\le H_0^{n-n_0-1}(\mu_\hbar) + H_0^{n_0-1}(\sigma_*^{-n}\mu_\hbar) + \mathcal{O}(\hbar^\infty)\\ &\le H_0^{n-n_0-1}(\mu_\hbar) + H_0^{n_0-1}(\mu_\hbar) + \mathcal{O}_{n_0}(\hbar^{\epsilon/3}).\end{aligned}$$

Here we used the fact that the function $\eta(s)$ satisfies $|\eta(s+s') - \eta(s)| \le \eta(|s'|)$. It is also easy to check that the "potential part" of the pressure $p(\mu_\hbar, w)$ satisfies this inequality:

$$\begin{aligned}&\sum_{|\boldsymbol\alpha|=n}\sum_{|\boldsymbol\beta|=n_0} \mu_\hbar([\cdot\boldsymbol{\alpha\beta}])\log w_{\boldsymbol{\alpha\beta}}\\ &\qquad = \sum_{|\boldsymbol\alpha|=n} \mu_\hbar([\cdot\boldsymbol\alpha])\log w_{\boldsymbol\alpha} + \sum_{|\boldsymbol\beta|=n_0} \sigma^{-n}\mu_\hbar([\cdot\boldsymbol\beta])\log w_{\boldsymbol\beta} + \mathcal{O}(\hbar^\infty)\\ &\qquad = \sum_{|\boldsymbol\alpha|=n} \mu_\hbar([\cdot\boldsymbol\alpha])\log w_{\boldsymbol\alpha} + \sum_{|\boldsymbol\beta|=n_0} \mu_\hbar([\cdot\boldsymbol\beta])\log w_{\boldsymbol\beta} + \mathcal{O}(\hbar^{\epsilon/2}),\end{aligned}$$

so we finally get the approximate pressure subadditivity

$$p_0^{n-1}(\mu_\hbar, w) \le p_0^{n-n_0-1}(\mu_\hbar, w) + p_0^{n_0-1}(\mu_\hbar, w) + \mathcal{O}_{n_0}(\hbar^{\epsilon/3}). \tag{3.37}$$

Taking the Euclidean quotient $n = qn_0 + r$, $r < n_0$, we can iterate this process q times and obtain:

$$p_0^{n+n_0-1}(\mu_\hbar, w) \le p_0^{r-1}(\mu_\hbar, w) + q\,p_0^{n_0-1}(\mu_\hbar, w) + \mathcal{O}_{n_0}(q\hbar^{\epsilon/3}).$$

A similar approximate subadditivity holds for the pressures $p(\tilde\mu_\hbar, v)$ associated with the "backwards" symbolic measure $\tilde\mu_\hbar$. We have thus obtained the following

Proposition 3.15. *Let $\mu_\hbar^u, \tilde\mu_\hbar^u$ be the associated forward and backwards symbolic measures associated with $(\psi_\hbar)$. Fix some $n_0 > 0$, and for $\hbar < \hbar_0$ split the*

Ehrenfest time $n = \lfloor 2T_{\epsilon,\hbar} \rfloor$ *into* $n = qn_0 + r$, $r \in [0, n_0)$. *Then, the pressures associated with these measures and the weights* $w_k = v_k = J^u(k)^{1/2}$ *satisfy the following lower bound:*

$$(3.38)\quad q\big(p_0^{n_0-1}(\mu_\hbar, w) + p_0^{n_0-1}(\tilde{\mu}_\hbar, v)\big) + p_0^{r-1}(\mu_\hbar, w) + p_0^{r-1}(\tilde{\mu}_\hbar, v) \\ \geq -(d-1+c\delta)|\log\hbar| + \mathcal{O}_{n_0}(\hbar^{\epsilon/4}).$$

Hence, approximate subadditivity has enabled us to transfer the bound on pressures for $n = \lfloor 2T_{\epsilon,\hbar} \rfloor$ into a bound on pressures at finite times n_0, r.

3.10. Back to the classical pressure. Using the fact that the pressures $p_0^{r-1}(\bullet)$ are uniformly bounded, we divide (3.38) by $qn_0 = 2T_{\epsilon,\hbar} - \mathcal{O}(1) = |\log\hbar| \times \big(1 + \mathcal{O}(\epsilon)\big)/\lambda_{\max}$. For $\hbar < \hbar_\epsilon$, we get

$$\frac{p_0^{n_0-1}(\mu_\hbar, w)}{n_0} + \frac{p_0^{n_0-1}(\tilde{\mu}_\hbar, v)}{n_0} \geq \frac{-(d-1+c\delta)|\log\hbar|}{qn_0} + \mathcal{O}_{n_0}(|\log\hbar|^{-1}) \\ \geq -(d-1+c\delta)\lambda_{\max}\big(1+\mathcal{O}(\epsilon)\big).$$

The right-hand side does not depend on $\hbar$, so it is still valid once we take the semiclassical limit of the left-hand side. From the properties of $\mu_\hbar$ and $\tilde{\mu}_\hbar$, the latter is equal to the ratio $2p_0^{n_0-1}(\mu_{\rm sc}, v, \mathcal{P}_{\rm sm})/n_0$, where we recall that $\mu_{\rm sc}$ is the semiclassical measure associated with $(\psi_\hbar)$, while $\mathcal{P}_{\rm sm}$ is the smoothed partition (3.7). We thus obtained the following lower bound on the classical pressure:

$$\frac{p_0^{n_0-1}(\mu_{\rm sc}, v, \mathcal{P}_{\rm sm})}{n_0} \geq -\frac{(d-1)\lambda_{\max}}{2} + \mathcal{O}(\delta, \epsilon).$$

Importantly, the implied constants do not depend on the degree of smoothness of the partition (that is, on the derivatives of the functions π_k), so we may send $\delta \to 0$ and get rid of the smoothing (the pressure is continuous in this limit, due to our assumption $\mu(\partial\mathcal{P}) = 0$). We thus obtain the following lower bound on the pressure associated with the *sharp* partition $\mathcal{P}$:

$$\frac{p_0^{n_0-1}(\mu_{\rm sc}, v, \mathcal{P})}{n_0} \geq -\frac{(d-1)\lambda_{\max}}{2} + \mathcal{O}(\epsilon).$$

We recall that ϵ majorizes the diameter of $\mathcal{P}$. The first part of the pressure is the entropy $H(\mu_{\rm sc}, \mathcal{P}^{\vee n_0})$, while the second part is given by the sum

$$-\sum_{|\boldsymbol{\alpha}|=n_0} \mu_{\rm sc}(E_{\boldsymbol{\alpha}}) \log J^u_{n_0}(\boldsymbol{\alpha}) = -n_0 \sum_{k=1}^{K} \mu_{\rm sc}(E_k) \log J^u(k),$$

where we remind that $J^u(k) = \min_{\rho\in E_k} J^u(\rho)$. We can now let $n_0 \to \infty$, and get

$$H_{\rm KS}(\mu_{\rm sc}, \mathcal{P}) \geq -\frac{(d-1)\lambda_{\max}}{2} + \sum_{k=1}^{K} \mu_{\rm sc}(E_k) \log J^u(k) + \mathcal{O}(\epsilon).$$

By taking finer and finer partitions (that is, $\epsilon \to 0$), we finally get the bound (1.8) for the semiclassical measure $\mu_{\rm sc}$. □

4. Proof of the hyperbolic dispersive estimate

4.1. Decomposition into adapted Lagrangian states. The proof of Proposition 3.6 starts from an arbitrary $\Psi \in L^2$ with $\|\Psi\|_{L^2} = 1$. The localized state $\psi_{\alpha_0} \stackrel{\text{def}}{=} \Pi_{\alpha_0}\chi^{(n)}\big(P(\hbar)\big)\Psi$ will then be decomposed into a linear combination of "nice" Lagrangian states. To construct these "nice states," we need to consider, on each neighbourhood $\widetilde{E}_k \supset E_k$, a coordinate chart $\{(y_i, \eta_i), i = 0, \ldots, d-1\}$ adapted to the dynamics of the geodesic flow. These coordinates are required to satisfy the following properties:

(1) $\widetilde{E}_k$ is contained in the polydisk $D(\epsilon,\epsilon) = \{(y,\eta), |y| \leq \epsilon, |\eta| \leq \epsilon\}$.
(2) the coordinate $\eta_0 = p_0(x,\xi)$, so that the energy shells are given by $\{\eta_0 = \text{const}\}$, and the conjugate variable y_0 represents the time along the trajectory.
(3) the planes $\{\eta = \text{const}\}$ are close to the local weak unstable manifolds W^{u0}_ϵ in $\widetilde{E}_k$. For this aim, we let the plane $\{\eta = 0\}$ coincide with the local unstable manifold $W^{u0}_\epsilon(\rho_k)$ for some arbitrary point $\rho_k \in E_k \cap \mathcal{E}$.

Definition 4.1. We say that a Lagrangian leaf $\Lambda \subset \widetilde{E}_k$ belongs to the *γ_1-cone* if it is represented, in the chart $\{(y,\eta)\}$, as

$$\Lambda = \big\{\big(y, \mathrm{d}S(y)\big), |y| \leq \epsilon\big\}, \qquad \text{with } \sup_{|y|\leq\epsilon} \|\mathrm{d}^2 S(y)\| \leq \gamma_1. \tag{4.1}$$

Fixing some $\gamma_1 > 0$, there exists $\epsilon_{\gamma_1} > 0$ such that, provided the diameters of the $\widetilde{E}_k$ are all smaller than ϵ_{γ_1}, then the γ_1-cone contains all the local unstable manifolds $W^{u0}_\epsilon(\rho)$, $\rho \in \widetilde{E}_k$, while all local stable manifolds $W^s_\epsilon(\rho)$ are uniformly transverse to this cone. We call such a cone an *unstable γ_1-cone*.

Let $\mathcal{U}_k$ be a semiclassical Fourier Integral Operator (FIO) associated with the change of coordinates

$$(x,\xi) \in T^*X \mapsto (y,\eta) \in \mathbb{R}^{2d}, \tag{4.2}$$

unitary microlocally near $\widetilde{E}_k \times D(\epsilon,\epsilon)$. This means that for any $\psi_k \in L^2(X)$, $\|\psi\| = \mathcal{O}(1)$, microlocalized inside $\widetilde{E}_k$, we have

$$\|\psi\|_{L^2(X)} = \|\mathcal{U}_k\psi\|_{L^2(\mathbb{R}^d)} + \mathcal{O}(\hbar^\infty),$$

and the function $\mathcal{U}_k\psi_k(y)$ is microlocalized inside $D(\epsilon,\epsilon)$. Hence, using the cutoff

$$\chi \in C_0^\infty(\{|y| \leq 2\epsilon\}), \qquad \chi = 1 \text{ for } |y| \leq \epsilon,$$

we may construct a family of "localized plane waves"

$$e_\eta(y) = \chi(y)\exp(\mathrm{i}\langle\eta, y\rangle/\hbar), \qquad (y,\eta) \in D(2\epsilon, 2\epsilon),$$

such that the function $\mathcal{U}_k\psi_k(y)$ can be Fourier expanded into

$$\mathcal{U}_k\psi_k = (2\pi\hbar)^{-d/2}\int_{|\eta|\leq 2\epsilon} e_\eta\hat{\psi}_k(\eta)\,\mathrm{d}\eta + \mathcal{O}_{L^2}(\hbar^\infty),$$

where $\hat{\psi}_k = \mathcal{F}_\hbar\psi_k$ is the $\hbar$-Fourier transform of $\mathcal{U}_k\psi_k(y)$. Each state e_η, $|\eta| \leq 2\epsilon$, is a Lagrangian state associated with the "horizontal" Lagrangian leave $\Lambda_\eta = \{(y,\eta), |y| \leq 2\epsilon\}$.

The change of coordinates (4.2) brings the energy layer $\{|p - \frac{1}{2}| \leq \hbar^{1-\delta}\}$ into the slice $\{|\eta_0| \leq \hbar^{1-\delta}\}$. As a result, the states ψ_k which are sharply localized in energy are easy to characterize.

Lemma 4.2. *Assume that for some integer $m \leq C_\delta |\log \hbar|$ the state ψ_k satisfies*

$$\chi^{(m)}\big(P(\hbar)\big)\psi_k = \psi_k + \mathcal{O}_{L^2}(\hbar^\infty). \tag{4.3}$$

In that case, the state ψ_k can be decomposed into

$$\psi_k = (2\pi\hbar)^{-d/2} \int_{|\eta_0| \leq e^{m\delta}\hbar^{1-\delta}, |\eta'| \leq 2\epsilon} \mathcal{U}_k^* e_\eta \hat{\psi}_k(\eta)\, d\eta + \mathcal{O}_{L^2}(\hbar^\infty). \tag{4.4}$$

Proof. The assumption (4.3) and the microlocalization of $\mathcal{U}_k\psi_k$ inside $D(2\epsilon, 2\epsilon)$ imply that its Fourier transform $\hat{\psi}_k$ satisfies $\hat{\psi}_k(\eta) = \mathcal{O}\big((\hbar/\langle\eta\rangle)^\infty\big)$ for η outside the strip $\{|\eta_0| \leq e^{m\delta}\hbar^{1-\delta}, |\eta'| \leq 2\epsilon\}$. □

Our aim is to prove (3.20). We consider a sequence $\boldsymbol{\alpha}$ of length n, and an arbitrary $\Psi \in L^2(X)$. We then apply the above decomposition to the case $k = \alpha_0$ and the state $\psi_{\alpha_0} = \Pi_{\alpha_0}\chi^{(n)}\big(P(\hbar)\big)\Psi$. By construction, this state satisfies the energy localization (4.3) if we take $m \geq n + 1$.

4.2. Evolution of individual Lagrangian states. Our strategy will consist in controlling the states $\Pi_{\boldsymbol{\alpha}}\mathcal{U}_{\alpha_0}^* e_\eta$ individually. For each $|\eta| \leq 2\epsilon$, the state

$$\psi^0 \stackrel{\text{def}}{=} \mathcal{U}_{\alpha_0}^* e_\eta \in L^2(X) \tag{4.5}$$

is a Lagrangian state associated with a certain Lagrangian leaf Λ^0 which belongs to some unstable γ_1-cone in $\widetilde{E}_{\alpha_0}$. The operator $\Pi_{\boldsymbol{\alpha}}$ is a succession of evolutions along the Schrödinger flow (U) and truncations by quasiprojectors Π_{α_i}. Each quasiprojector $\Pi_k = \mathrm{Op}_\hbar(\widetilde{\pi}_k)$ is a pseudodifferential operator, which transforms a Lagrangian state associated with some Lagrangian leaf Λ, into another Lagrangian state on the same Λ, by modifying its symbol. This modification will generally reduce the L^2 norm of the state. In turn, the propagator U is a unitary Fourier Integral Operator associated with the map g^1, which transforms a Lagrangian state on Λ into a Lagrangian state on $g(\Lambda)$, keeping the L^2-norm unchanged.

More precisely, the operator $\Pi_{\alpha_i}U$ acts as follows on Lagrangian states.

Proposition 4.3. *Consider a Lagrangian leaf $\Lambda^0 \subset \widetilde{E}_{\alpha_0}$ in some unstable γ_1-cone, and a Lagrangian state $\psi^0 \in L^2(X)$ localized on this leaf, of the form*

$$\mathcal{U}_{\alpha_0}\psi^0(y) = a^0(y)e^{iS^0(y)/\hbar}, \qquad a^0 \in C_0^\infty\big(D(\epsilon)\big).$$

Then, the state $\psi^1 = \Pi_{\alpha_1}U\psi^0$ is a Lagrangian state associated with the leaf $\Lambda^1 = g(\Lambda^0)$. It can be expressed (in the coordinates attached to $\widetilde{E}_{\alpha_1}$) as

$$\mathcal{U}_{\alpha_1}\psi^1(y) = a^1(y, \hbar)e^{iS^1(y)/\hbar},$$

where S^1 is a generating function for Λ^1. The symbol $a^1(y, \hbar)$ admits an expansion

$$a^1(y, \hbar) = \sum_{j=0}^{L-1} \hbar^j a_j^1(y) + \hbar^L r_l(y, \hbar). \tag{4.6}$$

The inverse flow $g^{-1}_{\restriction\Lambda^1} : \Lambda^1 \to \Lambda^0$ can be expressed in the coordinates y attached respectively to $\widetilde{E}_{\alpha_1}$ and $\widetilde{E}_{\alpha_0}$, through a map

$$y^1 \in \pi\Lambda^1 \subset D(2\epsilon) \mapsto y^0 = \pi g^{-1}\big(y^1, dS^1(y^1)\big) \in D(2\epsilon). \tag{4.7}$$

Then, the principal symbol $a_0^1(x)$ *in* (4.6) *reads*

$$a_0^1(y^1) = e^{i\beta^1} a^0(y^0) \left| \det \frac{\partial y^0}{\partial y^1} \right|^{1/2} \widetilde{\pi}_{\alpha_1}(y^1, dS^1(y^1)), \tag{4.8}$$

with β^1 *a constant phase. The higher-order symbols* a_j^1 *and the remainder* r_l *satisfy the following bounds, for any* $l \in \mathbb{N}$:

$$\|a_j^1\|_{C^l} \leq C_{l,j} \|a^0\|_{C^{l+2j}}, \qquad 0 \leq j \leq L-1, \tag{4.9}$$

$$\|r_l(\cdot, \hbar)\|_{C^l} \leq C_{l,L} \|a^0\|_{C^{l+2L+d}}, \tag{4.10}$$

$$r_l = \mathcal{O}\left(\left(\frac{\hbar}{\hbar + \mathrm{dist}(\bullet, \pi\widetilde{E}_{\alpha_1})} \right)^{\infty} \right) \quad \text{outside } \pi\widetilde{E}_{\alpha_1}.$$

The constants $C_{l,j}$ *depend on the Lagrangian* Λ^0.

The proximity of Λ^0 from the unstable manifold $\Lambda_{\alpha_0} \stackrel{\text{def}}{=} W^{u0}(\rho_{\alpha_0})$ has another consequence. The map $y^1 \mapsto y^0$ in (4.7), projection of $g^{-1}_{\restriction \Lambda^1}$ to the coordinates y, is close to the projection of $g^{-1}_{\restriction g(\Lambda_{\alpha_0})}$. For this reason, the Jacobian $\det(\partial y^0/\partial y^1)$ appearing in (4.13) is close to the Jacobian of $g^{-1}_{\restriction g(\Lambda_{\alpha_0})}$. Using our coarse-grained Jacobians (3.16), we find

$$\left| \det \frac{\partial y^0}{\partial y^1} \right| = J^u(\alpha_0)^{-1}\big(1 + \mathcal{O}(\gamma_1, \epsilon^{\gamma})\big), \tag{4.11}$$

where $\gamma \in (0,1]$ depends on the Hölder regularity of the unstable foliation.

4.3. *n*-steps evolution. The above proposition describes the 1-step evolution $\Pi_{\alpha_i} U$. We need to apply many ($n \sim \log(1/\hbar)$) similar steps. To control these many steps uniformly with respect to n, we first need to analyze the evolution of the Lagrangian leaf Λ^0 through the classical evolution corresponding to the operator $\Pi_{\boldsymbol{\alpha}}$: for $i = 0, \ldots, n-1$, the leaf Λ^{i+1} is obtained by truncating Λ^i on $\operatorname{supp} \widetilde{\pi}_{\alpha_i} \subset \widetilde{E}_{\alpha_i}$, and then evolving this truncated leaf through g^1. We will assume that the sequence $\boldsymbol{\alpha}$ is *admissible*, in the sense that Λ^i is nonempty for all $i = 1, \ldots, n$. Then, the Anosov structure of the geodesic flow induces the following *inclination lemma* [18], which describes the leaves Λ^i in the adapted coordinates (y, η) on $\widetilde{E}_{\alpha_i}$:

Lemma 4.4. *Assume the Lagrangian leave* $\Lambda^0 \subset \widetilde{E}_{\alpha_0}$ *belongs to a certain unstable* γ_1*-cone, as defined in Definition* 4.1. *Then, the Lagrangians* $\Lambda^i \subset \widetilde{E}_{\alpha_i}$, $i = 1, \ldots, n$, *also belong to the corresponding unstable* γ_1*-cones*:

$$\Lambda^i \subset \big\{ (y, dS^i(y)), |y| \leq \epsilon \big\}, \qquad \sup_{|y| \leq \epsilon} \|d^2 S^i(y)\| \leq \gamma_1.$$

We also have a uniform control on the higher derivatives of the generating functions S^i. *For any* $l > 1$, *there exists* $\gamma_l > 0$ *such that, assuming* dS^0 *is in the* γ_1*-cone and satisfies* $\|dS^0\|_{C^l} \leq \gamma_l$, *then the evolved Lagrangians also satisfy*

$$\|dS^i\|_{C^l} \leq \gamma_l, \qquad i = 1, \ldots, n.$$

The above lemma shows that the evolution $\Lambda^i \mapsto \Lambda^{i+1}$ remains uniformly under control at long times. Putting together the lemma with Proposition 4.3, we get the following

Proposition 4.5. *For $\hbar < \hbar_0$ and $n < C\log\hbar^{-1}$, take any sequence $\boldsymbol{\alpha} = \alpha_0 \cdots \alpha_n$ of length $n+1$. Consider a Lagrangian leaf $\Lambda^0 \subset E_{\alpha_0} \cap \mathcal{E}_{\eta_0}$ in some unstable γ_1-cone, and an associated Lagrangian state $\psi^0 \in L^2(X)$ localized on this leaf, of the form*

$$\mathcal{U}_{\alpha_0}\psi^0(y) = a^0(y)\mathrm{e}^{\mathrm{i}S^0(y)/\hbar}, \qquad a^0 \in C_c^\infty\big(D(\epsilon)\big).$$

We are interested in the evolved state

$$\psi^n \stackrel{\text{def}}{=} \Pi_{\alpha_n} U \Pi_{\alpha_{n-1}} U \cdots \Pi_{\alpha_1} U \psi^0.$$

Then,

(i) *If the manifold Λ^n (obtained from Λ^0 from the classical evolution) is empty, then $\|\psi^n\|_{L^2} = \mathcal{O}(\hbar^\infty)$.*

(ii) *Otherwise, ψ^n is a Lagrangian state associated with Λ^n. It reads*

$$\mathcal{U}_{\alpha_n}\psi^n(y) = a^n(y,\hbar)\mathrm{e}^{\mathrm{i}S^n(y)/\hbar} + \mathcal{O}_{\mathcal{S}}(\hbar^\infty),$$

where the symbol $a^n(\bullet,\hbar)$ is supported in $D(\epsilon)$ and satisfies the bound

$$\|a^n(\bullet,\hbar)\|_{C^0(D(\epsilon))} \leq CJ(\alpha_0\cdots\alpha_{n-1})^{-1/2}\|a^0\|_{C^0}.$$

As a consequence, we obtain the L^2 estimate

$$\|\psi^n\|_{L^2(X)} \leq CJ(\alpha_0\cdots\alpha_{n-1})^{-1/2}\|a^0\|_{C^0}. \tag{4.12}$$

The constant C is uniform when the function S^0 generating Λ^0 remains in a bounded set in the C^∞ topology.

Proof. This proposition is proved in [3, Lemma 3.5], but we will rather use the notations of a similar result valid in a more general setup in [23, Proposition 4.1]. The strategy consists in a tedious but straightforward bookkeeping of the properties of the symbols

$$a^i(y,\hbar) = \sum_{j=0}^{L-1} \hbar^j a^i_j(y) + r^i_l(y,\hbar)$$

associated with the intermediate states $\mathcal{U}_{\alpha_i}\psi^i(y)$. Namely, one manages to control the C^l norms of the symbols and of the remainder, using the equations (4.13), (4.9), (4.10).

The principal symbol a_0^n is given by the explicit formula

$$a_0^n(y^n) = \mathrm{e}^{\mathrm{i}\sum_{i=1}^n \beta^n} a^0(y^0)\prod_{i=1}^n \left|\det\frac{\partial y^{i-1}}{\partial y^i}\right|^{1/2} \widetilde{\pi}_{\alpha_i}\big(y^i, \mathrm{d}S^i(y^i)\big), \tag{4.13}$$

where each $y^{i-1} = \pi g_{S^i}^{-1}\big(y^i, \mathrm{d}S^i(y^i)\big)$ is the coordinate of the $(i-1)$th iterate of the point $\big(y^0, \mathrm{d}S^0(y^0)\big) \in \Lambda^0$. This formula shows that a_0^n results from a transport of the amplitude (or half-density) a^0 through the flow, and a multiplication by successive factors $|\widetilde{\pi}_{\alpha_i}| \leq 1 + \mathcal{O}(\hbar)$. From this expression and (4.11) we directly get the C^0 bound

$$\|a_0^n\|_{C^0(D(\epsilon))} \leq CJ(\alpha_0\cdots\alpha_{n-1})^{-1/2}\|a^0\|_{C^0}. \tag{4.14}$$

The derivatives $\partial^l a_0^n/\partial(y^n)^l$ are computed by applying the Leibnitz rule to the product (4.13) (which leads to $\mathcal{O}(n^l)$ terms), and then the chain rule $\partial f(y^i)/\partial y^n =$

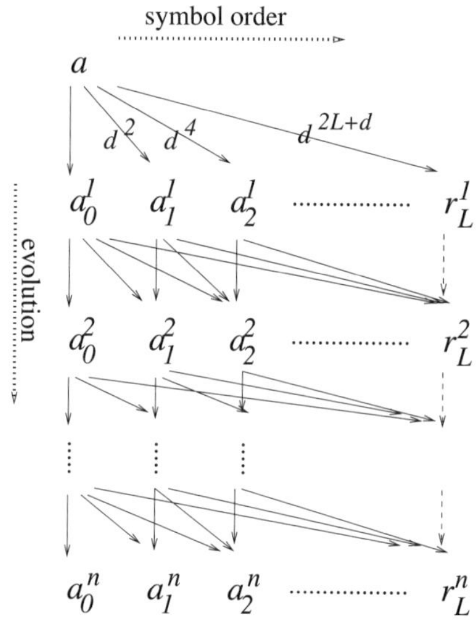

FIGURE 1. Each symbol a_j^i is linked to its direct "descendents." Vertical arrows represent operators of transport+multiplication, while oblique arrows include a certain number of differentiations, as given in (4.9), (4.10).

$(\partial f(y^i)/\partial y^i)(\partial y^i/\partial y^n)$. Using the fact that the Jacobian matrices $\partial y^i/\partial y^n$ are uniformly bounded, one obtains the bounds

$$\|a_0^n\|_{C^l} \leq C_l\, n^l\, J(\alpha_0 \cdots \alpha_{n-1})^{-1/2} \|a^0\|_{C^l}, \qquad l \geq 0.$$

The higher-order symbol a_j^i, $1 \leq j \leq L-1$, is obtained by the same transport-and-multiplication of the symbol a_j^{i-1}, but also by transporting and differentiating $2(j-j')$ times the symbols $a_{j'}^{i-1}$, $j' \leq j$. This procedure is sketched in Figure 1. On this figure, each symbol a_j^n results from the sum of $\mathcal{O}(n^j)$ paths starting from a^0, each path consisting in a succession of "long" vertical evolutions, and j "oblique" evolutions, involving altogether $2j$ differentiations performed at various stages. We have seen above that each differentiation leads, through Leibnitz's rule, to a factor $\mathcal{O}(n)$. Taking into account the number of paths, we obtain the bounds

$$\|a_j^n\|_{C^0} \leq C_j n^{3j} J(\alpha_0 \cdots \alpha_{n-1})^{-1/2} \|a^0\|_{C^{2j}}, \qquad 1 \leq j \leq L-1,$$

and l differentiations of the symbol a_j^n provide, for the same reasons as above, an additional factor n^l:

$$\|a_j^n\|_{C^l} \leq C_{j,l} n^{l+3j} J(\alpha_0 \cdots \alpha_{n-1})^{-1/2} \|a^0\|_{C^{l+2j}}, \qquad 1 \leq j \leq L-1.$$

At each stage, one also gets a remainder r_l^i, which results from the symbols a_j^{i-1} through transport, multiplication and differentiation, as well as from the *unitary* evolution of the previous remainder r_l^{i-1} (dashed vertical arrow). Taking the above bounds into account, one easily obtains the L^2 bound

$$\|r_l^n\|_{L^2} \leq C_l \|a^0\|_{C^{2L+d}}.$$

This last estimate shows that the full symbol $a^n(\bullet,\hbar)$ is dominated by the principal symbol a_0^n, so that the estimate (4.14) also applies to $\|a^n(\bullet,\hbar)\|_{C^0}$, and hence to $\|a^n(\bullet,\hbar)\|_{L^2} = \|\psi^n\|_{L^2} + \mathcal{O}(\hbar^\infty)$. $\square$

We now conclude the proof of Proposition 3.6. We use the decomposition (4.4) of the state ψ_{α_0}, and apply Proposition 4.5 to each state $\psi^0 = \mathcal{U}^*_{\alpha_0} e_\eta$, $|\eta| \leq 2\epsilon$. Notice that the manifolds Λ_η remain in a bounded cone in C^∞: a generating function for Λ_η is simply $S_\eta(y) = \langle \eta, y\rangle$. Therefore, the constant C in the estimate (4.12)) is uniform with respect to η. The triangle inequality then implies the bound

$$\|\Pi_{\alpha_n} U \Pi_{\alpha_{n-1}} \cdots \Pi_{\alpha_1} U \psi_{\alpha_0}\| \leq C\hbar^{-d/2} J(\alpha_0\cdots\alpha_{n-1})^{-1/2} \int |\hat{\psi}_{\alpha_0}(\eta)|\, \mathrm{d}\eta + \mathcal{O}_{L^2}(\hbar^\infty).$$

The right-hand side contains the L^1 bound for the Fourier transform $\hat{\psi}_{\alpha_0}(\eta)$. Since this function is $\mathcal{O}(\hbar^\infty)$ outside the set $\{|\eta_0| \leq \mathrm{e}^{m\delta}\hbar^{1-\delta}, |\eta'| \leq 2\epsilon\}$, the Cauchy–Schwarz inequality leads to

$$\|\hat{\psi}_{\alpha_0}\|_{L^1} \leq (C\epsilon \mathrm{e}^{m\delta}\hbar^{1-\delta})^{1/2}\, \|\hat{\psi}_{\alpha_0}\|_{L^2} + \mathcal{O}_{L^2}(\hbar^\infty).$$

Since $\|\hat{\psi}_{\alpha_0}\|_{L^2} = \|\psi_{\alpha_0}\|_{L^2} \leq 1 + \mathcal{O}(\hbar^\infty)$, we obtain (3.20). $\square$

5. Extension to Anosov toral diffeomorphisms

In this section we will give the few modifications necessary to prove Theorem 1.7(ii) dealing with Anosov toral diffeomorphisms. Although the strategy of proof is exactly the same as for the Laplacian eigenstates, the quantum setting is slightly different, and not so well-known as the one used above. One advantage of dealing with maps instead of flows is that we will not need any sharp energy cutoff: the torus will take the place of the energy shell $\mathcal{E}$, so there won't be any need to localize ourselves on a submanifold.

Let us mention that the bound (1.9) is not a new result in the case of *linear* hyperbolic symplectomorphisms on the 2-dimensional torus: as explained in Section 1.4, it is a consequence of Brooks' more precise result (Theorem 1.9). For linear symplectomorphisms in higher dimension, an improvement of the lower bound (1.9) has been recently obtained by G. Rivière [26].

5.1. Quantum mechanics on a torus phase space. We briefly recall the properties of quantum mechanics associated with the torus phase space. $\mathbb{T}^{2d} = \mathbb{R}^{2d}/\mathbb{Z}^{2d}$ is equipped with the symplectic form $\omega = \sum_{i=1}^d \mathrm{d}\xi_i \wedge \mathrm{d}x_i$. Quantum states are defined as the distributions $\psi \in \mathcal{S}'(\mathbb{R}^d)$ which are $\mathbb{Z}^d$-periodic, and the $\hbar$-transforms of which are also $\mathbb{Z}^d$-periodic:

(5.1)
$$\forall n \in \mathbb{Z}^d, \forall x \in \mathbb{R}^d, \quad \psi(x+n) = \psi(x), \qquad \forall \xi \in \mathbb{R}^d, \quad (\mathcal{F}_\hbar\psi)(\xi+n) = (\mathcal{F}_\hbar\psi)(\xi).$$

A simple calculation shows that such distributions can be nontrivial iff

$$\hbar = \hbar_N \stackrel{\text{def}}{=} (2\pi N)^{-1} \qquad \text{for some integer } N > 0.$$

Such values of $\hbar$ are called *admissible*. From now on we will only consider admissible values of $\hbar$. The distributions (5.1) then form a N^d-dimensional subspace of $\mathcal{S}'(\mathbb{R}^d)$, (denoted by $\mathcal{H}_N$), which is spanned by the following basis of "Dirac combs":

$$(5.2) \quad e_j(x) = N^{-d/2} \sum_{\nu\in\mathbb{Z}^d} \delta\Big(x - \frac{j}{N} - \nu\Big), \qquad j \in (\mathbb{Z}/N\mathbb{Z})^d \simeq \{0,\ldots,N-1\}^d.$$

It is natural to equip $\mathcal{H}_N$ with the Hermitian norm $\|\bullet\|_{\mathcal{H}_N}$ for which the basis $\{e_j, j \in (\mathbb{Z}/N\mathbb{Z})^d\}$ is orthonormal. One can construct the space $\mathcal{H}_N$ by "projecting" states $\psi \in \mathcal{S}(\mathbb{R}^d)$:

$$\forall \psi \in \mathcal{S}(\mathbb{R}^d), \qquad \Pi_N \psi(x) \stackrel{\text{def}}{=} \sum_{\mu,\nu \in \mathbb{Z}^d} \psi(x-\nu) e^{2i\pi \langle \mu, x\rangle}$$

belongs to $\mathcal{H}_N$, and the map Π_N is surjective. In general, there is no obvious link between the norms $\|\psi\|_{L^2}$ and $\|\Pi_N\psi\|_{\mathcal{H}_N}$. Yet, imposing some localization for ψ, we get the following relation:

Lemma 5.1. *Assume that $\psi \in \mathcal{S}(\mathbb{R}^d)$ is microlocalized inside a set $E \subset \mathbb{R}^{2d}$ of diameter$< \frac{1}{2}$. Then, its projection $\Pi_N \psi \in \mathcal{H}_N$ satisfies*

$$\|\Pi_N \psi\|_{\mathcal{H}_N} = \|\psi\|_{L^2} + \mathcal{O}(\hbar^\infty).$$

Let us now describe observables on $\mathbb{T}^{2d}$. Any smooth function on $\mathbb{T}^{2d}$ is also a $\mathbb{Z}^{2d}$-periodic function on $\mathbb{R}^{2d}$. It is natural to introduce symbol classes

$$S^k_\nu(\mathbb{T}^{2d}) = \{f(\hbar) \in C^\infty(\mathbb{T}^{2d}), |\partial_x^\alpha \partial_\xi^\beta f(\hbar)| \leq C_{\alpha,\beta} \hbar^{-k-\nu|\alpha+\beta|}\}, \qquad \nu \in [0, \tfrac{1}{2}), k \in \mathbb{R}$$

(due to periodicity, we cannot have growth or decay in the variable ξ as in the classes (2.8)).

Observables $f = (f(\hbar)) \in S^k_\nu(\mathbb{T}^{2d})$ can be Weyl-quantized as operators $\mathrm{Op}_\hbar(f)$ acting on $\mathcal{S}(\mathbb{R}^d)$, but also on $\mathcal{S}'(\mathbb{R}^d)$ by duality. We already know that any observable $f \in S^0_\nu(\mathbb{T}^{2d})$ satisfies

$$\|\mathrm{Op}_\hbar(f)\|_{L^2\to L^2} \leq \|f\|_\infty + \mathcal{O}(\hbar^{1-2\nu}).$$

Proposition 5.2 ([6]). *Take $f \in C^\infty(\mathbb{T}^{2d})$. For any admissible $\hbar = \hbar_N$, the operator $\mathrm{Op}_\hbar(f)$ leaves invariant the subspace $\mathcal{H}_N \subset \mathcal{S}'(\mathbb{R}^d)$. Let us call $\mathrm{Op}_N(f)$ its restriction on $\mathcal{H}_N$. These two operators satisfy*

$$\|\mathrm{Op}_N(f)\|_{\mathcal{H}_N \to \mathcal{H}_N} \leq \|\mathrm{Op}_\hbar(f)\|_{L^2\to L^2}. \tag{5.3}$$

This property allows to carry the pseudodifferential calculus on $\mathbb{R}^{2d}$ down to the torus. For instance, for any two observables $f, g \in S^0_0(\mathbb{T}^{2d})$, the product $\mathrm{Op}_N(f)\,\mathrm{Op}_N(g)$ is the restriction of $\mathrm{Op}_\hbar(f)\,\mathrm{Op}_\hbar(g) = \mathrm{Op}_\hbar(f \sharp g)$. One can check that $f \sharp g$ is a periodic function, and that each term in the expansion $f \sharp g = \sum_{j=0}^{L-1} \hbar^j (f \sharp g)_j + \hbar^L R_l(f, g, \hbar)$ (including the remainder) is also periodic. Hence, the estimates on $\|\mathrm{Op}_\hbar\big((f\sharp g)_j\big)\|_{L^2}$ and $\big\|\mathrm{Op}_\hbar\big(R_l(f,g,\hbar)\big)\big\|_{L^2}$ can be directly translated to estimates of their restrictions on $\mathcal{H}_N$.

5.2. Quantum maps on the torus. We now give a brief overview of what we mean by a quantum propagator associated with a smooth symplectic diffeomorphism $\kappa\colon \mathbb{T}^{2d} \circlearrowleft$. We will not try to provide a general recipe to "quantize" all possible κ, but only give some relevant examples.

(1) Consider the flow g^t generated by some Hamilton function $p \in C^\infty(\mathbb{T}^{2d})$. If we quantize p into $P(\hbar) = \mathrm{Op}_\hbar(p)$, then the quantum propagator quantizing g^t is naturally $U^t = e^{-itP(\hbar)/\hbar}$. Since $P(\hbar)$ leaves $\mathcal{H}_N$ invariant, so does the propagator. Hence, the map $\kappa \stackrel{\text{def}}{=} g^1$ is quantized by

$$U_N(\kappa) = e^{-iP(\hbar)/\hbar} \restriction \mathcal{H}_N = \exp\big(-i2\pi N\,\mathrm{Op}_N(p)\big),$$

which is unitary on $\mathcal{H}_N$. The Egorov property (2.13) can be directly translated to the torus setting:

$$U_N(\kappa)^{-1}\,\mathrm{Op}_N(f)U_N(\kappa) = \mathrm{Op}_N(f\circ\kappa) + \mathcal{O}_{\mathcal{H}_N\to\mathcal{H}_N}(\hbar). \tag{5.4}$$

(2) If κ is a linear symplectomorphism of $\mathbb{T}^{2d}$ associated with the symplectic matrix $S_\kappa = \left(\begin{smallmatrix} A & B \\ C & D\end{smallmatrix}\right) \in \mathrm{Sp}(2d,\mathbb{Z})$, then it can be quantized as a metaplectic transformation $U_\hbar(\kappa)$, which acts unitarily on $L^2(\mathbb{R}^d)$. Provided S_κ satisfies some "checkerboard conditions" [6], the extension of $U_\hbar(\kappa)$ to $\mathcal{S}'(\mathbb{R}^d)$ leaves $\mathcal{H}_N$ invariant; its restriction $U_N(\kappa) = U_\hbar(\kappa)\restriction \mathcal{H}_N$ is unitary on $\mathcal{H}_N$. If S_κ has no eigenvalues on the unit circle (meaning that the matrix is hyperbolic), the map κ is Anosov. Such a map is often called a "generalized quantum cat map," by reference to Arnold's "cat map", a specific hyperbolic matrix in $SL(2,\mathbb{Z})$. It satisfies an exact Egorov property: $U_N(\kappa)^{-1}\,\mathrm{Op}_N(f)U_N(\kappa) = \mathrm{Op}_N(f\circ\kappa)$.

(3) Let us combine these two types of maps, namely a linear symplectomorphism κ_0 (satisfying the checkerboard condition), and the flow g^t generated by some $p\in C^\infty(\mathbb{T}^{2d})$, to get

$$\kappa = g^1\circ\kappa_0. \tag{5.5}$$

If S_{κ_0} is hyperbolic and p is small enough in the C^2 topology, then it is known that κ still has the Anosov property (and it is topologically conjugate with κ_0). The quantum propagator can be defined as

$$U_N(\kappa) \stackrel{\mathrm{def}}{=} \exp\big(-\mathrm{i}2\pi N\,\mathrm{Op}_N(p)\big)\circ U_N(\kappa_0).$$

It obviously satisfies the Egorov property (5.4).

The long-time Egorov theorem (Proposition 3.4) can also be brought to the torus framework:

Proposition 5.3. *Choose $\epsilon > 0$ small and $\nu \in \big((1-\epsilon)/2, \frac12\big)$. Take $f\in S^0_0(\mathbb{T}^{2d})$. Then, for any admissible $\hbar = \hbar_N$ and any time $n = n(\hbar)$ in the range $|n|\le T_{\epsilon,\hbar}$, we have*

$$U_N^{-n}\,\mathrm{Op}_N(f)U_N^n = \mathrm{Op}_N(\tilde f_n) + \mathcal{O}(\hbar^\infty), \tag{5.6}$$
$$\textit{with } \tilde f_n\in S^0_\nu(\mathbb{T}^{2d}),\ \tilde f_n - f\circ\kappa^n \in S_\nu^{-(1+\epsilon)/2}(\mathbb{T}^{2d}).$$

5.3. Quantum partitions on the torus. Through the quantization $f\mapsto \mathrm{Op}_N(f)$, we can associate to any semiclassical sequence of normalized states $(u_N\in\mathcal{H}_N)_{N\to\infty}$ one or several semiclassical measures, as in Section 2.4. Starting from an Anosov map κ of the form (5.5), we consider sequences $(\psi_N\in\mathcal{H}_N)_{N\to\infty}$ where each ψ_N is an eigenstate of $U_N(\kappa)$. After possibly extracting a subsequence, this sequence admits a single semiclassical measure μ_{sc}, which is a probability κ-invariant measure on $\mathbb{T}^{2d}$. To analyze the KS entropy of this measure (defined as in Section 3.1 after replacing g^1 by κ), we set up a partition $\mathcal{P} = \bigsqcup_{k=1}^K E_k$ of $\mathbb{T}^{2d}$ of diameter ϵ, such that $\mu_{\mathrm{sc}}(\partial\mathcal{P}) = 0$. The partition $\mathcal{P}$ is smoothed into $\mathcal{P}_{\mathrm{sm}} = \{\pi_k\}$, where $\pi_k\in C^\infty(\mathbb{T}^{2d},[0,1])$, each π_k is supported near E_k, and $\sum_k \pi_k = 1$. The quantum partition is defined similarly as in Section 3.2, except that $\Pi_k = \mathrm{Op}_N(\widetilde\pi_k)$, $\widetilde\pi_k\in S(\mathbb{T}^{2d})$ satisfy

$$\sum_{k=1}^K \Pi_k^2 = \mathrm{Id}_{\mathcal{H}_N} + \mathcal{O}(\hbar^\infty). \tag{5.7}$$

The refined quasiprojectors $\Pi_{\boldsymbol{\alpha}}$ are defined similarly as in (3.9), after setting $U = U_N(\kappa)$. The symbolic measures μ_N, $\tilde{\mu}_N$ associated with ψ_N are defined as in (3.14), (3.34).

5.4. Hyperbolic dispersive estimate. One needs to adapt the proof of the hyperbolic dispersion estimate (5.8) to the torus setting. We can prove the following

Proposition 5.4. *Fix a constant $C_1 \gg 1$. Then there exists $C > 0$ such that for any $N \geq 1$, any $0 \leq n \leq C_1 \log N$ and any sequence $\boldsymbol{\alpha}$ of length n, the following estimate holds*:

$$\|\Pi_{\boldsymbol{\alpha}}\| \leq C N^{d/2} J_n^u(\boldsymbol{\alpha})^{-1/2}. \tag{5.8}$$

Here $J_n^u(\boldsymbol{\alpha}) = \prod_{j=0}^{n-1} J^u(\alpha_j)$ is the coarse-grained unstable Jacobian of the map κ. Notice that the power of $N = (2\pi\hbar)^{-1}$ is $d/2$ instead of $(d-1+\delta)$ in (3.20).

Proof of the proposition. The proof of this estimates proceeds along the same lines as in Section 4, that is by explicitly computing the action of the operator $\Pi_{\boldsymbol{\alpha}}$ on an arbitrary normalized state $\Psi \in \mathcal{H}_N$. To do this, we first expand each localized piece $\psi = \Pi_k \Psi$, $k = 1, \ldots, K$, into an orthonormal basis $\{f_j, j \in (\mathbb{Z}/N\mathbb{Z})^d\}$ obtained from the original basis (5.2) through a well-chosen quantized linear symplectomorphism $U(\tilde{\kappa}_k)$. Let us recall a few facts about linear symplectomorphisms of the torus. Any $\tilde{\kappa} \in \mathrm{Sp}(2d, \mathbb{R})$ acting on $\mathbb{R}^{2d}$ maps a "position Lagrangian" $\Lambda^{\mathbb{R}}_{x_0} = \{x = x_0, \xi \in \mathbb{R}\}, x_0 \in \mathbb{R}^d$ into another "linear" Lagrangian[7]

$$\tilde{\Lambda}^{\mathbb{R}}_{x_0} = \tilde{\kappa}(\Lambda^{\mathbb{R}}_{x_0}) = \{\xi = DB^{-1}x + (C - DB^{-1}A)x_0,\ x \in \mathbb{R}^d\}. \tag{5.9}$$

Each position Lagrangian $\Lambda^{\mathbb{R}}_{x_0}$ is associated with a "position eigenstate" $e^{\mathbb{R}}_{x_0}(x) = (2\pi\hbar)^{d/2}\delta(x - x_0)$. The metaplectic operator $U_\hbar(\tilde{\kappa})$ maps this position eigenstate into the Lagrangian state

$$\begin{aligned} &f^{\mathbb{R}}_{x_0} \stackrel{\text{def}}{=} U_\hbar(\tilde{\kappa}) e^{\mathbb{R}}_{x_0}, \\ &f^{\mathbb{R}}_{x_0}(x) = \frac{1}{\sqrt{\det(B)}} \mathrm{e}^{(\mathrm{i}/2\hbar)(\langle DB^{-1}x, x\rangle - 2\langle x, B^{-1}x_0\rangle + \langle B^{-1}Ax_0, x_0\rangle)}, \end{aligned} \tag{5.10}$$

which is associated with the Lagrangian $\tilde{\Lambda}^{\mathbb{R}}_{x_0}$. If $\tilde{\kappa}$ has integer coefficients and satisfies the "checkerboard condition", then this construction can be brought down to the torus. Each basis state e_j in (5.2) is associated with the Lagrangian $\Lambda^{\mathbb{T}}_{j/N}$, the projection of $\Lambda^{\mathbb{R}}_{j/N}$ on $\mathbb{T}^{2d}$. The state $f_j \stackrel{\text{def}}{=} U_N(\tilde{\kappa}) e_j \in \mathcal{H}_N$ is a "Lagrangian state on the torus" associated with the projected Lagrangian $\tilde{\Lambda}^{\mathbb{T}}_{j/N} = \tilde{\kappa}(\Lambda^{\mathbb{T}}_{j/N})$.

For each partition component E_k, we select an appropriate linear transformation $\tilde{\kappa}_k$. In each neighbourhood $\widetilde{E}_k$ of $\operatorname{supp} \pi_k$ (assumed to have diameter $\leq \epsilon$), we use an adapted coordinate chart $\{(y, \eta)\}$ as in Section 4.1. Using these coordinates, we consider the family of γ_1-cones defined in Definition 4.1.

Lemma 5.5. *Provided the diameter of $\widetilde{E}_k$ is small enough, we can choose the automorphism $\tilde{\kappa}_k$ such that each connected component of $\tilde{\Lambda}^{\mathbb{T}}_{x_0} \cap \widetilde{E}_k$ belongs to the unstable γ_1-cone.*

---

[7]We use the representation $S_{\tilde{\kappa}} = \begin{pmatrix} A & B \\ C & D \end{pmatrix}$ and assume for simplicity that B is nonsingular.

PROOF. The pieces of unstable manifolds inside $\widetilde{E}_k$ are then "almost flat" and "almost parallel," so they can be approached by a family of (local) parallel linear Lagrangians; the latter can be assumed to be rational, in the sense that they are parallel to $\tilde{\kappa}_k(\Lambda_0^{\mathbb{T}})$ for some well-chosen transformation $\tilde{\kappa}_k \in \mathrm{Sp}(2d, \mathbb{Z})$. □

For each $k = 1, \ldots, K$ we expand $\psi = \Pi_k \Psi$ in the orthonormal basis $\{f_j = U_N(\tilde{\kappa}_k) e_j, j \in (\mathbb{Z}/N\mathbb{Z})^d\}$. Using cutoffs $\mathbb{1}_{\widetilde{E}_k} \succ \pi_k^\sharp \succ \pi_k$ we can localize the Lagrangian states f_j inside $\widetilde{E}_k$:

$$\psi = \sum_{j \in (\mathbb{Z}/N\mathbb{Z})^d} \psi_j f_j = \sum_{j \in (\mathbb{Z}/N\mathbb{Z})^d} \psi_j \tilde{f}_j + \mathcal{O}(\hbar^\infty), \qquad \tilde{f}_j \stackrel{\text{def}}{=} \mathrm{Op}_N(\pi_k^\sharp) f_j.$$

We can thus proceed as in Section 4, namely compute separately each $\Pi_{\boldsymbol{\alpha}} \tilde{f}_j$. To be able to use Section 4, we will switch back from states in $\mathcal{H}_N$ to states in $\mathcal{S}(\mathbb{R}^d)$.

Lemma 5.6. *Each localized Lagrangian state $\tilde{f}_j = \mathrm{Op}_N(\pi_k^\sharp) f_j \in \mathcal{H}_N$ is equal (up to $\mathcal{O}_{\mathcal{H}_N}(\hbar^\infty)$) to the projection on $\mathcal{H}_N$ of finitely many Lagrangian states $\tilde{f}_{j,n}^{\mathbb{R}} \in \mathcal{S}(\mathbb{R}^d)$ microlocalized inside a single representative $\widetilde{E}_k^{\mathbb{R}}$ of $\widetilde{E}_k$ in $\mathbb{R}^{2d}$.*

PROOF. The localized Lagrangian state $\tilde{f}_j = \mathrm{Op}_N(\pi_k^\sharp) f_j$ is microlocalized on $\tilde{\Lambda}_{j/N}^{\mathbb{T}} \cap \widetilde{E}_k$, which is a finite collection of Lagrangian leaves. Each of these leaves is the projection on the torus of a leaf of the form $\tilde{\Lambda}_{j/N+n}^{\mathbb{R}} \cap \widetilde{E}_k^{\mathbb{R}}$, where $\tilde{\Lambda}_*^{\mathbb{R}}$ is given in (5.9), $n \in \mathbb{Z}^d$ and $\widetilde{E}_k^{\mathbb{R}}$ is a certain (arbitrary) representative of $\dot{E}_k$ in $\mathbb{R}^{2d}$. Accordingly, each $\tilde{f}_j$ can be split into a finite linear combination of Lagrangian states $\tilde{f}_{j,n} \in \mathcal{H}_N$ supported on these individual leaves. If we call $\pi_k^{\sharp\mathbb{R}} \in C_c^\infty(\mathbb{R}^{2d})$ the component of $\pi_k^\sharp$ supported on the representative $\widetilde{E}_k^{\mathbb{R}}$, then the Lagrangian state $\tilde{f}_{j,n}^{\mathbb{R}} = \mathrm{Op}_\hbar(\pi_k^{\sharp\mathbb{R}}) f_{j/N+n}^{\mathbb{R}}$ is microlocalized inside $\widetilde{E}_k^{\mathbb{R}}$. We claim that

$$\tilde{f}_{j,n} = \Pi_N \tilde{f}_{j,n}^{\mathbb{R}} + \mathcal{O}_{\mathcal{H}_N}(\hbar^\infty).$$

We will only give the proof in the case where f_j is a momentum state $\langle e_l, f_j \rangle = N^{-d/2} \mathrm{e}^{2\mathrm{i}\pi\langle l,j\rangle/N}$ associated with the momentum Lagrangian $\{\xi = \xi_j = j/N\}$, and the cutoff is the operator $\mathrm{Op}_N(\chi_1)\,\mathrm{Op}_N(\chi_2)$, where $\chi_1(x)$ (resp. $\chi_2(\xi)$) is obtained by periodizing $\chi_1^{\mathbb{R}} \in C_c^\infty((0,1))$ (resp. $\chi_2^{\mathbb{R}} \in C_c^\infty((0,1))$). In that case, the Lagrangian state $\tilde{f}_j = \mathrm{Op}_N(\chi_1)\,\mathrm{Op}_N(\chi_2) f_j$ admits the components $\langle e_l, \tilde{f}_j \rangle = N^{-d/2}\, \chi_1(l/N) \chi_2(j/N) \mathrm{e}^{2\mathrm{i}\pi\langle l,j\rangle/N}$. On the other hand, the state $\tilde{f}_j^{\mathbb{R}} = \mathrm{Op}_\hbar(\chi_1^{\mathbb{T}}) \times \mathrm{Op}_\hbar(\chi_2^{\mathbb{R}}) f_j^{\mathbb{R}}$ can be expressed as $\tilde{f}_j^{\mathbb{R}}(x) = \chi_1^{\mathbb{R}}(x)\, \chi_2^{\mathbb{R}}(\xi_j) \mathrm{e}^{\mathrm{i}\langle \xi_j, x\rangle/\hbar}$. The projection on $\mathcal{H}_N$ of that state gives

$$\begin{aligned}
\Pi_N \tilde{f}_j^{\mathbb{T}}(x) &= \chi_2^{\mathbb{R}}(\xi_j) \sum_{\nu,\mu} \chi_1^{\mathbb{R}}(x - \nu) \mathrm{e}^{2\mathrm{i}\pi N\langle \xi_j + \mu, x\rangle} \\
&= \chi_2^{\mathbb{R}}(\xi_j) \chi_1^{\mathbb{R}}([x]) \mathrm{e}^{2\mathrm{i}\pi N\langle \xi_j, x\rangle} \delta_{\mathbb{Z}^d}(Nx) \\
&= N^{-d/2} \chi_2^{\mathbb{R}}\left(\frac{j}{N}\right) \sum_{l \in (\mathbb{Z}/N\mathbb{Z})^d} e_l(x) \chi_1\left(\frac{l}{N}\right) \mathrm{e}^{2\mathrm{i}\pi\langle j,l\rangle} = \tilde{f}_j(x).
\end{aligned}$$

□

This lemma allows us to use our control of the evolution of the Lagrangian states $\tilde{f}^{\mathbb{R}} \stackrel{\text{def}}{=} \tilde{f}_{j,n}^{\mathbb{R}}$ through the sequence of operators $U_\hbar(\kappa)$ and $\Pi_k = \mathrm{Op}_\hbar(\widetilde{\pi}_k)$. We have to be a little careful when applying Propositions 4.3 and 4.5, because $\widetilde{\pi}_{\alpha_j}$ are now periodic symbols. Still, the Lagrangian state $\tilde{f}^{\mathbb{R}}$ is localized inside a single copy

$\widetilde{E}^{\mathbb{R}}_{\alpha_0}$ (of diameter$\leq \epsilon$), so its image $U_\hbar(\kappa)\tilde{f}^{\mathbb{R}}$ is a Lagrangian state microlocalized in a set of diameter $\mathcal{O}(\epsilon)$, which can intersect at most one copy $\widetilde{E}^{\mathbb{R}}_{\alpha_1} + n_1$. As a result, the state $\Pi_{\alpha_1} U_\hbar(\kappa)\tilde{f}^{\mathbb{R}}$ is also localized in this single copy, and is of the form of the state ψ_1 in Proposition 4.3. Importantly, the estimates we have on the remainder $r_l(\bullet, \hbar)$ in the expansion (4.6) show that $\mathrm{e}^{\mathrm{i}S_1/\hbar} r_l$ is microlocalized in the single copy $\widetilde{E}_{\alpha_1} + n_1$, and decays fast away from it. As a result, Lemma 5.1 implies that the projection on $\mathcal{H}_N$ of that remainder has a norm comparable with $\|r_l\|_{L^2(\mathbb{R}^d)}$.

Since the initial Lagrangian piece $\tilde{\Lambda}$ lies in some γ_1-unstable cone, we can iterate the evolution as in Proposition 4.5. At each step we get a Lagrangian state and some remainder $\mathrm{e}^{\mathrm{i}S_t(y)/\hbar} r_l^t(y,\hbar)$, which can be projected to $\mathcal{H}_N$ with a control on its norm. When we act on this remainder through operators $\Pi_{\alpha_j} U_n(\kappa)$, its norm can increase at most by a factor $\big(1 + \mathcal{O}(\hbar^\infty)\big)$. Finally, we obtain a Lagrangian state microlocalized in some representative $\widetilde{E}^{\mathbb{R}}_{\alpha_n} + n_n$, and a sum of remainders. The projection in $\mathcal{H}_N$ of that state satisfies the same bound as (4.12):

$$\|\Pi_{\boldsymbol{\alpha}} \tilde{f}_j\|_{\mathcal{H}_N} \leq C J^u(\boldsymbol{\alpha})^{-1/2}.$$

Summing over all the states $\tilde{f}_j$ and taking into account $\|\psi\|_{\mathcal{H}_N} = \sqrt{\sum_{i\in(\mathbb{Z}/N\mathbb{Z})^d} |\psi_i|^2} \leq 1$, we obtain (invoking Cauchy – Schwarz)

$$\|\Pi_{\boldsymbol{\alpha}} \Psi\|_{\mathcal{H}_N} \leq \sum_{j\in(\mathbb{Z}/N\mathbb{Z})^d} |\psi_j| C J^u(\boldsymbol{\alpha})^{-1/2} \leq C N^{d/2} J^u(\boldsymbol{\alpha})^{-1/2}. \qquad \square$$

5.5. EUP and subadditivity. The Hyperbolic dispersive estimate can now be injected in some form of Entropic Uncertainty Principle as in Section 3.8. Since we don't need any energy cutoff, the setting we need is actually simpler than Proposition 3.12: we can take $S_{c_1} = S_{c_2} = \mathrm{Id}_{\mathcal{H}_N}$. In the application, we also take $\varepsilon = \hbar^L$ for any large $L > 0$. We obtain a lower bound of the form (3.33) on the pressures associated with the symbolic measures $\mu_N, \tilde{\mu}_N$ and the Ehrenfest time $n = \lfloor 2T_{\epsilon,\hbar} \rfloor$, with a constant $C_{\mathrm{cone}}(\hbar) = C\hbar^{-d/2}$. Using the Egorov theorem up to $T_{\epsilon,\hbar}$, one shows that the quantum pressures satisfy an approximate subadditivity property similar with (3.37), which allows to prove

$$\frac{p_0^{n_0-1}(\mu_{\mathrm{sc}}, v, \mathcal{P})}{n_0} \geq -\frac{d\lambda_{\max}}{2} + \mathcal{O}(\epsilon).$$

The rest of the proof is unchanged.

References

1. N. Anantharaman, *Entropy and the localization of eigenfunctions*, Ann. of Math. (2) **168** (2008), no. 2, 435–475.
2. N. Anantharaman and S. Nonnenmacher, *Entropy of semiclassical measures of the Walsh-quantized baker's map*, Ann. Henri Poincaré **8** (2007), no. 1, 37–74.
3. ———, *Half-delocalization of eigenfunctions for the Laplacian on an Anosov manifold*, Ann. Inst. Fourier (Grenoble) **57** (2007), no. 7, 2465–2523.
4. N. Anantharaman, H. Koch, and S. Nonnenmacher, *Entropy of eigenfunctions*, Proceedings of the International Congress on Mathematical Physics (Rio de Janeiro, 2006), available at arXiv:0704.1564.
5. J. Bourgain and E. Lindenstrauss, *Entropy of quantum limits*, Comm. Math. Phys. **233** (2003), no. 1, 153–171; corrigendum available at http://www.math.princeton.edu/~elonl/Publications/.
6. A. Bouzouina and S. De Bièvre, *Equipartition of the eigenfunctions of quantized ergodic maps on the torus*, Comm. Math. Phys. **178** (1996), no. 1, 83–105.

7. A. Bouzouina and D. Robert, *Uniform semiclassical estimates for the propagation of quantum observables*, Duke Math. J. **111** (2002), no. 2, 223–252.
8. S. Brooks, *On the entropy of quantum limits for 2-dimensional cat maps*, Comm. Math. Phys. **293** (2010), no. 1, 231–255.
9. Y. Colin de Verdière, *Ergodicité et fonctions propres du laplacien*, Comm. Math. Phys. **102** (1985), no. 3, 497–502.
10. H. Donnelly, *Quantum unique ergodicity*, Proc. Amer. Math. Soc. **131** (2003), no. 9, 2945–2951.
11. L. C. Evans and M. Zworski, *Lectures on semiclassical analysis*, available at http://math.berkeley.edu/~zworski/.
12. F. Faure and S. Nonnenmacher, *On the maximal scarring for quantum cat map eigenstates*, Comm. Math. Phys. **245** (2004), no. 1, 201–214.
13. F. Faure, S. Nonnenmacher, and S. De Bièvre, *Scarred eigenstates for quantum cat maps of minimal periods*, Comm. Math. Phys. **239** (2003), no. 3, 449–492.
14. P. Gérard and É. Leichtnam, *Ergodic properties of eigenfunctions for the Dirichlet problem*, Duke Math. J. **71** (1993), no. 2, 559–607.
15. B. Gutkin, *Entropic bounds on semiclassical measures for quantized one-dimensional maps*, Comm. Math. Phys. **294** (2010), no. 2, 303–342.
16. B. Helffer, A. Martinez, and D. Robert, *Ergodicité et limite semi-classique*, Comm. Math. Phys. **109** (1987), no. 2, 313–326.
17. E. J. Heller, *Bound-state eigenfunctions of classically chaotic Hamiltonian systems: scars of periodic orbits*, Phys. Rev. Lett. **53** (1984), no. 16, 1515–1518.
18. A. Katok and B. Hasselblatt, *Introduction to the modern theory of dynamical systems*, Encyclopedia Math. Appl., vol. 54, Cambridge Univ. Press, Cambridge, 1995.
19. D. Kelmer, *Scarring on invariant manifolds for perturbed quantized hyperbolic toral automorphisms*, Comm. Math. Phys. **276** (2007), no. 2, 381–395.
20. ———, *Arithmetic quantum unique ergodicity for symplectic linear maps of the multidimensional torus*, Ann. of Math. **171** (2010), no. 2, 815–879.
21. E. Lindenstrauss, *Invariant measures and arithmetic quantum unique ergodicity*, Ann. of Math. (2) **163** (2006), no. 1, 165–219.
22. H. Maassen and J. B. M. Uffink, *Generalized entropic uncertainty relations*, Phys. Rev. Lett. **60** (1988), no. 12, 1103–1106.
23. S. Nonnenmacher and M. Zworski, *Quantum decay rates in chaotic scattering*, Acta Math. **203** (2009), no. 2, 149–233.
24. G. Rivière, *Entropy of semiclassical measures in dimension* 2, Duke Math. J., to appear, available at arXiv:0809.0230.
25. ———, *Entropy of semiclassical measures for nonpositively curved surface*, preprint, available at arXiv:0911.1840.
26. ———, *Entropy of quantum limits for symplectic linear maps of the multidimensional torus*, preprint, available at arXiv:1003.5061.
27. Z. Rudnick and P. Sarnak, *The behaviour of eigenstates of arithmetic hyperbolic manifolds*, Comm. Math. Phys. **161** (1994), no. 1, 195–213.
28. A. Schnirelman, *Ergodic properties of eigenfunctions*, Uspehi Mat. Nauk **29** (1974), no. 6(180), 181–182 (Russian).
29. J. Sjöstrand and M. Zworski, *Asymptotic distribution of resonances for convex obstacles*, Acta Math. **183** (1999), no. 2, 191–253.
30. S. Zelditch, *Uniform distribution of eigenfunctions on compact hyperbolic surfaces*, Duke Math. J. **55** (1987), no. 4, 919–941.
31. ———, *On the rate of quantum ergodicity.* I: *Upper bounds*, Comm. Math. Phys. **160** (1994), no. 1, 81–92.
32. S. Zelditch and M. Zworski, *Ergodicity of eigenfunctions for ergodic billiards*, Comm. Math. Phys. **175** (1996), no. 3, 673–682.

Institut de Physique Théorique, CEA-Saclay, 91191 Gif-sur-Yvette, France
E-mail address: snonnenmacher@cea.fr

Centre de Recherches Mathématiques
CRM Proceedings and Lecture Notes
Volume **52**, 2010

Geometry of the High Energy Limit of Differential Operators on Vector Bundles

Alexander Strohmaier

ABSTRACT. At high energies relativistic quantum systems describing scalar particles behave classically. This observation plays an important role in the investigation of eigenfunctions of the Laplace operator on manifolds for large energies and allows to establish relations to the dynamics of the corresponding classical system. Relativistic quantum systems describing particles with spin such as the Dirac equation do not behave classically at high energies. Nonetheless, the dynamical properties of the classical frame flow determine the behavior of eigensections of the corresponding operator for large energies. We review what a high energy limit is and how it can be described for geometric operators.

1. Introduction

A quantum physical system is usually described by an algebra $\mathcal{A}$ of operators on a Hilbert space $\mathcal{H}$ and the time evolution, which is a one-parameter group $\mathrm{U}(t)$ of unitary operators on $\mathcal{H}$. An important example of such a system is the one describing the motion of a quantum particle on a compact Riemannian manifold M. In this case the Hilbert space is $L^2(M, \mu_g)$, where μ_g is the Riemannian measure. The time evolution is described by the Schrödinger equation, which means that the unitary one-parameter group $\mathrm{U}(t)$ is given by

$$\mathrm{U}(t) = \mathrm{e}^{-itH},$$

where H is the Schrödinger operator. For nonrelativistic quantum systems H would typically be $\Delta + V$, where Δ is the metric Laplace operator on M and $V \in C^\infty(M, \mathbb{R})$ is a potential. As we are interested in the high energy limit we will consider the Klein – Gordon time evolution which describes relativistic particles. So we take

$$H = (\Delta + m^2 + V)^{1/2}$$

where m is a positive real number, the mass of the particle. We will assume here that $m^2 + V(x) \geq 0$ so that H is a positive first order pseudodifferential operator

2010 *Mathematics Subject Classification.* Primary 81Q50; Secondary 35P20, 37D30, 58J50, 81Q005.

This is the final form of the paper.

with principal symbol $\sigma_H(\xi) = \|\xi\|$. The above group is defined by spectral calculus as

$$\Delta + m^2 + V\colon H^2(M) \to L^2(M)$$

is a self-adjoint operator with the Sobolev space $H^2(M)$ as its domain. The algebra of observables would be a unital $*$-subalgebra of $\mathcal{B}(\mathcal{H})$, the algebra of all bounded operators on $\mathcal{H}$. A state on $\mathcal{A}$ is a linear functional $\omega\colon \mathcal{A} \to \mathbb{C}$ such that

(i) ω is complex linear,
(ii) ω is positive, i.e., $\omega(A^*A) \geq 0$ for all $A \in \mathcal{A}$,
(iii) $\omega(1) = 1$.

The physical interpretation is that a state is a state of the system and assign to each observable its expectation value. An example is

$$\omega(A) = \langle \psi, A\psi \rangle,$$

where ψ is a vector of unit length in $\mathcal{H}$. More generally, any trace class operator ρ with $\rho \geq 0$ and $\mathrm{Tr}(\rho) = 1$ defines a state by

$$\omega_\rho(A) = \mathrm{Tr}(A\rho).$$

These states are often referred to as normal. If A is a self-adjoint element in $\mathcal{A}$ then A has a spectral decomposition

$$A = \int_{\mathbb{R}} \lambda \, \mathrm{d}E_\lambda.$$

If the system is in the normal state ω_ρ then the probability that a measurement of the observable A yields a value in the Borel measurable set $\mathcal{O} \subset \mathbb{R}$ is given by

$$\omega_\rho(E_{\mathcal{O}}) = \mathrm{tr}(\rho E_{\mathcal{O}}),$$

where $E_{\mathcal{O}}$ is the spectral projection onto $\mathcal{O}$

$$E_{\mathcal{O}} = \int_{\mathcal{O}} \mathrm{d}E_\lambda.$$

This is the probabilistic interpretation of quantum mechanics. For example if $\mathcal{U}$ is a subset of M then the characteristic function $\chi_{\mathcal{U}}$ corresponds to the binary experiment that measures whether the particle is in the region $\mathcal{U}$. The probability of finding the particle in $\mathcal{U}$ is therefore given by $\omega_\rho(\chi_{\mathcal{U}})$ or if the state is of the form ω_ψ with $\psi \in L^2(M)$ then of course we get for the probability of finding the particle in $\mathcal{U}$

$$\int_{\mathcal{U}} |\psi(x)|^2 \, \mathrm{d}x.$$

Whereas in the literature $\mathcal{B}(\mathcal{H})$ itself is often chosen as the algebra of observables, this choice is sometimes not very convenient for practical purposes. On the physical side it is impossible to build a detector that measures $\chi_{\mathcal{O}}$. The reason is that such a measurement would involve a detector that near the boundary of $\mathcal{O}$ had an arbitrary high resolution. If we wanted to be more realistic we would restrict ourselves to algebras that contain functions that are only smooth or continuous. On the mathematical side it is much easier to specify a state on a smaller algebra rather than on the full algebra of bounded operators. Knowledge of the state on the smaller algebra is often sufficient to extend it uniquely to a larger subalgebra. So which algebra to choose for the particle on the manifold?

The state of a classical particle is completely determined by its momentum and its position. We expect the same to be true for quantum particles. In order to measure the position we take the algebra $C^\infty(M)$. Measurement of the momentum involves unbounded operators of the form $\mathrm{i}X$, where X is a vector field. Thus, we need an algebra of operators that contains enough bounded functions of X so that we can approximate the spectral projections of $\mathrm{i}X$ by elements in our algebra. If we choose a classical symbol p of order 0 on $\mathbb{R}$ then $p(X)$ is a classical pseudodifferential operator of order 0. If we choose the algebra of pseudodifferential operators $\Psi\mathrm{DO}^0_{\mathrm{cl}}(M)$ on M this algebra contains enough observables to measure the location and the momentum of our particle up to arbitrary small errors. And, indeed, the restriction of a normal state ω_ρ to the algebra $\mathcal{A}$ determines this state completely. To see this note that normal states are continuous in the weak-$*$-topology on $\mathcal{B}(\mathcal{H})$ and $\Psi\mathrm{DO}^0_{\mathrm{cl}}(M)$ is weak-$*$-dense in $\mathcal{B}(\mathcal{H})$. Since states are automatically norm continuous any state can by continuity be uniquely extended to a state on the norm closure. It is therefore reasonable to use as the algebra of observables the norm closure of the algebra of pseudodifferential operators, that is

$$\mathcal{A} = \overline{\Psi\mathrm{DO}^0_{\mathrm{cl}}(M)}.$$

The high energy limit is now the limit of the quantum mechanical system for particles of high energies. Measuring the energy of a particle corresponds to the unbounded operator

$$(\Delta + m^2 + V)^{1/2}$$

which generates the unitary one-parameter group $\mathrm{U}(t)$. So we should think of a normal state ω_ρ as a state with energy larger than $(\lambda + m^2)^{1/2}$ if

$$\omega_\rho(E_{[-\infty,\lambda)}) = 0$$

where E is the spectral projection of $\Delta + V$.

Note that the set of states is weak-$*$-compact and the algebra $\mathcal{A}$ is separable. Therefore, the set of states is sequentially compact in the weak-$*$-topology. That is every sequence of normal states has weak-$*$ limit points. Suppose ω_{ρ_n} is a sequence of normal states that converges to a not necessarily normal state ω_∞ in the weak-$*$-topology. Then we think of ω_∞ as a high energy limit if

$$\omega_{\rho_n}(E_{[-\infty,\lambda)}) \to 0$$

for all $\lambda > 0$. Using spectral calculus one finds that this is equivalent to

$$\omega_\infty\big((|\Delta + V| + 1)^{-1}\big) = 0.$$

Now suppose that K is a pseudodifferential operator of order -1. Then, $K \times \sqrt{|\Delta + V| + 1}$ is a pseudodifferential operator of order 0 and is therefore bounded. By the Cauchy – Schwarz inequality

$$|\omega_\infty(K)|^2 \leq \omega_\infty\Big(\big|K\sqrt{|\Delta + V| + 1}\big|^2\Big) \cdot \omega_\infty\Big(\big(\sqrt{|\Delta + V| + 1}\big)^{-2}\Big)$$

and therefore

$$\omega_\infty(K) = 0$$

for any high energy limit ω_∞. Hence, ω_∞ vanishes on the algebra of pseudodifferential operators of order -1 and since it is continuous it also vanishes on the ideal of compact operators $\mathcal{K}$. Consequently, at high energies the states become states on the quotient algebra

$$\mathcal{A}/\mathcal{K}.$$

Now it is well known (e.g., [20, Theorem 11.1]) that the principal symbol map

$$\sigma\colon \Psi\mathrm{DO}^0_{\mathrm{cl}}(M) \to C^\infty(S^*M)$$

extends continuously to a map

$$\hat\sigma\colon \mathcal{A} \to C(S^*M)$$

and $\ker\hat\sigma = \mathcal{K}$. This means that

$$\mathcal{A}/\mathcal{K}$$

is naturally isomorphic to the commutative algebra $C(S^*M)$. States on $C(S^*M)$ are by the Riesz representation theorem in one to one correspondence to regular Borel probability measures on S^*M. So for every high energy limit ω_∞ there exists a unique probability measure μ on S^*M such that

$$\omega_\infty(A) = \int_{S^*M} \hat\sigma_A(\xi)\,\mathrm{d}\mu(\xi).$$

The fact that the high energy limit states are actually states on an abelian algebra can be interpreted as the passage from quantum to classical mechanics. The system behaves classically for very large energies. That the quantum mechanical time evolution becomes the classical motion along geodesics is now a consequence of Egorov's theorem. Namely, if $A \in \mathcal{A}$ then also

$$A(t) = \mathrm{U}(-t)A\,\mathrm{U}(t) \in \mathcal{A}$$

and

$$\hat\sigma_{A(t)} = G_t^*(\hat\sigma_A),$$

where G_t is the geodesic flow on S^*M and G_t^* is its pull-back acting on functions. This means that the group of $*$-automorphisms

$$\alpha_t(A) = \mathrm{U}(-t)A\,\mathrm{U}(t)$$

which describes the quantum mechanical time evolution on the level of observables (the so-called Heisenberg picture) factors to $\mathcal{A}/\mathcal{K} \cong C(S^*M)$ and becomes there the geodesic flow. This is a very concise way of saying that in the limit of high energy the quantum system becomes classical and the time-evolution becomes the motion along geodesics with constant speed.

Note that in the high energy limit the potential does not play a role any more. From the physical point of view this is expected as particles with high energy do not "feel" a potential and move at the speed of light along light-like geodesics.

Interestingly some high energy limit can be computed explicitly. Let ϕ_j be a complete orthonormal sequence of eigenfunctions of Δ such that

$$\Delta\phi_j = \lambda_j^2\phi_j, \qquad 0 = \lambda_1 \le \lambda_2 \le \lambda_3 \le \cdots.$$

Then, the sequence of normal states ω_N defined by

$$\omega_N(A) = \frac{1}{N}\sum_{j=1}^N \langle\phi_j, A\phi_j\rangle$$

has a high energy limit. It follows from the classical Tauberian theorem of Karamata that the limit of this sequence is given by

$$\omega_\infty(A) = \lim_{t\to 0+}\frac{\mathrm{Tr}(A\mathrm{e}^{-t(\Delta+V)})}{\mathrm{Tr}(\mathrm{e}^{-t(\Delta+V)})}.$$

The (microlocal) heat kernel expansion then shows that

$$\lim_{t\to 0+} \frac{\mathrm{Tr}(A\mathrm{e}^{-t(\Delta+V)})}{\mathrm{Tr}(\mathrm{e}^{-t(\Delta+V)})} - \int_{S^*} \sigma_A(\xi)\, \mathrm{d}\mu_L(\xi),$$

where μ_L is the normalized Liouville measure on S^*M.

The state ω_t defined by

$$\omega_t(A) = \frac{\mathrm{Tr}(A\mathrm{e}^{-t(\Delta+V)})}{\mathrm{Tr}(\mathrm{e}^{-t(\Delta+V)})}$$

is the KMS-state with temperature t^{-1} describing a quantum system at temperature t^{-1} in thermal equilibrium. In the limit as the temperature goes to infinity the state converges to the Liouville measure on the unit-cotangent bundle.

The sequence of eigenstates $\langle \phi_j, \cdot \phi_j \rangle$ is a sequence of invariant states and any weak-$*$-limit point is a therefore an invariant high energy limit. The above says that on average these states converge to the Liouville measure. If the Liouville measure is ergodic with respect to the geodesic flow this means that the tracial state ω_∞ defined above is ergodic. In other words there is no non-trivial decomposition of ω_∞ into a convex combination of invariant states. From this one can conclude that any subsequence of $\langle \phi_j, \cdot \phi_j \rangle$ that does not have the state ω_∞ as a weak-$*$-limit has to have counting density zero, as otherwise it would give rise to a decomposition of ω_∞ into invariant states. A more careful argument along these lines (see [14, 21, 22, 24]) shows that in fact there is a subsequence of counting density one of eigenfunctions $\phi_{j(k)}$ such that

$$\lim_{k\to\infty} \langle \phi_{j(k)}, A\phi_{j(k)} \rangle = \omega_\infty(A),$$

for all $A \in \mathcal{A}$. This is usually referred to as *quantum ergodicity*.

2. The Dirac equation and Laplace type operators

Relativistic quantum systems that describe particles with spin like electrons, neutrinos are not described by the Klein – Gordon equation. As the particles have an internal degree of freedom, the spin, the Hilbert space will consist of vector valued functions and the observable algebra needs to include operators that detect these internal degrees of freedom. This is appropriately described by the following construction. Suppose that $E \to M$ is a complex Hermitian vector bundle over M we take as the Hilbert space the space of square integrable sections of E

$$\mathcal{H} = L^2(M; E)$$

and as an algebra of observables we take the norm closure of the space of zero order classical pseudodifferential operators acting on sections of this vector bundle

$$\mathcal{A} = \overline{\Psi\mathrm{DO}^0_{\mathrm{cl}}(M; E)}.$$

Now a second order differential operator P is said to be of Laplace-type if in local coordinates it has the form

$$P = -\sum_{ij} g^{ij} \frac{\partial^2}{\partial x^i \partial x^j} + B,$$

where B has order 1, or in other words if and only if

$$\sigma_P(\xi) = g(\xi, \xi).$$

Similarly, a first order differential operator D is said to be of Dirac type if and only if

$$\sigma_D(\xi)^2 = g(\xi, \xi).$$

Of course a first order operator is of Dirac type if and only if its square is a Laplace type operator.

As we saw in the previous section the Klein – Gordon operator $\Delta + V + m^2$ is a Laplace-type operator acting on the sections of the trivial vector bundle. The time evolution in this case is generated by the square root of a Laplace type operator. To describe electrons one typically chooses a spin structure on M and then the complex vector bundle is the associated spinor bundle S. The algebra of observables in this case is the algebra of zero order classical pseudodifferential operators acting on sections of the spinor bundle. The time-evolution is described by the Dirac operator D acting on sections of the spinor bundle. This operator will however not be positive any more, which is a typical feature of relativistic quantum theory. The system describes electrons and positrons at the same time. Pure electron states are states that are supported in the positive spectral subspace of D. Since on this subspace the operators D and $|D|$ coincide the time evolution of such states may as well be described by the operator $H = |D|$. Whereas from the viewpoint of a one-particle theory it might seem strange to take the operator $|D|$ as the generator of the time evolution this is perfectly justified in a fully quantized theory. The generator of the time-evolution of the fully quantized free electron-positron field restricted to the one-particle subspace is given by $|D|$ rather than D. The apparent violation of Einstein causality by the infinite propagation speed of the operator $\exp(-it|D|)$ is resolved in the fully quantized theory and is not causing a problem there (see, e.g., [23] for details).

Note that if one chooses the group generated by D instead of $|D|$ the time evolution does not leave the space of operators invariant. Instead of passing to the $|D|$ one can also restrict the algebra of observables. This approach is favored in [15] and also used in [2–6] in order to investigate the semi-classical limit of the Dirac operator.

Spin 1 particles like photons and mesons are described by Maxwell's equation or the Proca equation. For example the quantum system describing photons is given as follows. The vector bundle is the complexified cotangent bundle $\Lambda^1 M = T^*M$. The Hilbert space is the closure of the space of coclosed 1-forms in $L^2(M; E)$ and the space of observables is the algebra

$$\mathcal{A} = \overline{P\,\Psi\mathrm{DO}^0_{\mathrm{cl}}(M; \Lambda^1 M) P},$$

where $\Psi\mathrm{DO}^0_{\mathrm{cl}}(M; \Lambda^1 M)$ is the algebra of pseudodifferential operators and P is the orthogonal projection onto the space of coclosed 1-forms. The relativistic time evolution is given by the one parameter group generated by the Laplace – Beltrami operator Δ_1 acting on one-forms.

As we can see the time evolution is in these examples given by the root of a Laplace-type operator acting on the sections of a vector bundle. The algebra of observables is either the full algebra of pseudodifferential operators or an appropriate subalgebra that is invariant under the time-evolution. The symbol map σ is now a map

$$\Psi\mathrm{DO}^0_{\mathrm{cl}}(M; E) \to C^\infty\big(S^*M; \pi^* \operatorname{End}(E)\big)$$

from the pseudodifferential operators of order 0 to the smooth functions on S^*M with values in the bundle $\mathrm{End}(E)$. Here π^*E denotes the pull back of the bundle E on M under the projection $\pi\colon S^*M \to M$. As in the scalar case the symbol map has a continuous extension $\hat{\sigma}$ to a map

$$\mathcal{A} \to C\big(S^*M; \pi^* \mathrm{End}(E)\big)$$

where $\mathcal{A} = \overline{\Psi\mathrm{DO}^0_{\mathrm{cl}}(M;E)}$ is the norm closure of the space of pseudodifferential operators acting on $L^2(M;E)$.

In contrast to the Egorov theorem for scalar pseudodifferential operators the Egorov theorem for matrix valued pseudodifferential operators involves terms of the time-evolution that are of lower order ([16,17]). Let us give some invariant meaning to this. If Δ_E is a self-adjoint Laplace type operator acting on the sections of some Hermitian vector bundle E, then there exists a unique connection ∇_E on E and a unique potential $V \in C^\infty\big(M, \mathrm{End}(E)\big)$ such that

$$\Delta_E = \nabla_E^* \nabla_E + V.$$

The locally defined connection-1-form can be interpreted as the subprincipal symbol of Δ_E (see [18]). The connection ∇_E of course defines a connection $\nabla_{\mathrm{End}(E)}$ on $\mathrm{End}(E)$. This connection can be used to extend the geodesic flow on S^*M to a flow on $\pi^*\big(\mathrm{End}(E)\big)$ by parallel translation. We will denote the induced action on the sections of $\pi^*\big(\mathrm{End}(E)\big)$ by β_t. It is easy to check that β_t is a one-parameter group of $*$-automorphisms on

$$C\left(S^*M; \pi^*\big(\mathrm{End}(E)\big)\right).$$

The analog of Egorov's theorem can now be stated as follows. Let $\mathcal{A}$ be the algebra $\overline{\Psi\mathrm{DO}^0_{\mathrm{cl}}(M;E)}$. Then, if $\mathrm{U}(t) = \mathrm{e}^{-\mathrm{i}t\sqrt{\Delta_E}}$ and $A \in \mathcal{A}$ we have

$$A_t = \mathrm{U}(-t)A\,\mathrm{U}(t) \in \mathcal{A}, \qquad \hat{\sigma}_{A_t} = \beta_t(\sigma_A).$$

In other words β_t is the high energy limit of the quantum time evolution. A proof can be found in [13,18].

3. Geometric operators and the frame flow

Most geometric operators like the Dirac operator and the Laplace–Beltrami operator are acting on sections of vector bundles that are constructed in a geometric way from the manifold. We assume here that M is oriented and that FM is the bundle of oriented orthonormal frames. We will show that in many cases the bundle $\mathrm{End}(E)$ as a Hermitian vector bundle with connection is isomorphic to an induced bundle of the frame bundle by some representation $\rho\colon\ \mathrm{SO}(n) \to \mathrm{Aut}\big(\mathrm{End}(\mathbb{C}^m)\big)$ of $\mathrm{SO}(n)$ by $*$-automorphisms of $\mathrm{End}(\mathbb{C}^m)$

$$\mathrm{End}(E) = FM \times_\rho \mathrm{End}(\mathbb{C}^m)$$

with connection induced by the Levi-Civita connection on FM.

Example 1 (Dirac operators). Suppose that D is the Dirac operator associated with a spin-structure or spin^{c}-structure acting on the sections of the associated spinor bundle S. Then the action of the complex Clifford algebra bundle $\mathrm{Cl}(TM)$ on S is irreducible and therefore $\mathrm{End}(S)$ is a quotient of the bundle $\mathrm{Cl}(TM)$. The connection on S is compatible with the Clifford action and therefore, the induced connection on $\mathrm{End}(S)$ is compatible with the Clifford connection on $\mathrm{Cl}(TM)$. But the Clifford algebra bundle is as a Hermitian vector bundle with connection obtained as an associated bundle

$$\mathrm{Cl}(TM) = FM \times_\rho \mathrm{Cl}(\mathbb{R}^n),$$

where ρ is the canonical representation of $\mathrm{SO}(n)$ on $\mathrm{Cl}(\mathbb{R}^n)$. Note that the spinor bundle itself is not an associated bundle of FM, but $\mathrm{End}(S)$ nevertheless is. By the Bochner – Lichnerowicz – Weitzenböck – Schrödinger formula

$$D^2 = \nabla^*\nabla + V,$$

where V is some potential (for example $\frac{1}{4}R$ in the case of a spin structure). Thus, D^2 is a Laplace-type operator and the corresponding connection on S and on $\mathrm{End}(S)$ is the Levi-Civita connection.

Example 2. The bundle $\Lambda^p M$ is an associated bundle of the frame bundle

$$\Lambda^p M = FM \times_\sigma \Lambda^p \mathbb{C}^n$$

where σ is the canonical representation of $\mathrm{SO}(n)$ on $\Lambda^p\mathbb{C}^n$. This of course induces a connection on $\Lambda^p M$ which is the Levi-Civita connection on forms. The Hodge – Laplace operator on p-forms Δ_p is then defined by

$$\Delta_p = d\delta + \delta d,$$

where $d\colon C^\infty(\Lambda^p M) \to C^\infty(\Lambda^{p+1} M)$ is the exterior differential and $\delta\colon C^\infty(\Lambda^{p+1}M) \to C^\infty(\Lambda^p M)$ its formal adjoint. Again, by the Bochner – Weitzenböck formula the

$$\Delta_p = \nabla^*\nabla + V,$$

where V involves curvature terms and ∇ is the Levi-Civita connection. Note that

$$\mathrm{End}(\Lambda^p M) = FM \times_\rho \mathrm{End}(\Lambda^p\mathbb{C}^n),$$

where $\rho = \sigma \otimes \sigma^*$.

These examples show that in geometric situations the bundle $\mathrm{End}(E)$ is an associated bundle of FM. Consequently, the bundle $\pi^*\,\mathrm{End}(E) \to S^*M$ is an associated bundle of $FM \to S^*M$, where the map $\pi\colon FM \to S^*M$ is defined by projecting onto the first vector in the frame and identifying vectors and covectors using the metric. So, if we view FM as an $\mathrm{SO}(n-1)$-principal bundle over S^*M then we can think of $\pi^*\,\mathrm{End}(E)$ as the associated bundle

$$\pi^*\,\mathrm{End}(E) = FM \times_{\hat\rho} \mathrm{End}(\mathbb{C}^m),$$

where $\hat\rho$ is the restriction of $\rho\colon\ \mathrm{SO}(n) \to \mathrm{Aut}\bigl(\mathrm{End}(\mathbb{C}^m)\bigr)$ to the subgroup $\mathrm{SO}(n-1)$ which we think of as the subgroup that fixes the first vector in the standard representation on $\mathbb{R}^n$. Therefore, sections of $\pi^*\,\mathrm{End}(E)$ can be identified with functions f on FM with values in $\mathrm{End}(\mathbb{C}^m)$ satisfying the transformation property

$$f(x\cdot g) = \hat\rho^{-1}(g) f(x), \tag{1}$$

for all $g \in \mathrm{SO}(n-1)$ and $x \in FM$. The frame flow on FM is the extension of the geodesic flow on S^*M by parallel translation to the space FM. More explicitly, if $(e_1,\dots,e_n) \in FM$ is an orthonormal frame, then the frame flow $\gamma_t(e_1,\dots,e_n) = \bigl(e_1(t),\dots,e_n(t)\bigr)$ can be defined as follows. The vector $e_1(t)$ is the tangent of the endpoint of the unique geodesic of length t with starting tangent vector e_1. In other words $e_1(t) = G_t(e_1)$. The rest of the frame $(e_2,\dots,e_n)$ is parallel transported along this geodesic using the Levi-Civita connection to give the orthonormal basis $\bigl(e_2(t),\dots,e_n(t)\bigr)$ in the orthogonal complement of $e_1(t)$. It is important here that the Levi-Civita connection preserves angles so that the parallel transport of the frame yields a frame. The frame flow gives rise to a flow on the space of functions on FM by pull-back. By construction γ_t commutes with the right action of $\mathrm{SO}(n-1)$ on FM and therefore the space of functions satisfying the transformation property (1) is left invariant. Since all constructions are compatible it turns out that the flow β_t originally constructed from the connection on E coincides with this flow. That is

$$(\beta_t f)(x) = f(\gamma_{-t}x),$$

where γ_t is the frame flow.

4. The high energy limit for geometric operators

Suppose that ϕ_j is a complete orthonormal sequence of eigensections of a positive Laplace-type operator Δ_E such that

$$\Delta_E \phi_j = \lambda_j^2 \phi_j, \qquad 0 = \lambda_1 \le \lambda_2 \le \lambda_3 \le \cdots .$$

Then, the sequence of normal states ω_N defined by

$$\omega_N(A) = \frac{1}{N}\sum_{j=1}^{N} \langle \phi_j, A\phi_j\rangle$$

has a high energy limit ω_∞. As in the scalar case we have

$$\omega_\infty(A) = \lim_{t\to 0^+} \frac{\mathrm{Tr}(A\mathrm{e}^{-t\Delta_E})}{\mathrm{Tr}(\mathrm{e}^{-t\Delta_E})}$$

and

$$\lim_{t\to 0^+} \frac{\mathrm{Tr}(A\mathrm{e}^{-t\Delta_E})}{\mathrm{Tr}(\mathrm{e}^{-t\Delta_E}))} = \frac{1}{\mathrm{rk}\, E}\int_{S^*} \mathrm{Tr}\big(\hat{\sigma}_A(\xi)\big)\, \mathrm{d}\mu_L(\xi),$$

where μ_L is the normalized Liouville measure on S^*M and Tr is the trace. Indeed this state is obviously invariant under the classical time evolution β_t.

In order to understand which quantum limit can be obtained from subsequences of nonzero counting density one needs to decompose the state ω_∞ into ergodic states with respect to the action β_t.

Indeed, one can prove the following theorem ([18, 19, 25])

Theorem 3. *Suppose that $p_1, \ldots, p_r$ are projections in $\mathcal{A}$ which commute with Δ_E. Suppose furthermore that $\sum_{i=1}^r p_i = \mathrm{id}$ and that the decomposition*

$$\omega_\infty(\cdot) = \sum_{i=1}^{r} \omega_\infty(p_i \cdot)$$

is a decomposition into ergodic states

$$\omega_i(A) = \frac{1}{\omega_\infty(p_i)}\omega_\infty(p_i A).$$

Then Shnirelman's theorem holds in the subspaces onto which p_i project. More precisely, if ϕ_j is an orthonormal sequence of eigensections of Δ_E such that

$$\Delta_E \phi_j = \lambda_j^2 \phi_j, \qquad \lambda_1 \le \lambda_2 \le \lambda_3 \le \cdots, \qquad p_i \phi_j = \phi_j,$$

and such that ϕ_j span the range of p_i. Then there is a subsequence of eigensections $\phi_{k(j)}$ of counting density one such that

$$\lim_{j\to\infty} \langle \phi_{k(j)}, A\phi_{k(j)}\rangle = \omega_i(A).$$

5. Ergodic decomposition of the tracial state

Let $\tau\colon \mathrm{SO}(n-1) \to \mathrm{Aut}\big(\mathrm{End}(\mathbb{C}^k)\big)$ be a representation of $\mathrm{SO}(n-1)$ by $*$-automorphisms. Then there is a unique projective unitary representation

$$\rho\colon \mathrm{SO}(n-1) \to \mathrm{PU}(k)$$

such that

$$\tau(g)(x) = \rho(g)^{-1} x \rho(g).$$

As before the frame flow induces a flow β_t on the space of sections of

$$F = FM \times_\tau \mathrm{End}(\mathbb{C}^k)$$

by $*$-automorphisms. The following theorem is proved in ([18, 19]).

Theorem 4. *Suppose that the frame flow on FM is ergodic and that ρ is irreducible. Then the tracial state*

$$\omega(A) = \frac{1}{k} \int_{S^*M} \mathrm{Tr}\big(\hat{\sigma}_A(\xi)\big)\, \mathrm{d}\mu_L(\xi),$$

on $C(S^*M; F)$ is ergodic.

This gives us a strategy to decompose the tracial state on $C\big(S^*M, \mathrm{End}(E)\big)$ into ergodic states assuming that $\mathrm{End}(E)$ is an associated bundle of the frame bundle. Namely, suppose that $\rho : \mathrm{SO}(n-1) \to \mathrm{PU}(k)$ is a projective unitary representation which gives rise to a representation $\rho = \rho \otimes \rho^*$ on $\mathrm{End}(\mathbb{C}^k)$ such that

$$\mathrm{End}(E) = FM \times_\tau \mathrm{End}(\mathbb{C}^k).$$

Then decompose $\mathbb{C}^k$ into invariant subspaces for ρ

$$\mathbb{C}^k = V_1 \oplus \cdots \oplus V_r.$$

The orthogonal projection p_i onto V_i is a matrix in $\mathrm{End}(\mathbb{C}^k)$ and we have $\sum_{i=1}^r p_i = \mathrm{id}$. This matrix is invariant under the action of ρ and therefore, the constant function $p_i \in C^\infty(FM)$ can be understood as a section in $\mathrm{End}(E)$ as it satisfies the transformation rule. By the above theorem the state

$$\omega_i(\cdot) = \frac{1}{\omega(p_i)} \omega(p_i A)$$

is ergodic and we have constructed an ergodic decomposition of the tracial state. If there are pseudodifferential operators P_i in $\Psi\mathrm{DO}^0_{\mathrm{cl}}(M, E)$ that are mutually commuting and commute with Δ_E such that $\sigma_{P_i} = p_i$ we can then apply the theorem above and conclude that quantum ergodicity holds in subspaces onto which P_i projects.

6. Examples

6.1. Dirac operators. As before assume that S is the spinor bundle of some Spin structure or Spin^c structure. Let D be the Dirac operator and $\mathrm{sign}(D)$ defined by spectral calculus. Then $\mathrm{sign}(D)$ is a zero-order pseudodifferential operator and its principal symbol is given by

$$\sigma_{\mathrm{sign}(D)(\xi)} = \gamma_\xi,$$

where γ_ξ denotes Clifford multiplication with ξ. Then the projections

$$P_\pm = \frac{1}{2}\big(1 \pm \mathrm{sign}(D)\big)$$

are pseudodifferential operators that commute with $|D|$. Their symbols are elements in $C^\infty\big(S^*X, \pi^* \operatorname{End}(S)\big)$ that are invariant under the flow β_t. If we identify sections of this bundle with functions on FM with values in the $\operatorname{End}(\mathbb{C}^{2^{[n/2]}})$ then this function corresponds to Clifford multiplication with first vector in the $\mathrm{Cl}(\mathbb{R}^n)$-module $\mathbb{C}^{2^{[n/2]}}$. This projects onto an irreducible subspace of the projective representation

$$\mathrm{SO}(n-1) \to \mathrm{PU}\big(2^{[n/2]}\big).$$

If the frame flow is ergodic quantum ergodicity holds in the positive spectral subspace and negative spectral subspace respectively.

6.2. Laplace-Beltrami operator on p-forms. Since the bundles $\Lambda^p T^* M \to M$ are associated bundles of the representation $\Lambda^p \rho$, if $\rho\colon \mathrm{SO}(n) \to \mathbb{C}^n$ the bundle $\pi^* \Lambda^p M \to S^* M$ is associated with the restriction of this representation to the subgroup $\mathrm{SO}(n-1)$. Note that whereas the representations $\Lambda^p \rho$ are irreducible for $p \neq n/2$ the restriction to the group $\mathrm{SO}(n-1)$ is not irreducible unless $p = 0$ or $p = n$. The reason is that since $\mathbb{C}^n = \mathbb{C} \oplus \mathbb{C}^{n-1}$ is a decomposition into invariant subspaces of the $\mathrm{SO}(n-1)$-action also

$$\Lambda^p \mathbb{C}^n = \Lambda^p \mathbb{C}^{n-1} \oplus \Lambda^{p-1} \mathbb{C}^{n-1}$$

is a decomposition into invariant subspaces. This decomposition is into irreducible subspaces unless $p = (n-1)/2$ in which case the first summand is not irreducible or $p = (n+1)/2$ in which case the second summand is not irreducible.

As in the case the Dirac operator one can find pseudodifferential operators of order zero that have symbols that project onto these irreducible subspaces. Namely, define Δ_p^{-1} as the inverse of Δ_p on $(\ker \Delta_p)^\perp$ and to be zero otherwise. Then

$$P = \Delta_p^{-1} \delta d, \qquad Q = \Delta_p^{-1} d \delta.$$

are projections that commute with Δ_p. Their principal symbols are invariant elements in $C^\infty\big(S^* M, \pi^* \operatorname{End}(\Lambda^p)\big)$ that give rise to a decomposition of the tracial state as

$$\sigma_P + \sigma_Q = 1.$$

Suppose now that the frame flow on FM is ergodic. If $p \neq (n-1)/2$ then the state

$$\sigma_P(\cdot) = \frac{1}{\omega(P)} \omega(P \cdot)$$

is ergodic as σ_P corresponds to the projection onto the first summand in. If $p \neq (n-1)/2$ then the state

$$\sigma_Q(\cdot) = \frac{1}{\omega(Q)} \omega(Q \cdot)$$

is ergodic for the analogous reason. Thus, quantum ergodicity holds in the subspaces onto which P and Q project onto, which are the subspaces of co-exact and exact p-forms.

In the case $p = (n-1)/2$ the state ω_P is not ergodic. There is a further pseudodifferential operator commuting with Δ_p and with P, namely the operator

$$R = \Delta_p^{-1/2} * d,$$

where $*$ is the Hodge star operator. Note that $R^2 = 1$ on $\operatorname{rg}(P)$ and the decomposition of the state ω_P into $+1$ and -1 eigenspaces is ergodic in case the frame flow is ergodic. The eigenvalues of R correspond to polarized forms. For example the case of $n = 1$, $p = 1$ and co-closed 1-forms corresponds to electrodynamics in dimension

3. It is well known that electromagnetic waves can be decomposed into circular polarized waves and that this decomposition is invariant under the time-evolution determined by the Maxwell equation. Sequences of differently polarized eigensections give rise to different quantum limits as the observable R that measures the polarization gives rise to an observable in the high energy limit, namely σ_R, that distinguishes them.

7. Operators on Kähler manifolds

The two examples of the previous section may also be discussed in the category of Kähler manifolds or other special geometries. A Kähler manifold can be thought of as a Riemannian manifold of dimension $2m$ such that the frame bundle can be reduced to a $\mathrm{U}(m)$-principal bundle in such a way that the parallel transport preserves the $\mathrm{U}(m)$-structure. This is equivalent to the existence of a covariantly constant complex structure. On Kähler manifolds the frame flow is not ergodic as the complex structure is preserved and gives rise to conservation laws. It is much more natural however to consider the $\mathrm{U}(m)$-bundle UM of unitary frames instead and the restriction of the frame flow to it. Again, a lot of geometric constructions in the category of Kähler manifolds can be understood as associated bundle constructions starting from the unitary frame bundle. The bundles of (p,q)-forms are associated bundles of UM and natural geometric operators to consider are the Dolbeault Dirac operator and the Dolbeault Laplace operator. Under the assumption that the unitary frame flow is ergodic the ergodic decomposition of the tracial state can be found explicitly. It is closely related to the action of a certain Lie-superalgebra on the space of exterior differential forms of a Kähler manifold. This action can be seen as the quantum counterpart of the classical symmetry that prevents the frame flow on FM to be ergodic. A detailed discussion of this and its implications can be found in [19].

8. Conclusions and further remarks

It was already found by [2–6] for the Dirac operator in $\mathbb{R}^n$ that the semi-classical limit for this operator can be described by a suitable extension of the classical flow and that ergodicity of that flow implies quantum ergodicity. The geometric framework discussed in the present article was introduced in [18] and further developed in [19]. It deals with high energy limits rather than with the semiclassical limit so that only the non-trivial geometry of space contributes to the classical dynamics. A representation theoretic framework generalizing the representation theoretic lift on locally symmetric spaces to induced bundles over locally symmetric spaces can be found in [13]. This article also contains some discussion of quantum ergodicity questions for vector bundles.

As advertised [25] the language of states and ergodicity of states over C^*-dynamical systems is the appropriate one to describe the high energy limit or quantum ergodicity of quantum systems. Its application to quantum systems with spin or more mathematically to geometric operators acting on vector bundles naturally leads to associated bundle constructions over the frame bundle. The underlying dynamics being the frame flow. Quantum ergodicity questions translate into questions about the frame flow. In particular ergodicity of the frame flow has strong implications for quantum systems with spin. It implies quantum ergodicity on certain natural subspaces that can be found more or less constructively from our method. Finally we would like to mention that the frame flow was already considered by Arnol'd in [1]. In negative curvature, it was studied by Brin, together with Gromov, Karcher and Pesin, in a series of papers [7–12] independently of any connection to spectral theory for operators on vector bundles or quantum ergodicity questions. Many examples of manifolds with ergodic frame flow are known (for example manifolds of constant negative curvature to name only the simplest ones) and much progress has been made towards its understanding. We would like to refer the reader to the above mentioned literature on frame flows for further details.

Acknowledgements. The author would like to thank the CRM (Montréal) and its Analysis lab for their kind hospitality during his stay in summer 2008.

References

1. V. I. Arnol'd, *Some remarks on flows of line elements and frames*, Dokl. Akad. Nauk SSSR **138** (1961), 255–257 (Russian).
2. J. Bolte, *Semiclassical expectation values for relativistic particles with spin* 1/2, Found. Phys. **31** (2001), no. 2, 423–444.
3. J. Bolte and R. Glaser, *Zitterbewegung and semiclassical observables for the Dirac equation*, J. Phys. A **37** (2004), no. 24, 6359–6373.
4. ———, *A semiclassical Egorov theorem and quantum ergodicity for matrix valued operators*, Comm. Math. Phys. **247** (2004), no. 2, 391–419.
5. J. Bolte and S. Keppeler, *Semiclassical time evolution and trace formula for relativistic spin-1/2 particles*, Phys. Rev. Lett. **81** (1998), no. 10, 1987–1991.
6. ———, *A semiclassical approach to the Dirac equation*, Ann. Physics **274** (1999), no. 1, 125–162.
7. M. Brin, *Topological transitivity of a certain class of dynamical systems, and flows of frames on manifolds of negative curvature*, Funkcional. Anal. i Priložen. **9** (1975), no. 1, 9–19 (Russian); English transl., Functional Anal. Appl. **9** (1975), 8–16.
8. ———, *The topology of group extensions of C-systems*, Mat. Zametki **18** (1975), no. 3, 453–465 (Russian); English transl., Math. Notes **18** (1975), no. 3-4, 858–864.
9. ———, *Ergodic theory of frame flows*, Ergodic Theory and Dynamical Systems. II (College Park, MD, 1979/1980), Progr. Math., vol. 21, Birkhäuser, Boston, MA, 1982, pp. 163–183.
10. M. Brin and M. Gromov, *On the ergodicity of frame flows*, Invent. Math. **60** (1980), no. 1, 1–7.
11. M. Brin and H. Karcher, *Frame flows on manifolds with pinched negative curvature*, Compositio Math. **52** (1984), no. 3, 275–297.
12. M. Brin and Ya. Pesin, *Partially hyperbolic dynamical systems*, Izv. Akad. Nauk SSSR Ser. Mat. **38** (1974), 170–212 (Russian).
13. U. Bunke and M. Olbrich, *On quantum ergodicity for vector bundles*, Acta Appl. Math. **90** (2006), no. 1-2, 19–41.
14. Y. Colin de Verdière, *Ergodicité et fonctions propres du laplacien*, Comm. Math. Phys. **102** (1985), no. 3, 497–502.
15. H. O. Cordes, *A pseudo-algebra of observables for the Dirac equation*, Manuscripta Math. **45** (1983), no. 1, 77–105.
16. N. Dencker, *On the propagation of polarization sets for systems of real principal type*, J. Funct. Anal. **46** (1982), no. 3, 351–372.
17. C. Emmrich and A. Weinstein, *Geometry of the transport equation in multicomponent WKB approximations*, Comm. Math. Phys. **176** (1996), no. 3, 701–711.
18. D. Jakobson and Alexander Strohmaier, *High energy limits of Laplace-type and Dirac-type eigenfunctions and frame flows*, Comm. Math. Phys. **270** (2007), no. 3, 813–833.
19. D. Jakobson, A. Strohmaier, and S. Zelditch, *On the spectrum of geometric operators on Kähler manifolds*, J. Mod. Dyn. **2** (2008), no. 4, 701–718.
20. R. T. Seeley, *Integro-differential operators on vector bundles*, Trans. Amer. Math. Soc. **117** (1965), 167–204.

21. A. I. Shnirelman, *Ergodic properties of eigenfunctions*, Uspehi Mat. Nauk **29** (1974), no. 6(180), 181–182 (Russian).
22. ———, *On the asymptotic properties of eigenfunctions in the regions of chaotic motion*, KAM Theory and Semiclassical Approximations to Eigenfunctions, Ergeb. Math. Grenzgeb. (3), vol. 24, Springer, Berlin, 1993, pp. 313–337.
23. B. Thaller, *The Dirac equation*, Texts Monogr. Phys., Springer, Berlin, 1992.
24. S. Zelditch, *Uniform distribution of eigenfunctions on compact hyperbolic surfaces*, Duke Math. J. **55** (1987), no. 4, 919–941.
25. ———, *Quantum ergodicity of C^* dynamical systems*, Comm. Math. Phys. **177** (1996), no. 2, 507–528.

School of Mathematics, Loughborough University, Loughborough, Leicestershire, LE11 3TU, UK

E-mail address: a.strohmaier@lboro.ac.uk

Centre de Recherches Mathématiques
CRM Proceedings and Lecture Notes
Volume **52**, 2010

Classical and Quantum Dynamics in Transverse Geometry of Riemannian Foliations

Yuri A. Kordyukov

ABSTRACT. First, we survey some results on classical and quantum dynamical systems associated with transverse Dirac operators on Riemannian foliations. Then we illustrate these results by two examples of Riemannian foliations: a foliation given by the fibers of a fibration and a linear foliation on the two-dimensional torus.

Introduction

In classical mechanics, a particle, moving on a compact manifold M, is described by a point of the phase space, which is the cotangent bundle T^*M of M. The evolution of the particle in the phase space is governed by the Hamilton equations of motion. In particular, a Riemannian metric g considered as a function on T^*M is the Hamiltonian of a free particle on M, and the corresponding motion is given by the geodesic flow on T^*M.

In quantum mechanics, a particle on a compact manifold M is described by a function in the Hilbert space $L^2(M)$ called the wave function. The evolution of the quantum particle is determined by the Schrödinger equation. The Hamiltonian of a free quantum particle, moving on a Riemannian manifold (M, g), is the Laplace–Beltrami operator Δ_g associated with the Riemannian metric g.

It would be more convenient for us to consider Heisenberg's picture of quantum mechanics, which deals with observable quantities and their dynamics. For a free particle on a compact manifold M, classical observables are real-valued functions on the phase space T^*M, and quantum observables are pseudodifferential operators on M considered as (unbounded) self-adjoint operators in $L^2(M)$. The quantum evolution of pseudodifferential operators is described by the Heisenberg equations of motion. The classical evolution of their principal symbols is induced by the action of the geodesic flow. The quantum and the classical evolutions are related by the Egorov theorem.

The purpose of this paper is to discuss classical and quantum dynamical systems for a particular class of Riemannian singular spaces, namely, for the leaf space

---

2010 *Mathematics Subject Classification.* 58J40, 58J42, 58B34.
Supported by the Russian Foundation of Basic Research, grant 07-01-00081-a.
This is the final form of the paper.

$M/\mathcal{F}$ of a Riemannian foliation $\mathcal{F}$ on a compact manifold M. In this case, it is very natural to use ideas and notions of noncommutative geometry by A. Connes [3]. The conormal bundle $N^*\mathcal{F}$ of $\mathcal{F}$ carries a natural foliation, $\mathcal{F}_N$, so that the leaf space $N^*\mathcal{F}/\mathcal{F}_N$ of this foliation can be naturally considered as the cotangent bundle of $M/\mathcal{F}$. Using constructions of noncommutative geometry, one can associate to the singular space $N^*\mathcal{F}/\mathcal{F}_N$ some noncommutative algebra, denoted in the paper by $C^\infty_{\text{prop}}(G_{\mathcal{F}_N}, |T\mathcal{G}_N|^{1/2})$, which plays the role of an algebra of smooth functions on this space, that is, of the algebra of classical observables. The algebra of quantum observables is given by the algebra of transverse pseudodifferential operators $\Psi^{0,-\infty}(M,\mathcal{F})$ on M. We consider quantum Hamiltonians defined by transverse Dirac operators. Then the associated classical dynamics on $C^\infty_{\text{prop}}(G_{\mathcal{F}_N}, |T\mathcal{G}_N|^{1/2})$ is induced by the action of the transverse geodesic flow on $N^*\mathcal{F}$, and the quantum evolution of transverse pseudodifferential operators is related with the classical evolution by a version of Egorov's theorem for transversally elliptic operators stated in [12, 13].

These results are closely related to the reduction theory for quantum Hamiltonian systems with symmetry (see, for instance, [7–9, 22, 24] and references therein).

We believe that the results discussed in the paper will play some important role in further investigations of spectral theory of transversally elliptic operators on foliated manifolds as well as in the study of asymptotic spectral problems for elliptic operators on foliated manifolds, such as adiabatic limits.

The paper is organized as follows. In Section 1, we review the results on classical and quantum dynamics in transverse geometry of Riemannian foliations mentioned above. In Section 2, we discuss two examples of Riemannian foliations: a foliation given by the fibers of a fibration and a linear foliation on the two-dimensional torus.

1. Definitions and main results

Throughout in the paper, $(M,\mathcal{F})$ is a compact foliated manifold, $\dim M = n$, $\dim \mathcal{F} = p$, $p + q = n$. We will consider foliated charts $\varkappa\colon U \subset M \to I^p \times I^q$ on M with coordinates $(x, y) \in I^p \times I^q$ (I is the open interval $(0, 1)$) such that the restriction of $\mathcal{F}$ to U is given by the sets $y = \text{const}$. We will also use the following notation: $T\mathcal{F}$ is the tangent bundle of $\mathcal{F}$; $Q = TM/T\mathcal{F}$ is the normal bundle of $\mathcal{F}$; $N^*\mathcal{F}$ is the conormal bundle of $\mathcal{F}$.

1.1. Classical phase space. To define the cotangent bundle of the leaf space $M/\mathcal{F}$ of the foliation $(M,\mathcal{F})$, one can proceed as follows. The conormal bundle $N^*\mathcal{F}$ carries a natural foliation, $\mathcal{F}_N$, called the linearized or the lifted foliation. One can define it by constructing its foliated atlas. Given a foliated chart $\varkappa\colon U \subset M \to I^p \times I^q$ on M with coordinates $(x, y) \in I^p \times I^q$, there is the corresponding chart in T^*M with coordinates written as $(x, y, \xi, \eta) \in I^p \times I^q \times \mathbb{R}^p \times \mathbb{R}^q$. In these coordinates, the restriction of the conormal bundle $N^*\mathcal{F}$ to U is given by the equation $\xi = 0$. So we have a chart $\varkappa_n\colon U_1 \subset N^*\mathcal{F} \xrightarrow{\sim} I^p \times I^q \times \mathbb{R}^q$ on $N^*\mathcal{F}$ with the coordinates $(x, y, \eta) \in I^p \times I^q \times \mathbb{R}^q$. This chart is a foliated chart on $N^*\mathcal{F}$ for the linearized foliation $\mathcal{F}_N$, and the restriction of $\mathcal{F}_N$ to U_1 is given by the level sets $y = \text{const}$, $\eta = \text{const}$.

If the foliation $\mathcal{F}$ is given by the fibers of a fibration $f\colon M \to B$ over a compact manifold B, then the leaf space $N^*\mathcal{F}/\mathcal{F}_N$ of the linearized foliation $\mathcal{F}_N$ coincides with the cotangent bundle T^*B of B. Therefore, for an arbitrary foliation $\mathcal{F}$, the leaf space $N^*\mathcal{F}/\mathcal{F}_N$ can be naturally considered as the cotangent bundle of the leaf space $M/\mathcal{F}$.

Remark 1. This construction can be viewed a particular example of the symplectic reduction in the sense of [16, Chapter III, Section 14] (see also [17, 18]).

Let (X, ω) be a symplectic manifold, and Y a submanifold of X such that the 2-form ω_Y induced by ω on Y is of constant rank. Let $\mathcal{F}_Y$ be the characteristic foliation of Y relative to ω_Y (that is, $T\mathcal{F}_Y$ is the skew-orthogonal complement of TY in TX). If the foliation $\mathcal{F}_Y$ is simple, that is, it is given by the fibers of a surjective submersion p of Y to a smooth manifold Z, then Z has a unique symplectic form ω_Z such that $p^*\omega_Z = \omega_Y$. The symplectic manifold (Z, ω_Z) is said to be the reduced symplectic manifold associated with Y. In a particular case when the submanifold Y is the preimage of a point under the momentum map associated with the Hamiltonian action of a Lie group, the symplectic reduction associated with Y is the Marsden–Weinstein symplectic reduction [19].

In the case under consideration, one can consider the symplectic reduction associated with the coisotropic submanifold $Y = N^*\mathcal{F}$ in the symplectic manifold $X = T^*M$. The corresponding characteristic foliation $\mathcal{F}_Y$ is the linearized foliation $\mathcal{F}_N$, and the leaf space $N^*\mathcal{F}/\mathcal{F}_N$ plays a role of the reduced phase space Z.

In the general case, the leaf space $N^*\mathcal{F}/\mathcal{F}_N$ is not a smooth manifold. We will use the ideas of the noncommutative geometry in the sense of A. Connes [3] and introduce a noncommutative algebra, $C^\infty_{\text{prop}}(G_{\mathcal{F}_N}, |T\mathcal{G}_N|^{1/2})$, as a noncommutative analogue of an algebra of smooth functions on $N^*\mathcal{F}/\mathcal{F}_N$.

First, we recall several notions (for more details, see, e.g., [14] and references therein). Let $\gamma\colon [0.1] \to M$ be a continuous leafwise path in M with the initial point $x = \gamma(0)$ and the final point $y = \gamma(1)$ and T_0 and T_1 arbitrary smooth submanifolds (possibly, with boundary), transversal to the foliation, such that $x \in T_0$ and $y \in T_1$. Sliding along the leaves of the foliation $\mathcal{F}$ determines a diffeomorphism $H_{T_0T_1}(\gamma)$ of a neighborhood of x in T_0 to a neighborhood of y in T_1, called the holonomy map along γ. The differential of $H_{T_0T_1}(\gamma)$ at x gives rise to a well-defined linear map $Q_x \to Q_y$, which is independent of the choice of transversals T_0 and T_1. This map is called the linear holonomy map and denoted by dh_γ. The adjoint of dh_γ yields a linear map $dh^*_\gamma\colon N^*\mathcal{F}_y \to N^*\mathcal{F}_x$.

The holonomy groupoid G of $\mathcal{F}$ consists of $\sim_h$-equivalence classes of continuous leafwise paths in M, where we set $\gamma_1 \sim_h \gamma_2$, if γ_1 and γ_2 have the same initial and final points and the same holonomy maps. The set of units $G^{(0)}$ is the manifold M. The multiplication in G is given by the product of paths. The corresponding source map $s\colon G \to M$ and range map $r\colon G \to M$ are given by $s(\gamma) = \gamma(0)$ and $r(\gamma) = \gamma(1)$. Finally, the diagonal map $\Delta\colon M \to G$ takes any $x \in M$ to the element in G given by the constant path $\gamma(t) = x, t \in [0, 1]$. To simplify the notation, we will identify $x \in M$ with $\Delta(x) \in G$. For any $x \in M$ the map s maps G^x on the leaf L_x through x. The group G^x_x coincides with the holonomy group of L_x. The map $s\colon G^x \to L_x$ is the covering map associated with the group G^x_x, called the holonomy covering.

Let us introduce the groupoid $G_{\mathcal{F}_N}$ as the set of all $(\gamma, \nu) \in G \times N^*\mathcal{F}$ such that $r(\gamma) = \pi(\nu)$. The source map $s_N\colon G_{\mathcal{F}_N} \to N^*\mathcal{F}$ and the range map $r_N\colon G_{\mathcal{F}_N} \to N^*\mathcal{F}$ are defined as $s_N(\gamma, \nu) = dh^*_\gamma(\nu)$ and $r_N(\gamma, \nu) = \nu$. We have a map $\pi_G\colon G_{\mathcal{F}_N} \to G$ given by $\pi_G(\gamma, \nu) = \gamma$. If $\mathcal{F}$ is a Riemannian foliation, $G_{\mathcal{F}_N}$ coincides with the holonomy groupoid of the linearized foliation $\mathcal{F}_N$, but, for a general foliation, these two groupoids may be different. Observe also that the leaf of $\mathcal{F}_N$ through $\nu \in N^*\mathcal{F}$ can be described as the set of all points $dh^*_\gamma(\nu) \in N^*\mathcal{F}$, where $\gamma \in G, r(\gamma) = \pi(\nu)$ (here $\pi\colon T^*M \to M$ is the bundle map).

The groupoid $G_{\mathcal{F}_N}$ carries a natural codimension q foliation $\mathcal{G}_N$. The leaf of $\mathcal{G}_N$ through a point $(\gamma, \nu) \in G_{\mathcal{F}_N}$ is the set of all $(\gamma', \nu') \in G_{\mathcal{F}_N}$ such that ν and ν' lie in the same leaf in $\mathcal{F}_N$. For any vector bundle V on M, denote by $|V|^{1/2}$ the associated half-density vector bundle. Let $|T\mathcal{G}_N|^{1/2}$ be the line bundle of leafwise half-densities on $G_{\mathcal{F}_N}$ with respect to the foliation $\mathcal{G}_N$. It is easy to see that

$$|T\mathcal{G}_N|^{1/2} = r_N^*(|T\mathcal{F}_N|^{1/2}) \otimes s_N^*(|T\mathcal{F}_N|^{1/2}),$$

where $s_N^*(|T\mathcal{F}_N|^{1/2})$ and $r_N^*(|T\mathcal{F}_N|^{1/2})$ denote the lifts of the line bundle $|T\mathcal{F}_N|^{1/2}$ of leafwise half-densities on $N^*\mathcal{F}$ via the source and the range mappings s_N and r_N respectively.

Let $\varkappa_0\colon U \to I^p \times I^q, \varkappa_1\colon U' \to I^p \times I^q$ be two foliated charts, let $T_0 = \varkappa_0^{-1}(\{0\} \times I^q)$, $T_1 = \varkappa_1^{-1}(\{0\} \times I^q)$ be the corresponding local transversals, and let $\pi_0\colon T_0 \to I^q, \pi_1\colon T_1 \to I^q$ be the corresponding diffeomorphisms. The foliated charts $\varkappa_0$ and $\varkappa_1$ are called compatible, if, for any $m \in U$ and $m' \in U'$ with $\pi_0(m) = \pi_1(m')$, there is a leafwise path γ from m to m' such that the corresponding holonomy diffeomorphism $H_{T_0T_1}(\gamma)$ considered as a local diffeomorphism of I^q is the identity in a neighborhood of $\pi_0(m) = \pi_1(m')$.

For any pair of compatible foliated charts $\varkappa_0$ and $\varkappa_1$ denote by $W(\varkappa_0, \varkappa_1)$ the subset in G, consisting of all $\gamma \in G$ such that $s(\gamma) = \varkappa_0^{-1}(x, y) \in U$ and $r(\gamma) = \varkappa_1^{-1}(x', y) \in U'$ for some $(x, x', y) \in I^p \times I^p \times I^q$ and the corresponding holonomy diffeomorphism $H_{T_0T_1}(\gamma)$ considered as a local diffeomorphism in I^q is the identity in a neighborhood of $y \in I^q$. There is a coordinate map $\Gamma\colon W(\varkappa_0, \varkappa_1) \to I^p \times I^p \times I^q$, which takes an element $\gamma \in W(\varkappa_0, \varkappa_1)$ as above to the corresponding triple $(x, x', y) \in I^p \times I^p \times I^q$. As shown in [2], the coordinate neighborhoods $W(\varkappa_0, \varkappa_1)$ form an atlas of a $(2p + q)$-dimensional manifold (in general, non-Hausdorff and non-paracompact) on G. Moreover, the groupoid G is a smooth groupoid.

Now let $\varkappa\colon U \subset M \to I^p \times I^q, \varkappa_1\colon U' \subset M \to I^p \times I^q$, be two compatible foliated charts on M. Then the corresponding foliated charts $\varkappa_n\colon U_1 \subset N^*\mathcal{F} \to I^p \times I^q \times \mathbb{R}^q, (\varkappa_1)_n\colon U_1' \subset N^*\mathcal{F} \to I^p \times I^q \times \mathbb{R}^q$, are compatible with respect to the foliation $\mathcal{F}_N$. So they define a foliated chart V on the foliated manifold $(G_{\mathcal{F}_N}, \mathcal{G}_N)$ with the coordinates $(x, x', y, \eta) \in I^p \times I^p \times I^q \times \mathbb{R}^q$, and the restriction of $\mathcal{G}_N$ to V is given by the level sets $y = \text{const}$, $\eta = \text{const}$.

A section $\sigma \in C^\infty(G_{\mathcal{F}_N}, |T\mathcal{G}_N|^{1/2})$ is said to be properly supported, if the restriction of the map $r_N : G_{\mathcal{F}_N} \to N^*\mathcal{F}$ to $\operatorname{supp}\sigma$ is a proper map. One can introduce the structure of involutive algebra on the space $C^\infty_{\text{prop}}(G_{\mathcal{F}_N}, |T\mathcal{G}_N|^{1/2})$ of smooth, properly supported sections of $|T\mathcal{G}_N|^{1/2}$ by the formulas

$$\begin{aligned}
\sigma_1 * \sigma_2(\gamma, \nu) &= \int_{(\gamma_1,\nu_1)(\gamma_2,\nu_2)=(\gamma,\nu)} \sigma_1(\gamma_1, \nu_1)\sigma_2(\gamma_2, \nu_2), \qquad &(\gamma, \nu) \in G_{\mathcal{F}_N}, \\
\sigma^*(\gamma, \nu) &= \overline{\sigma((\gamma, \nu)^{-1})}, \qquad &(\gamma, \nu) \in G_{\mathcal{F}_N},
\end{aligned} \tag{1}$$

where $\sigma, \sigma_1, \sigma_2 \in C^\infty_{\text{prop}}(G_{\mathcal{F}_N}, |T\mathcal{G}_N|^{1/2})$. The formula for $\sigma_1 * \sigma_2$ should be interpreted in the following way. The composition $(\gamma, \nu) = (\gamma_1, \nu_1)(\gamma_2, \nu_2)$ is defined if $\nu_2 = dh^*_{\gamma_1}(\nu_1)$ and equals $(\gamma, \nu) = (\gamma_1\gamma_2, \nu_1)$. Then we have

$$\sigma_1(\gamma_1,\nu_1)\sigma_2(\gamma_2,\nu_2) \in |T_{\nu_1}\mathcal{F}_N|^{1/2} \otimes |T_{dh^*_{\gamma_1}(\nu_1)}\mathcal{F}_N|^{1/2} \otimes |T_{\nu_2}\mathcal{F}_N|^{1/2} \otimes |T_{dh^*_{\gamma_2}(\nu_2)}\mathcal{F}_N|^{1/2}$$
$$\cong |T_{\nu_1}\mathcal{F}_N|^{1/2} \otimes |T_{\nu_2}\mathcal{F}_N|^{1} \otimes |T_{dh^*_{\gamma_1\gamma_2}(\nu_1)}\mathcal{F}_N|^{1/2},$$

and, integrating the $|T_{\nu_2}\mathcal{F}_N|^1$-component of $\sigma_1(\gamma_1,\nu_1)\sigma_2(\gamma_2,\nu_2)$ over the set of all $\nu_2 \in N^*\mathcal{F}$ of the form $\nu_2 = dh^*_{\gamma_1}(\nu)$, which coincides with the leaf of $\mathcal{F}_N$ through ν, we get a well-defined section of the bundle $r_N^*(|T\mathcal{F}_N|^{1/2}) \otimes s_N^*(|T\mathcal{F}_N|^{1/2}) = |T\mathcal{G}_N|^{1/2}$.

As mentioned above, the algebra $C^\infty_{\text{prop}}(G_{\mathcal{F}_N}, |T\mathcal{G}_N|^{1/2})$ plays a role of the noncommutative analogue of an algebra of functions on the leaf space $N^*\mathcal{F}/\mathcal{F}_N$.

1.2. Transversal pseudodifferential operators. Our quantum phase space is a certain completion of the algebra $\Psi^{*,-\infty}(M,\mathcal{F},E)$ of transversal pseudodifferential operators introduced in [11]. So we will briefly recall the definition of this algebra. Let E be a complex vector bundle over M of rank r. We will consider pseudodifferential operators, acting on half-densities. Let $C^\infty(M,E)$ denote the space of smooth sections of the vector bundle $E \otimes |TM|^{1/2}$, $L^2(M,E)$ the Hilbert space of square integrable sections of $E \otimes |TM|^{1/2}$, $\mathcal{D}'(M,E)$ the space of distributional sections of $E \otimes |TM|^{1/2}$, $\mathcal{D}'(M,E) = C^\infty(M,E)'$. Finally, let $\Psi^m(M,E)$ denote the standard classes of pseudodifferential operators, acting on $C^\infty(M,E)$.

For any $k_A \in S^m(I^p \times I^p \times I^q \times \mathbb{R}^q, \mathcal{L}(\mathbb{C}^r))$, define an operator $A\colon C^\infty_c(I^n,\mathbb{C}^r) \to C^\infty(I^n,\mathbb{C}^r)$ by the formula

$$Au(x,y) = (2\pi)^{-q} \int e^{i(y-y')\eta} k_A(x,x',y,\eta) u(x',y')\, dx'\, dy'\, d\eta, \tag{2}$$

where $u \in C^\infty_c(I^n,\mathbb{C}^r)$, $x \in I^p$, $y \in I^q$. The function k_A is called the complete symbol of A. As usual, we will consider only classical (or polyhomogeneous) symbols, that is, those symbols, which can be represented as an asymptotic sum of homogeneous (in η) components.

Let $\varkappa\colon U \subset M \to I^p \times I^q$, $\varkappa_1\colon U' \subset M \to I^p \times I^q$ be a pair of compatible foliated charts on M equipped with trivializations of the bundle E over them. Any operator A of the form (2) with the Schwartz kernel, compactly supported in $I^n \times I^n$, determines an operator $A\colon C^\infty_c(U, E|_U) \to C^\infty_c(U', E|_{U'})$, which extends to an operator in $C^\infty(M,E)$ in a trivial way. The resulting operator is called an elementary operator of class $\Psi^{m,-\infty}(M,\mathcal{F},E)$.

The class $\Psi^{m,-\infty}(M,\mathcal{F},E)$ consists of all operators A in $C^\infty(M,E)$, which can be represented in the form

$$A = \sum_i A_i + K,$$

where A_i are elementary operators of class $\Psi^{m,-\infty}(M,\mathcal{F},E)$, corresponding to a pair $\varkappa_i, \varkappa_i'$ of compatible foliated charts, $K \in \Psi^{-\infty}(M,E)$.

The principal symbol of the operator A given by (2) is the leafwise half-density

$$\sigma(A)(x,x',y,\eta) = k_{A,m}(x,x',y,\eta)|dx|^{1/2}|dx'|^{1/2}, \tag{3}$$

where $k_{A,m}$ is the degree m homogeneous component of the complete symbol k_A.

The global definition of the principal symbol is given as follows. Let π^*E denote the lift of the vector bundle E to $\widetilde{T}^*M = T^*M\backslash 0$ via the bundle map $\pi\colon \widetilde{T}^*M \to M$. Put $\widetilde{N}^*\mathcal{F} = N^*\mathcal{F}\backslash 0$. Denote by $\widetilde{\mathcal{F}}_N$ the restriction of $\mathcal{F}_N$ and by $G_{\widetilde{\mathcal{F}}_N}$ the groupoid associated with $\widetilde{\mathcal{F}}_N$. Let $\mathcal{L}(\pi^*E)$ be the vector bundle on $G_{\widetilde{\mathcal{F}}_N}$, whose fiber at a point $(\gamma,\nu) \in G_{\widetilde{\mathcal{F}}_N}$ is the space $\mathcal{L}\big((\pi^*E)_{s_N(\gamma,\nu)}, (\pi^*E)_{r_N(\gamma,\nu)}\big)$ of linear maps from $(\pi^*E)_{s_N(\gamma,\nu)}$ to $(\pi^*E)_{r_N(\gamma,\nu)}$. Consider the space $C^\infty_{\text{prop}}(G_{\widetilde{\mathcal{F}}_N}, \mathcal{L}(\pi^*E) \otimes |T\mathcal{G}_N|^{1/2})$ of smooth, properly supported sections of $\mathcal{L}(\pi^*E) \otimes |T\mathcal{G}_N|^{1/2}$. One can introduce the structure of involutive algebra on $C^\infty_{\text{prop}}(G_{\widetilde{\mathcal{F}}_N}, \mathcal{L}(\pi^*E) \otimes |T\mathcal{G}_N|^{1/2})$ by formulas similar to (1). Let $S^m(G_{\mathcal{F}_N}, \mathcal{L}(\pi^*E) \otimes |T\mathcal{G}_N|^{1/2})$ be the space of all $s \in C^\infty_{\text{prop}}(G_{\widetilde{\mathcal{F}}_N}, \mathcal{L}(\pi^*E) \otimes |T\mathcal{G}_N|^{1/2})$ homogeneous of degree m with respect to the action of $\mathbb{R}$ given by the multiplication in the fibers of the vector bundle $\pi_G\colon G_{\mathcal{F}_N} \to G$.

The principal symbol $\sigma(A)$ of an operator $A \in \Psi^{m,-\infty}(M,\mathcal{F},E)$ given in local coordinates by the formula (3) is globally defined as an element of the space $S^m(G_{\mathcal{F}_N}, \mathcal{L}(\pi^*E) \otimes |T\mathcal{G}_N|^{1/2})$. Thus, we have the half-density principal symbol mapping

$$\sigma\colon \Psi^{m,-\infty}(M,\mathcal{F},E) \to S^m(G_{\mathcal{F}_N}, \mathcal{L}(\pi^*E) \otimes |T\mathcal{G}_N|^{1/2}),$$

which satisfies

$$\sigma(AB) = \sigma(A)\sigma(B), \qquad \sigma(A^*) = \sigma(A)^*$$

for any $A \in \Psi^{m_1,-\infty}(M,\mathcal{F},E)$ and $B \in \Psi^{m_2,-\infty}(M,\mathcal{F},E)$.

1.3. Riemannian foliations. We will consider a particular class of Hamiltonians associated with a naive analogue of a Riemannian metric on the leaf space $M/\mathcal{F}$. It only exists for a particular class of foliations, called Riemannian foliations.

Let M be a compact manifold equipped with a foliation $\mathcal{F}$, g_M a Riemannian metric on M. Let $H = T\mathcal{F}^\perp$ be the orthogonal complement of the tangent bundle of $\mathcal{F}$. So we have the direct sum decomposition

$$TM = T\mathcal{F} \oplus H. \tag{4}$$

There are natural isomorphisms $H \cong Q$ and $H^* \cong Q^* \cong N^*\mathcal{F}$. One can also decompose the metric g_M into the sum of its leafwise and transverse components:

$$g_M = g_F + g_H. \tag{5}$$

In a foliated chart with coordinates $(x,y) \in I^p \times I^q$ the transverse part g_H can be written as $g_H = \sum_{\alpha\beta} g_{\alpha\beta}(x,y)\theta^\alpha\theta^\beta$, where $\theta^\alpha \in H^*$ is the (unique) lift of dy^α under the projection $I^p \times I^q \to I^q$. The metric g_M is called bundle-like, if in any foliated chart the components $g_{\alpha\beta}$ are independent of x, $g_{\alpha\beta}(x,y) = g_{\alpha\beta}(y)$. Equivalently, one can say that g_M is bundle-like, if, for any leafwise continuous path γ from x to y, the corresponding linear holonomy map $dh_\gamma\colon Q_x \to Q_y$ is an isometry (see, for instance, [20, 21] for more details). The foliation $\mathcal{F}$ is called Riemannian, if there exists a bundle-like metric on M. In the sequel, we will always assume that $\mathcal{F}$ is a Riemannian foliation and g_M is a bundle-like metric.

Let P_F (resp. P_H) denotes the orthogonal projection operator of TM onto $T\mathcal{F}$ (resp. H). Denote by ∇^L the Levi-Civita connection on TM defined by g_M. The following formulas define a connection ∇ in H:

$$\begin{aligned} &\nabla_X N = P_H[X,N], && X \in C^\infty(M,T\mathcal{F}), N \in C^\infty(M,H)\\ &\nabla_X N = P_H \nabla^L_X N, && X \in C^\infty(M,H), N \in C^\infty(M,H). \end{aligned} \tag{6}$$

Remark that the first identity in (6) yields a canonical flat connection in H, defined along the leaves of $\mathcal{F}$, which exists for an arbitrary foliation and is called the Bott connection. It turns out that ∇ depends only on the transverse part of the metric

g_M and preserves the inner product of H. It will be called the transverse Levi-Civita connection.

Let $\omega_{\mathcal{F}}$ denote the leafwise Riemannian volume form of $\mathcal{F}$. Let $f \in H_x$ and let $\tilde{f} \in C^\infty(M,H)$ denote any infinitesimal transformation of $\mathcal{F}$, which coincides with f at x. By the definition of infinitesimal transformation, the local flow generated by $\tilde{f}$ preserves the foliation and gives rise to a well-defined action on $\Lambda^p T^*\mathcal{F}$. The mean curvature vector field $\tau \in C^\infty(M,H)$ of $\mathcal{F}$ is defined by the identity

$$L_{\tilde{f}}\omega_{\mathcal{F}} = g_M(\tau, \tilde{f})\omega_{\mathcal{F}}.$$

If $e_1, e_2, \dots, e_p$ is a local orthonormal frame in $T\mathcal{F}$, then

$$\tau = \sum_{i=1}^{p} P_H(\nabla^L_{e_i} e_i).$$

1.4. Transverse Dirac operators. For any $x \in M$, denote by $\mathrm{Cl}(Q_x)$ the complex Clifford algebra of Q_x. Recall that, relative to an orthonormal basis $\{f_1, f_2, \dots, f_q\}$ of Q_x, $\mathrm{Cl}(Q_x)$ is the complex algebra generated by 1 and $f_1, f_2, \dots,$ f_q, satisfying the relations

$$f_\alpha f_\beta + f_\beta f_\alpha = -2\delta_{\alpha\beta}, \qquad \alpha, \beta = 1, 2, \dots, q.$$

Consider the $\mathbb{Z}_2$-graded vector bundle $\mathrm{Cl}(Q)$ over M whose fiber at $x \in M$ is $\mathrm{Cl}(Q_x)$. This bundle is associated to the principal $\mathrm{SO}(q)$-bundle $O(Q)$ of oriented orthonormal frames in Q, $\mathrm{Cl}(Q) = O(Q) \times_{O(q)} \mathrm{Cl}(\mathbb{R}^q)$. Therefore, the transverse Levi-Civita connection ∇ induces a natural leafwise flat connection $\nabla^{\mathrm{Cl}(Q)}$ on $\mathrm{Cl}(Q)$ which is compatible with the multiplication and preserves the $\mathbb{Z}_2$-grading on $\mathrm{Cl}(Q)$. If $\{f_1, f_2, \dots, f_q\}$ is a local orthonormal frame in T^HM, and $\omega^\gamma_{\alpha\beta}$ is the coefficients of the connection ∇: $\nabla_{f_\alpha} f_\beta = \sum_\gamma \omega^\gamma_{\alpha\beta} f_\gamma$, then

$$\nabla^{\mathrm{Cl}(Q)}_{f_\alpha} = f_\alpha + \frac{1}{4}\sum_{\gamma=1}^{q} \omega^\gamma_{\alpha\beta} c(f_\beta)c(f_\gamma), \tag{7}$$

where $c(a)$ denotes the action of an element $a \in C^\infty\big(M, \mathrm{Cl}(Q)\big)$ on $C^\infty\big(M, \mathrm{Cl}(Q)\big)$ by pointwise left multiplication.

A transverse Clifford module is a complex vector bundle $\mathcal{E}$ on M endowed with a fiberwise action of the bundle $\mathrm{Cl}(Q)$. We will denote the action of $a \in \mathrm{Cl}(Q_x)$ on $s \in \mathcal{E}_x$ as $c(a)s \in \mathcal{E}_x$. A transverse Clifford module $\mathcal{E}$ is called self-adjoint if it endowed with a Hermitian metric such that the operator $c(f)\colon \mathcal{E}_x \to \mathcal{E}_x$ is skew-adjoint for any $x \in M$ and $f \in Q_x$. Any transverse Clifford module $\mathcal{E}$ carries a natural $\mathbb{Z}_2$-grading $\mathcal{E} = \mathcal{E}_+ \oplus \mathcal{E}_-$.

A Hermitian connection $\nabla^{\mathcal{E}}$ on a transverse Clifford module $\mathcal{E}$ is called a Clifford connection if it is compatible with the Clifford action, that is, for any $f \in C^\infty(M,H)$ and $a \in C^\infty\big(M, \mathrm{Cl}(Q)\big)$,

$$[\nabla^{\mathcal{E}}_f, c(a)] = c(\nabla^{\mathrm{Cl}(Q)}_f a).$$

A self-adjoint transverse Clifford module $\mathcal{E}$ equipped with a leafwise flat Clifford connection $\nabla^{\mathcal{E}}$ is called a transverse Clifford bundle. For any transverse Clifford bundle $(\mathcal{E}, \nabla^{\mathcal{E}})$, we define the operator $D'_{\mathcal{E}}$ acting on the sections of $\mathcal{E}$ as the composition

$$C^\infty(M,\mathcal{E}) \xrightarrow{\nabla^{\mathcal{E}}} C^\infty(M, Q^* \otimes \mathcal{E}) = C^\infty(M, Q \otimes \mathcal{E}) \xrightarrow{c} C^\infty(M, \mathcal{E}).$$

Here we identify the bundle Q and Q^* by means of the metric g_M. The transverse Dirac operator $D_{\mathcal{E}}$ is defined as

$$D_{\mathcal{E}} = D'_{\mathcal{E}} - \tfrac{1}{2}c(\tau).$$

If $f_1, \dots, f_q$ is a local orthonormal frame for H, then

$$D_{\mathcal{E}} = \sum_{\alpha=1}^{q} c(f_\alpha)\big(\nabla^{\mathcal{E}}_{f_\alpha} - \tfrac{1}{2}g_M(\tau, f_\alpha)\big).$$

This operator is odd with respect to the natural $\mathbb{Z}_2$-grading on $\mathcal{E}$.

Denote by $(\cdot,\cdot)_x$ the inner product in the fiber $\mathcal{E}_x$ over $x \in M$. Then the inner product in $L^2(M,\mathcal{E})$ is given by the formula

$$(s_1, s_2) = \int_M \big(s_1(x), s_2(x)\big)_x \omega_M, \qquad s_1, s_2 \in L^2(M,\mathcal{E}),$$

where $\omega_M = \sqrt{\det g}\, dx$ denotes the Riemannian volume form on M. As shown in [13], the transverse Dirac operator $D_{\mathcal{E}}$ is formally self-adjoint in $L^2(M,\mathcal{E})$.

The transverse Dirac operators were introduced in [4,5] (see [13,15] for further references).

We will use the Riemannian volume form ω_M to identify the half-densities bundle with the trivial one. So the action of $D_{\mathcal{E}}$ on half-densities is defined by

$$D_{\mathcal{E}}(u|\omega_M|^{1/2}) = (D_{\mathcal{E}}u)|\omega_M|^{1/2}, \qquad u \in C^\infty(M,\mathcal{E}).$$

Example 2. Assume that $\mathcal{F}$ is transversely oriented and the normal bundle Q is spin. Thus the $\mathrm{SO}(q)$ bundle $O(Q)$ of oriented orthonormal frames in Q can be lifted to a $\mathrm{Spin}(q)$ bundle $O'(Q)$ so that the projection $O'(Q) \to O(Q)$ induces the covering projection $\mathrm{Spin}(q) \to \mathrm{SO}(q)$ on each fiber.

Let $F(Q), F_+(Q), F_-(Q)$ be the bundles of spinors

$$F(Q) = O'(Q) \times_{\mathrm{Spin}(q)} S, \qquad F_\pm(Q) = O'(Q) \times_{\mathrm{Spin}(q)} S_\pm.$$

Since $\dim Q = q$ is even $\operatorname{End} F(Q)$ is as a bundle of algebras over M isomorphic to the Clifford bundle $\mathrm{Cl}(Q)$. The transverse Levi-Civita connection ∇ lifts to a leafwise flat Clifford connection $\nabla^{F(Q)}$ on $F(Q)$. So $F(Q)$ is a transverse Clifford bundle.

More generally, one can take a Hermitian vector bundle $\mathcal{W}$ equipped with a leafwise flat Hermitian connection $\nabla^{\mathcal{W}}$. Then $F(Q)\otimes\mathcal{W}$ is a transverse Clifford bundle: the action of $a \in C^\infty\big(M, \mathrm{Cl}(Q)\big)$ on $C^\infty(M, F(Q)\otimes\mathcal{W})$ is given by $c(a)\otimes 1$ ($c(a)$ denotes the action of a on $C^\infty\big(M, F(Q)\big)$), and the product connection $\nabla^{F(Q)\otimes\mathcal{W}} = \nabla^{F(Q)}\otimes 1 + 1\otimes\nabla^{\mathcal{W}}$ on $F(Q)\otimes\mathcal{W}$ is a Clifford connection.

Example 3. Another example of a transverse Clifford bundle associated with a transverse almost complex structure on $(M,\mathcal{F})$, a transverse Clifford module $\Lambda^{0,*}$, is described in [15].

Example 4. The decomposition (4) induces a bigrading on ΛT^*M:

$$\Lambda T^*M = \bigoplus_{i,j=1}^{n} \Lambda^{i,j}T^*M, \qquad \Lambda^{i,j}T^*M = \Lambda^i T\mathcal{F}^* \otimes \Lambda^j H^*.$$

In this bigrading, the de Rham differential d can be written as

$$d = d_F + d_H + \theta, \tag{8}$$

where d_F and d_H are first order differential operators (called the tangential de Rham differential and the transversal de Rham differential accordingly), and θ is a zero-order differential operator.

By definition, the transverse signature operator is a first-order differential operator in $C^\infty(M,\Lambda H^*)$ given by

$$D_H = d_H + d_H^*.$$

Consider the vector bundle $\mathcal{E} = \Lambda H^* \otimes \mathbb{C}$ equipped with a natural structure of a transverse Clifford bundle. As shown in [13], for the associated transverse Dirac operator $D_\mathcal{E}$, we have

$$D_\mathcal{E} = d_H + d_H^* - \frac{1}{2}(\varepsilon_{\tau^*} + i_\tau),$$

where ε_{τ^*} denotes the exterior product by the covector $\tau^* \in Q_x^*$ dual to τ, i_τ the interior product by τ. So we see that the transverse signature operator D_H coincides with $D_\mathcal{E}$ if and only if $\tau = 0$, that is, all the leaves are minimal submanifolds.

1.5. Quantum dynamics. Let $D_\mathcal{E}$ be a transverse Dirac operator associated with a transverse Clifford bundle $(\mathcal{E},\nabla^\mathcal{E})$. The operator $D_\mathcal{E}$ is essentially self-adjoint with initial domain $C^\infty(M,\mathcal{E})$. Define an unbounded linear operator $\langle D_\mathcal{E}\rangle$ in the space $L^2(M,\mathcal{E})$ as

$$\langle D_\mathcal{E}\rangle = (D_\mathcal{E}^2 + I)^{1/2}.$$

By the spectral theorem, the operator $\langle D_\mathcal{E}\rangle$ is well-defined as a positive, self-adjoint operator in $L^2(M,\mathcal{E})$. It can be shown that the Sobolev space $H^1(M,\mathcal{E})$ is contained in the domain of $\langle D_\mathcal{E}\rangle$ in $L^2(M,\mathcal{E})$.

By the spectral theorem, the operator $\langle D_\mathcal{E}\rangle$ defines a strongly continuous group $e^{it\langle D_\mathcal{E}\rangle}$ of bounded operators in $L^2(M,\mathcal{E})$. Consider a one-parameter group Φ_t of $*$-automorphisms of the algebra $\mathcal{L}\big(L^2(M,\mathcal{E})\big)$ defined by

$$\Phi_t(T) = \mathrm{e}^{\mathrm{i}t\langle D_\mathcal{E}\rangle} T \mathrm{e}^{-\mathrm{i}t\langle D_\mathcal{E}\rangle}, \qquad T \in \mathcal{L}\big(L^2(M,\mathcal{E})\big), t \in \mathbb{R}.$$

By the spectral theorem, the operator $\langle D_\mathcal{E}\rangle^s = (D_\mathcal{E}^2 + I)^{s/2}$ is a well-defined positive self-adjoint operator in $\mathcal{H} = L^2(M,\mathcal{E})$ for any $s \in \mathbb{R}$, which is unbounded if $s > 0$. For any $s \geq 0$, denote by $\mathcal{H}^s$ the domain of $\langle D_\mathcal{E}\rangle^s$, and, for $s < 0$, $\mathcal{H}^s = (\mathcal{H}^{-s})^*$. Put also $\mathcal{H}^\infty = \bigcap_{s\geq 0}\mathcal{H}^s$, $\mathcal{H}^{-\infty} = (\mathcal{H}^\infty)^*$. It is clear that $H^s(M,\mathcal{E}) \subset \mathcal{H}^s$ for any $s \geqslant 0$ and $\mathcal{H}^s \subset H^s(M,\mathcal{E})$ for any $s < 0$. In particular, $C^\infty(M,\mathcal{E}) \subset \mathcal{H}^s$ for any s.

We say that a bounded operator A acting on $\mathcal{H}^\infty$ belongs to $\mathcal{L}(\mathcal{H}^{-\infty},\mathcal{H}^\infty)$ (resp. $\mathcal{K}(\mathcal{H}^{-\infty},\mathcal{H}^\infty)$), if, for any s and r, it extends to a bounded (resp. compact) operator from $\mathcal{H}^s$ to $\mathcal{H}^r$, or, equivalently, the operator $\langle D_\mathcal{E}\rangle^r A \langle D_\mathcal{E}\rangle^{-s}$ extends to a bounded (resp. compact) operator on $L^2(M,\mathcal{E})$. It is easy to see that $\mathcal{L}(\mathcal{H}^{-\infty},\mathcal{H}^\infty)$ is an involutive subalgebra in $\mathcal{L}(\mathcal{H})$ and $\mathcal{K}(\mathcal{H}^{-\infty},\mathcal{H}^\infty)$ is its ideal. We also introduce the class $\mathcal{L}^1(\mathcal{H}^{-\infty},\mathcal{H}^\infty)$, which consists of all operators from $\mathcal{K}(\mathcal{H}^{-\infty},\mathcal{H}^\infty)$ such that, for any s and r, the operator $\langle D_\mathcal{E}\rangle^r A \langle D_\mathcal{E}\rangle^{-s}$ is a trace class operator on $L^2(M,\mathcal{E})$. It should be noted that any operator K with smooth kernel belongs to $\mathcal{L}^1(\mathcal{H}^{-\infty},\mathcal{H}^\infty)$.

Theorem 5 ([13]). *Let $D_\mathcal{E}$ be a transverse Dirac operator associated with a transverse Clifford bundle $(\mathcal{E},\nabla^\mathcal{E})$. For any $K \in \Psi^{m,-\infty}(M,\mathcal{F},\mathcal{E})$, there exists an operator $K(t) \in \Psi^{m,-\infty}(M,\mathcal{F},\mathcal{E})$ such that $\Phi_t(K) - K(t), t \in \mathbb{R}$, is a smooth family of operators of class $\mathcal{L}^1(\mathcal{H}^{-\infty},\mathcal{H}^\infty)$.*

By this theorem, we have a well-defined one-parameter automorphism group

$$\Phi_t\colon \hat{\Psi}^{0,-\infty}(M,\mathcal{F},\mathcal{E}) \to \hat{\Psi}^{0,-\infty}(M,\mathcal{F},\mathcal{E}), \qquad t\in\mathbb{R},$$

of the completed algebra of transverse pseudodifferential operators defined by

$$\hat{\Psi}^{0,-\infty}(M,\mathcal{F},\mathcal{E}) = \Psi^{0,-\infty}(M,\mathcal{F},\mathcal{E}) + \mathcal{L}^1(\mathcal{H}^{-\infty},\mathcal{H}^{\infty}).$$

It describes the quantum dynamics associated with the quantum Hamiltonian $\langle D_{\mathcal{E}}\rangle$.

1.6. Classical dynamics. In this section, we give a definition of the transverse geodesic flow associated with a bundle-like metric g on a compact foliated manifold $(M,\mathcal{F})$. This notion is a particular case of the bicharacteristic flow associated with a first order transversally elliptic operator (see [12, 13]).

Denote by f_t the geodesic flow of the Riemannian metric g on T^*M. Since g is bundle-like, the flow f_t preserves the conormal bundle $N^*\mathcal{F}$, and its restriction to $N^*\mathcal{F}$ (denoted also by f_t) preserves the foliation $\mathcal{F}_N$, that is, takes any leaf of $\mathcal{F}_N$ to a leaf. Moreover, one can show existence of a flow F_t on $G_{\mathcal{F}_N}$ such that $s_N\circ F_t = f_t\circ s_N$, $r_N\circ F_t = f_t\circ r_N$, which preserves the foliation $\mathcal{G}_N$.

Let X be the generator of the geodesic flow f_t. Then it is tangent to $N^*\mathcal{F}$ and determines a vector field on $N^*\mathcal{F}$, denoted also by X. Since f_t preserves the foliation $\mathcal{F}_N$, X is an infinitesimal transformation of $\mathcal{F}_N$, and there exists a vector field $\mathcal{H}$ on $G_{\mathcal{F}_N}$ such that $ds_N(\mathcal{H}) = X$ and $dr_N(\mathcal{H}) = X$. Then the vector field $\mathcal{H}$ is the generator of the flow F_t.

In a foliation chart, X is given by

$$X = \sum_{j=1}^{p}\left(\partial_{\xi_j}\sqrt{g}\frac{\partial}{\partial x_j} - \partial_{x_j}\sqrt{g}\frac{\partial}{\partial \xi_j}\right) + \sum_{k=1}^{q}\left(\partial_{\eta_k}\sqrt{g}\frac{\partial}{\partial y_k} - \partial_{y_k}\sqrt{g}\frac{\partial}{\partial \eta_k}\right),$$

and $\mathcal{H}$ is given by

$$\begin{aligned}\mathcal{H}(x,x',y,\eta) = \sum_{j=1}^{p}\partial_{\xi_j}\sqrt{g}(x,y,0,\eta)\frac{\partial}{\partial x_j} + \sum_{j=1}^{p}\partial_{\xi_j}\sqrt{g}(x',y,0,\eta)\frac{\partial}{\partial x'_j}\\ + \sum_{k=1}^{q}\left(\partial_{\eta_k}\sqrt{g}(y,\eta)\frac{\partial}{\partial y_k} - \partial_{y_k}\sqrt{g}(y,\eta)\frac{\partial}{\partial \eta_k}\right),\\ (x,x',y,\eta)\in I^p\times I^p\times I^q\times\mathbb{R}^q.\end{aligned}$$

The flow F_t^* on $C^\infty_{\text{prop}}(G_{\mathcal{F}_N}, |T\mathcal{G}_N|^{1/2})$ induced by F_t is called the transverse geodesic flow.

Now consider a transverse Clifford bundle $(\mathcal{E},\nabla^{\mathcal{E}})$. The pull-back of the connection form of $\nabla^{\mathcal{E}}$ by the projection $\pi\colon N^*\mathcal{F}\to M$ yields a connection $\nabla^{\pi^*\mathcal{E}}$ on $\pi^*\mathcal{E}$. The parallel transport along the orbits of the transverse geodesic flow f_t on $N^*\mathcal{F}$ associated with the connection $\nabla^{\pi^*\mathcal{E}}$ defines a one-parameter automorphism group α_t of the vector bundle $\pi^*\mathcal{E}$, covering the flow f_t, and the induced flow α_t^* on $C^\infty(N^*\mathcal{F},\pi^*\mathcal{E})$. In its turn, this flow induces the flow $\mathrm{Ad}(\alpha_t)^*$ on the space $C^\infty_{\text{prop}}(G_{\mathcal{F}_N},\mathcal{L}(\pi^*\mathcal{E})\otimes|T\mathcal{G}_N|^{1/2})$, which satisfies

$$\frac{\mathrm{d}}{\mathrm{d}t}\mathrm{Ad}(\alpha_t)^*\sigma = \nabla_{\mathcal{H}}\sigma, \qquad \sigma\in C^\infty_{\text{prop}}(G_{\mathcal{F}_N},\mathcal{L}(\pi^*\mathcal{E})\otimes|T\mathcal{G}_N|^{1/2}).$$

This flow will be called the transverse parallel transport. One can show that

$$\mathrm{Ad}(\alpha_t)^*\circ s_N^* = s_N^*\circ\alpha_t^*, \qquad \mathrm{Ad}(\alpha_t)^*\circ r_N^* = r_N^*\circ\alpha_t^*.$$

The following theorem proved in [13] (for scalar operators, see [12]) relates the quantum dynamics defined by a transverse Dirac operator with the corresponding classical dynamics.

Theorem 6. *Let $D_{\mathcal{E}}$ be a transverse Dirac operator associated with a transverse Clifford bundle $(\mathcal{E}, \nabla^{\mathcal{E}})$. For any $K \in \Psi^{m,-\infty}(M, \mathcal{F}, \mathcal{E})$ with the principal symbol $\sigma \in S^m(G_{\mathcal{F}_N}, \mathcal{L}(\pi^*\mathcal{E}) \otimes |T\mathcal{G}_N|^{1/2})$, the principal symbol $\sigma_t \in S^m(G_{\mathcal{F}_N}, \mathcal{L}(\pi^*\mathcal{E}) \otimes |T\mathcal{G}_N|^{1/2})$ of the operator $K(t)$ is given by*

$$\sigma_t = \operatorname{Ad}(\alpha_t)^*\sigma.$$

Remark 7. The construction of the transversal geodesic flow provides an example of what can be called noncommutative symplectic (or, maybe, better, Poisson) reduction in the setting discussed above.

As in Remark 1, let (X, ω) be a symplectic manifold, and Y a submanifold of X such that the 2-form ω_Y induced by ω on Y is of constant rank. Let $\mathcal{F}_Y$ be the characteristic foliation of Y relative to ω_Y. Suppose that the foliation $\mathcal{F}_Y$ is simple and (B, ω_B) is the associated reduced symplectic manifold. If Y is invariant under the Hamiltonian flow of a Hamiltonian $H \in C^\infty(X)$ (this is equivalent to the fact that $(dH)\big|_Y$ is constant along the leaves of the characteristic foliation $\mathcal{F}_Y$), there exists a unique function $\widehat{H} \in C^\infty(B)$, called the reduced Hamiltonian, such that $H\big|_Y = \widehat{H} \circ p$. Furthermore, the map p projects the restriction of the Hamiltonian flow of H to Y to the reduced Hamiltonian flow on B defined by the reduced Hamiltonian $\widehat{H}$ (see, for instance, [16, Chapter III, Theorem 14.6]).

In the case under consideration, one can consider as above the symplectic reduction associated with the coisotropic submanifold $Y = N^*\mathcal{F}$. The Hamiltonian of the geodesic flow on T^*M determined by a bundle-like metric g_M on M satisfies the above invariance condition, and, if the foliation $\mathcal{F}$ is simple, one can apply the symplectic reduction procedure mentioned above, that gives us the geodesic flow on the cotangent bundle of the base as the corresponding reduced Hamiltonian flow (see also Section 2.1). For a general foliation, following the ideas of noncommutative geometry, one can consider the corresponding reduced Hamiltonian flow as the flow F_t^* on the algebra $C^\infty_{\text{prop}}(G_{\mathcal{F}_N}, |T\mathcal{G}_N|^{1/2})$. Following the ideas of [1, 23], one can interpret the algebra $C^\infty_{\text{prop}}(G_{\mathcal{F}_N}, |T\mathcal{G}_N|^{1/2})$ as a noncommutative Poisson manifold and the flow F_t^* as a noncommutative Hamiltonian flow.

2. Examples

2.1. Riemannian submersions. Suppose that a foliation $\mathcal{F}$ on a compact manifold M is given by the fibers of a fibration $p\colon M \to B$ over a compact manifold B. Then, for any $x \in M$, $N_x^*\mathcal{F}$ coincides with the image of the cotangent map $dp(x)^*\colon T^*_{p(x)}B \to T^*_xM$. The inverse maps $(dp(x)^*)^{-1}\colon N^*_x\mathcal{F} \to T^*_{p(x)}B$ determine a fibration $N^*\mathcal{F} \to T^*B$ whose fibers are the leaves of the linearized foliation $\mathcal{F}_N$. Thus, $N^*\mathcal{F}$ is diffeomorphic to the fiber product

$$M \times_B T^*B = \{(x, \xi) \in M \times T^*B : p(x) = \pi(\xi)\}$$

with a diffeomorphism $M \times_B T^*B \xrightarrow{\cong} N^*\mathcal{F}$, given by

$$(x, \xi) \in M \times_B T^*B \mapsto dp(x)^*(\xi) \in N^*_x\mathcal{F}. \tag{9}$$

The holonomy groupoid G of $\mathcal{F}$ is the fiber product

$$M \times_B M = \{(x,y) \in M \times M : p(x) = p(y)\},$$

where $s(x,y) = y, r(x,y) = x$. Similarly, the holonomy groupoid $G_{\mathcal{F}_N}$ is the fiber product $N^*\mathcal{F} \times_{T^*B} N^*\mathcal{F}$, which consists of all $(\nu_x, \nu_y) \in N^*_x\mathcal{F} \times N^*_y\mathcal{F}$ such that $(x,y) \in M \times_B M$ and $(dp(x)^*)^{-1}(\nu_x) = (dp(x)^*)^{-1}(\nu_y)$, with $s_N(\nu_x,\nu_y) = \nu_y$, $r_N(\nu_x,\nu_y) = \nu_x$. On the other hand, $N^*\mathcal{F} \times_{T^*B} N^*\mathcal{F}$ is also diffeomorphic to the fiber product

$$G \times_B T^*B = \{(x,y,\xi) \in M \times M \times T^*B : p(x) = p(y) = \pi_B(\xi)\}.$$

A diffeomorphism $G \times_B T^*B \xrightarrow{\cong} N^*\mathcal{F} \times_{T^*B} N^*\mathcal{F}$ can be defined as

$$(x,y,\xi) \in G \times_B T^*B \mapsto \big(dp(x)^*(\xi), dp(y)^*(\xi)\big) \in N^*\mathcal{F} \times_{T^*B} N^*\mathcal{F}.$$

The foliation $\mathcal{G}_N$ is given by the fibers of the fibration

$$(x,y,\xi) \in G_{\mathcal{F}_N} \cong G \times_B T^*B \mapsto \xi \in T^*B.$$

As in Remark 1, consider the cotangent bundle T^*M as a symplectic manifold equipped with the canonical symplectic structure, and $N^*\mathcal{F}$ as its closed coisotropic submanifold. Then the linearized foliation $\mathcal{F}_N$ coincides with the characteristic foliation of this coisotropic submanifold. It is well-known (see, for instance, [6]) that the fiber product $N^*\mathcal{F} \times_{T^*B} N^*\mathcal{F}$ is a canonical relation in T^*M, which is often called the flowout of the coisotropic submanifold $N^*\mathcal{F}$. For a complex vector bundle E on M, the algebra of Fourier integral operators in $C^\infty(M,E)$ associated with this canonical relation coincides with the algebra $\Psi^{*,-\infty}(M,\mathcal{F},E)$. It is a particular case of a general construction described in [6].

In this case, the corresponding classes of symbols $S^m(G_{\mathcal{F}_N}, \mathcal{L}(\pi^*E) \otimes |T\mathcal{G}_N|^{1/2})$ can be described as follows. For any $\xi \in T^*B$, let $\Psi^{-\infty}\big((N^*\mathcal{F})_\xi, (\pi^*E)_\xi\big)$ be the involutive algebra of all smoothing operators, acting on $C^\infty\big((N^*\mathcal{F})_\xi, (\pi^*E)_\xi\big)$, where $(N^*\mathcal{F})_\xi$ is the fiber of the fibration $N^*\mathcal{F} \to T^*B$ at ξ and $(\pi^*E)_\xi$ is the restriction of π^*E to $(N^*\mathcal{F})_\xi$. Consider a field $\Psi^{-\infty}(N^*\mathcal{F}, \pi^*E)$ of involutive algebras on T^*B whose fiber at $\xi \in T^*B$ is $\Psi^{-\infty}\big((N^*\mathcal{F})_\xi, (\pi^*E)_\xi\big)$. For any section K of the field $\Psi^{-\infty}(N^*\mathcal{F}, \pi^*E)$, the Schwartz kernels of the operators K_ξ in $C^\infty\big((N^*\mathcal{F})_\xi, (\pi^*E)_\xi\big)$ determine a well-defined section σ_K of $\mathcal{L}(\pi^*E) \otimes |T\mathcal{G}_N|^{1/2}$ over $G_{\mathcal{F}_N} \cong N^*\mathcal{F} \times_{T^*B} N^*\mathcal{F}$. We say that the section K is smooth, if the corresponding section σ_K is smooth. This defines an algebra isomorphism of $\Psi^{-\infty}(N^*\mathcal{F}, \pi^*E)$ with $C^\infty(G_{\mathcal{F}_N}, \mathcal{L}(\pi^*E) \otimes |T\mathcal{G}_N|^{1/2})$.

A Riemannian metric g_M on M is bundle-like if and only if there exists a Riemannian metric g_B on B such that, for any $x \in M$, the tangent map dp induces an isometry from (H_x, g_H) to $(T_{p(x)}B, g_B)$, or, equivalently, $p : (M, g_M) \to (B, g_B)$ is a Riemannian submersion. Then the transverse geodesic flow f_t of g_M projects under the map $N^*\mathcal{F} \to T^*B$ to the geodesic flow f_t^B of g_B that implies commutativity of the following diagram

$$\begin{array}{ccc} N^*\mathcal{F} & \xrightarrow{f_t} & N^*\mathcal{F} \\ \downarrow & & \downarrow \\ T^*B & \xrightarrow{f_t^B} & T^*B \end{array}$$

Commutativity of this diagram allows us to lift the flow f_t to the flow F_t on the holonomy groupoid $G_{\mathcal{F}_N} \cong N^*\mathcal{F} \times_{T^*B} N^*\mathcal{F}$ in the following way:

$$F_t(\nu_x, \nu_y) = \big(f_t(\nu_x), f_t(\nu_y)\big), \qquad (\nu_x, \nu_y) \in N^*\mathcal{F} \times_{T^*B} N^*\mathcal{F}.$$

If W is a Clifford bundle on B equipped with a Clifford connection ∇^W and D_W is the associated Dirac operator acting on $C^\infty(B, W)$, then its lift p^*W to M has a natural structure of a transverse Clifford bundle, the pull-back of the connection ∇^W is a leafwise flat, transverse Clifford connection on p^*W, and the associated transverse Dirac operator can be considered as a natural lift of D_W to an operator in $C^\infty(M, p^*W)$.

Remark 8. We refer the reader to [13] for a discussion of a particular case when the fibration $p\colon M \to B$ is a principal K-bundle with a compact group K.

Remark 9. Since any compact reduced orbifold is diffeomorphic to the quotient of a compact manifold by an action of a compact Lie group with finite isotropy groups, Theorems 5 and 6 imply the Egorov theorem for Dirac-type operators on compact reduced Riemannian orbifolds (see [13] for more details).

2.2. Linear foliation on the two-torus. Consider the linear foliation $\mathcal{F}_\theta$ on the two-dimensional torus $\mathbb{T}^2 = \mathbb{R}^2/\mathbb{Z}^2$, whose leaves are the images of the lines $\tilde{L}_{(u_0,v_0)} = \{(u_0 + \tau, v_0 + \theta\tau) \in \mathbb{R}^2 : \tau \in \mathbb{R}\}$, $(u_0, v_0) \in \mathbb{R}^2$, under the projection $\mathbb{R}^2 \to \mathbb{T}^2$.

The conormal bundle $N^*\mathcal{F}_\theta$ is described as

$$N^*\mathcal{F}_\theta = \{(u, v, p_u, p_v) \in T^*\mathbb{T}^2 \cong \mathbb{T}^2 \times \mathbb{R}^2 : p_u + \theta p_v = 0\},$$

that is diffeomorphic to $\mathbb{T}^2 \times \mathbb{R}$ with coordinates (u, v, p_v). The leaves of the lifted foliation $\mathcal{F}_N$ are the images of the lines $\tilde{L}_{(u_0,v_0,p_v^0)} = \{(u_0 + \tau, v_0 + \theta\tau, p_v^0) \in \mathbb{R}^2 \times \mathbb{R} : \tau \in \mathbb{R}\}$, $(u_0, v_0, , p_v^0) \in \mathbb{R}^2 \times \mathbb{R}$, under the natural projection $\mathbb{R}^2 \times \mathbb{R} \to \mathbb{T}^2 \times \mathbb{R}$.

If $\theta \in \mathbb{Q}$, then all the leaves of $\mathcal{F}_\theta$ are closed and diffeomorphic to the circle S^1, and the foliation $\mathcal{F}_\theta$ is determined by the fibers of a fibration $\mathbb{T}^2 \to S^1$. So this is a particular case of the situation discussed in Section 2.1. The reduced phase space can be described as $T^*S^1 \cong S^1 \times \mathbb{R}$ with the coordinates $y = v - \theta u \in S^1$ and $\eta = p_v \in \mathbb{R}$, and the reduced symplectic structure is given by the two-form $\omega = \mathrm{d}y \wedge \mathrm{d}\eta$.

Consider a Riemannian metric g on $\mathbb{T}^2$:

$$g = a\,\mathrm{d}u^2 + 2b\,\mathrm{d}u\,\mathrm{d}v + c\,\mathrm{d}v^2, \qquad a, b, c \in C^\infty(\mathbb{T}^2), ac - b^2 > 0.$$

Let $\omega_{\mathbb{T}^2}$ be the Riemannian volume form on $\mathbb{T}^2$:

$$\omega_{\mathbb{T}^2} = \sqrt{\det g_{\mathbb{T}^2}}\,\mathrm{d}u\,\mathrm{d}v. \qquad \det g_{\mathbb{T}^2} = ac - b^2.$$

The tangent space $T_{(u,v)}\mathcal{F}_\theta \subset T_{(u,v)}\mathbb{T}^2 = \mathbb{R}^2$ of $\mathcal{F}_\theta$ is spanned by $(1, \theta) \in \mathbb{R}^2$, and its orthogonal complement $H_{(u,v)}$ is spanned by $(-(b(u,v)+c(u,v)\theta), a(u,v)+b(u,v)\theta)$ $\in \mathbb{R}^2$. The decomposition (5) holds with

$$g_F = \frac{\big((a + b\theta)\,\mathrm{d}u + (b + c\theta)\,\mathrm{d}v\big)^2}{a + 2b\theta + c\theta^2}, \qquad g_H = \frac{ac - b^2}{a + 2b\theta + c\theta^2}(\mathrm{d}v - \theta\,\mathrm{d}u)^2.$$

Therefore, the metric g is bundle-like iff

$$\left(\frac{\partial}{\partial u} + \theta\frac{\partial}{\partial v}\right)a_H = 0,$$

where a_H is the positive smooth function on $\mathbb{T}^2$ defined by

$$a_H^2 = \frac{ac - b^2}{a + 2b\theta + c\theta^2}.$$

In particular, if θ is irrational, then g is bundle-like iff $a_H \equiv \text{const}$.

There is an orthonormal base $\{e_F, e_H\}$ in $T\mathbb{T}^2$ given by

$$e_F = \frac{1}{\sqrt{a + 2b\theta + c\theta^2}} \left(\frac{\partial}{\partial u} + \theta \frac{\partial}{\partial v} \right) \in F,$$

$$e_H = \frac{1}{\sqrt{a + 2b\theta + c\theta^2}} \frac{-(b + c\theta)\partial/\partial u + (a + b\theta)\partial/\partial v}{\sqrt{ac - b^2}} \in H.$$

The dual orthonormal base $\{E_F, E_H\}$ in $T^*\mathbb{T}^2$ is given by

$$E_F = \frac{1}{\sqrt{a + 2b\theta + c\theta^2}} \big((a + b\theta)\, \mathrm{d}u + (b + c\theta)\, \mathrm{d}v \big),$$

$$E_H = \frac{\sqrt{ac - b^2}}{\sqrt{a + 2b\theta + c\theta^2}} (-\theta\, \mathrm{d}u + \mathrm{d}v).$$

The decomposition (8) holds with

$$d_F f = e_F(f) \cdot E_F, \quad d_H f = e_H(f) \cdot E_H, \quad \theta(f) = 0, \qquad f \in C^\infty(\mathbb{T}^2).$$

The adjoint d_H^* of the operator $d_H \colon C^\infty(\mathbb{T}^2) \to C^\infty(\mathbb{T}^2, H^*)$ is given by

$$d_H^*(g \cdot E_H) = e_H^*(g) = -e_H(g) + F \cdot g,$$

where $F \in C^\infty(\mathbb{T}^2)$ is given by

$$F = \frac{1}{\sqrt{ac - b^2}} \left(\frac{\partial}{\partial u} \left(\frac{b + c\theta}{\sqrt{a + 2b\theta + c\theta^2}} \right) - \frac{\partial}{\partial v} \left(\frac{a + b\theta}{\sqrt{a + 2b\theta + c\theta^2}} \right) \right). \tag{10}$$

The function F is closely related with the mean curvature vector τ:

$$\tau = F e_H.$$

For any $(u, v) \in \mathbb{T}^2$, the complex Clifford algebra $\mathrm{Cl}(Q_{(u,v)})$ is the complex algebra generated by 1 and e_H with the relation $e_H^2 = -1$. Thus, the vector bundle $\mathrm{Cl}(Q)$ over $\mathbb{T}^2$ is trivial, and the sections $1 \in C^\infty(\mathbb{T}^2)$ and $e_H \in C^\infty(\mathbb{T}^2, Q)$ determine a natural trivialization of this bundle. We also have $\nabla_{e_H} e_H = 0$, and

$$\nabla_{e_H}^{Cl(Q)} = e_H.$$

Consider the transverse Clifford bundle $\mathcal{E}$ defined in Example 4. Thus, we have $\mathcal{E} = \Lambda H^* \otimes \mathbb{C} \cong \mathbb{T}^2 \times \mathbb{C}^2$, It is equipped with the trivial connection $\nabla^{\mathcal{E}}$, and the action of $\mathrm{Cl}(Q_{(u,v)})$ on $\mathcal{E}_{(u,v)} \cong \mathbb{C}^2$ is given by

$$c(e_H(u, v)) = \begin{pmatrix} 0 & -1 \\ 1 & 0 \end{pmatrix}.$$

The transverse Dirac operator $D_{\mathcal{E}}$ associated with $\mathcal{E}$ is a matrix-valued first-order differential operator acting on $C^\infty(\mathbb{T}^2, \mathbb{T}^2 \times \mathbb{C}^2)$:

$$D_{\mathcal{E}} = \begin{pmatrix} 0 & -e_H + \frac{1}{2}F \\ e_H - \frac{1}{2}F & 0 \end{pmatrix}. \tag{11}$$

The geodesic flow f_t on $T^*\mathbb{T}^2$ is the Hamiltonian flow of the function

$$h(u, v, p_u, p_v) = \big(G_{(u,v)}(p_u, p_v) \big)^{1/2},$$

where G is the induced metric on $T^*\mathbb{T}^2$ given by

$$G_{(u,v)}(p_u,p_v) = \frac{1}{ac-b^2}(cp_u^2 - 2bp_up_v + ap_v^2), \qquad (p_u,p_v)\in T^*_{(u,v)}\mathbb{T}^2.$$

The infinitesimal generator X of f_t is given by

$$\begin{aligned} X = \big(G_{(u,v)}(p_u,p_v)\big)^{-1/2}\frac{cp_u-bp_v}{ac-b^2}\frac{\partial}{\partial u} &+ \big(G_{(u,v)}(p_u,p_v)\big)^{-1/2}\frac{-bp_u+ap_v}{ac-b^2}\frac{\partial}{\partial v}\\ &-\frac{1}{2}\big(G_{(u,v)}(p_u,p_v)\big)^{-1/2}\frac{\partial G_{(u,v)}(p_u,p_v)}{\partial u}\frac{\partial}{\partial p_u}\\ &-\frac{1}{2}\big(G_{(u,v)}(p_u,p_v)\big)^{-1/2}\frac{\partial G_{(u,v)}(p_u,p_v)}{\partial v}\frac{\partial}{\partial p_v}.\end{aligned}$$

The restriction of G to $N^*\mathcal{F}_\theta$ is given by

$$G_{(u,v)}(-\theta p_v,p_v) = \frac{1}{a_H^2}p_v^2, \qquad p_v\in\mathbb{R}.$$

It follows that

$$X(p_u+\theta p_v) = -\frac{1}{2}\big(G_{(u,v)}(p_u,p_v)\big)^{-1/2}\left(\frac{\partial}{\partial u}+\theta\frac{\partial}{\partial v}\right)G_{(u,v)}(p_u,p_v),$$

that immediately implies that, if the metric is bundle-like, then the flow f_t preserves $N^*\mathcal{F}_\theta$. We also see that in this case the restriction of X to $N^*\mathcal{F}_\theta$ is given by

$$X = -a_H\frac{b+c\theta}{ac-b^2}\operatorname{sign} p_v\frac{\partial}{\partial u} + a_H\frac{a+b\theta}{ac-b^2}\operatorname{sign} p_v\frac{\partial}{\partial v} + a_H^{-2}\frac{\partial a_H}{\partial v}|p_v|\frac{\partial}{\partial p_v}.$$

We have

$$X(v-\theta u) = -\frac{1}{a_H}\operatorname{sign} p_v,$$

that is constant along the leaves of $\mathcal{F}_N$, because a_H depends only on $v-\theta u$. It follows that the flow f_t takes each leaf of $\mathcal{F}_N$ to a leaf.

So if $\theta\in\mathbb{Q}$, the reduced dynamic on the reduced phase space T^*S^1 with coordinates $y = v-\theta u$ and $\eta = p_v$ is determined by the reduced Hamiltonian $h(y,\eta) = \big(1/a_H(y)\big)|\eta|$:

$$\frac{\mathrm{d}y}{\mathrm{d}t} = -\frac{1}{a_H(y)}\operatorname{sign}\eta, \qquad \frac{\mathrm{d}\eta}{\mathrm{d}t} = a_H^{-2}\frac{\partial a_H}{\partial y}|\eta|.$$

If $\theta\notin\mathbb{Q}$, the foliation $\mathcal{F}_\theta$ is given by the orbits of a free action of $\mathbb{R}$ on $\mathbb{T}^2$, and its holonomy groupoid is described as follows: $G=\mathbb{T}^2\times\mathbb{R}$, $G^{(0)}=\mathbb{T}^2$, $s(u,v,\tau)=(u-t,v-\theta\tau)$, $r(u,v,\tau)=(u,v)$, $(u,v)\in\mathbb{T}^2$, $\tau\in\mathbb{R}$, and the product of (u_1,v_1,τ_1) and (u_2,v_2,τ_2) is defined if $u_2=u_1-\tau_1, v_2=v_1-\theta\tau_1$ and equals

$$(u_1,v_1,\tau_1)(u_2,v_2,\tau_2) = (u_1,v_1,\tau_1+\tau_2).$$

Remark 10. If $\theta\in\mathbb{Q}$, then the linear foliation on $\mathbb{T}^2$ is given by the orbits of a free group action of S^1 on $\mathbb{T}^2$, and its holonomy groupoid coincides with the crossed product groupoid $G=\mathbb{T}^2\rtimes S^1$.

The holonomy groupoid $G_{\mathcal{F}_N}$ is described as follows: $G_{\mathcal{F}_N}=\mathbb{T}^2\times\mathbb{R}^2$, $G^{(0)}_{\mathcal{F}_N}=\mathbb{T}^2\times\mathbb{R}$, $s_N(u,v,p_v,\tau)=(u-t,v-\theta\tau,p_v)$, $r_N(u,v,p_v,\tau)=(u,v,p_v)$, $(u,v,p_v)\in\mathbb{T}^2\times\mathbb{R}$, $\tau\in\mathbb{R}$, and the product of $(u_1,v_1,p_{v,1},\tau_1)$ and $(u_2,v_2,p_{v,2},\tau_2)$ is defined if $u_2=u_1-\tau_1, v_2=v_1-\theta\tau_1, p_{v,1}=p_{v,2}(=p_v)$, and equals

$$(u_1,v_1,p_v,\tau_1)(u_2,v_2,p_v,\tau_2) = (u_1,v_1,p_v,\tau_1+\tau_2).$$

There is a natural trivialization of the line bundle $T\mathcal{F}_N$ denoted by $d\tau$. Let $\omega_F \in C^\infty(\mathbb{T}^2, T\mathcal{F}_N)$ be the leafwise Riemannian volume form:

$$\omega_F(u,v,p_v) = \sqrt{\det g_F(u,v)}\,d\tau, \qquad \det g_F = a + 2b\theta + c\theta^2.$$

An arbitrary section $\sigma \in C^\infty(G_{\mathcal{F}_N}, |T\mathcal{G}_N|^{1/2})$ can be written as

$$\sigma(u,v,p_v,\tau) = k(u,v,p_v,\tau)|\omega_F(u,v,p_v)|^{1/2}|\omega_F(u-\tau, v-\theta\tau, p_v)|^{1/2}, \quad (u,v)\in\mathbb{T}^2, p_v\in\mathbb{R}, \tau\in\mathbb{R}, \tag{12}$$

It is properly supported if and only if k has compact support as a function of τ for any (u,v,p_v).

For any $\sigma_j \in C^\infty_{\text{prop}}(G_{\mathcal{F}_N}, |T\mathcal{G}_N|^{1/2})$, of the form (12) with some k_j, $j=1,2$, their product $\sigma = \sigma_1 * \sigma_2$ is written in the form (12) with

$$\begin{aligned}(k_1 * k_2)(u,v,p_v,\tau) &= \int_{-\infty}^{\infty} k_1(u,v,p_v,\tau_1)k_2(u-\tau_1, v-\theta\tau_1, p_v, \tau-\tau_1)\sqrt{\det g_F(u-\tau_1, v-\theta\tau_1)}\,d\tau_1,\\ &\qquad (u,v)\in\mathbb{T}^2, p_v\in\mathbb{R}, \tau\in\mathbb{R}.\end{aligned}$$

The infinitesimal generator $\mathcal{H}$ of the induced flow F_t on $G_{\mathcal{F}_N}$ is given by

$$\begin{aligned}\mathcal{H} = &-a_H\frac{b(u,v)+c(u,v)\theta}{a(u,v)c(u,v)-b(u,v)^2}\operatorname{sign}p_v\frac{\partial}{\partial u}\\ &+a_H\frac{a(u,v)+b(u,v)\theta}{a(u,v)c(u,v)-b(u,v)^2}\operatorname{sign}p_v\frac{\partial}{\partial v}\\ &+a_H\left(\frac{b(u-\tau,v-\theta\tau)+c(u-\tau,v-\theta\tau)\theta}{a(u-\tau,v-\theta\tau)c(u-\tau,v-\theta\tau)-b(u-\tau,v-\theta\tau)^2}\right.\\ &\qquad\left.-\frac{b(u,v)+c(u,v)\theta}{a(u,v)c(u,v)-b(u,v)^2}\right)\operatorname{sign}p_v\frac{\partial}{\partial\tau}.\end{aligned}$$

We have

$$\mathcal{L}_X|\omega_F|^{1/2}(u,v,p_v) = \tfrac{1}{2}F(u,v)|\omega_F|^{1/2}(u,v,p_v),$$

where the mean curvature F is given by (10). Thus, for any $\sigma \in C^\infty(G_{\mathcal{F}_N}, |T\mathcal{G}_N|^{1/2})$ of the form (12), we have

$$\begin{aligned}\mathcal{L}_\mathcal{H}\sigma(u,v,p_v,\tau) &= \big(\mathcal{H} + \tfrac{1}{2}F(u,v) + \tfrac{1}{2}F(u-\tau, v-\theta\tau)\big)k(u,v,p_v,\tau)\\ &\quad\times|\omega_F(u,v,p_v)|^{1/2}|\omega_F(u-\tau,v-\theta\tau,p_v)|^{1/2},\\ &\qquad (u,v)\in\mathbb{T}^2, p_v\in\mathbb{R}, \tau\in\mathbb{R}.\end{aligned}\tag{13}$$

The bundle $\pi^*\mathcal{E}$ is a trivial two-dimensional complex bundle over $N^*\mathcal{F}_\theta \cong \mathbb{T}^2\times\mathbb{R}$. Therefore, an element $\sigma \in C^\infty(G_{\mathcal{F}_N}, \mathcal{L}(\pi^*\mathcal{E})\otimes|T\mathcal{G}_N|^{1/2})$ can be written as

$$\begin{aligned}\sigma(u,v,p_v,\tau) &= k(u,v,p_v,\tau)|\omega_F(u,v,p_v)|^{1/2}|\omega_F(u-\tau,v-\theta\tau,p_v)|^{1/2},\\ &\qquad (u,v,p_v,\tau)\in G_{\mathcal{F}_N}\cong\mathbb{T}^2\times\mathbb{R}\times\mathbb{R},\end{aligned}$$

where k is a smooth function with values in the space $M_2(\mathbb{C})$ of complex 2×2 matrices. Since the Clifford connection $\nabla^\mathcal{E}$ is trivial, the action of $\mathcal{H}$ on such a σ is defined by the formula (13) with

$$(\mathcal{H}k)_{\alpha\beta} = \mathcal{H}(k_{\alpha\beta}), \qquad \alpha,\beta = 1,2.$$

The class $\Psi^{d,-\infty}(\mathbb{T}^2,\mathcal{F}_\theta,\mathbb{T}^2\times\mathbb{C}^2)$, $d\in\mathbb{R}$, can be described as follows. Suppose that a polyhomogeneous symbol $k\in S^d\big(\mathbb{R}^4\times\mathbb{R},M_2(\mathbb{C})\big)$ satisfies the conditions:

(1) for any $(m,n)\in\mathbb{Z}^2$,

$$k(u+m,v+n,u'+m,v'+n,\eta)=k(u,v,u',v',\eta),\qquad (u,v,u',v',\eta)\in\mathbb{R}^4\times\mathbb{R}; \tag{14}$$

(2) there exists $R>0$ such that

$$k(u,v,u',v',\eta)=0,\qquad (u,v,u',v',\eta)\in\mathbb{R}^4\times\mathbb{R},\ (u-u')^2+(v-v')^2>R^2. \tag{15}$$

Define an operator $K: C^\infty(\mathbb{R}^2,\mathbb{R}^2\times\mathbb{C}^2)\to C^\infty(\mathbb{R}^2,\mathbb{R}^2\times\mathbb{C}^2)$ by the formula

$$Kf(u,v)=\int \mathrm{e}^{\mathrm{i}(v-v'-\theta(u-u'))\eta}k(u,v,u',v',\eta)\,f(u',v') \times\sqrt{\det g_F(u',v')}\,du'\,dv'\,d\eta. \tag{16}$$

By (14) and (15), the operator K takes any $\mathbb{Z}^2$-invariant function from $C^\infty(\mathbb{R}^2,\mathbb{R}^2\times\mathbb{C}^2)$ to a $\mathbb{Z}^2$-invariant function from $C^\infty(\mathbb{R}^2,\mathbb{R}^2\times\mathbb{C}^2)$, and determines an operator, acting on $C^\infty(\mathbb{T}^2,\mathbb{T}^2\times\mathbb{C}^2)$.

It is clear that any operator from $\Psi^{d,-\infty}(\mathbb{T}^2,\mathcal{F}_\theta,\mathbb{T}^2\times\mathbb{C}^2)$ can be written in this form. On the other hand, one can show that any operator K defined by (16) belongs to $\Psi^{d,-\infty}(\mathbb{T}^2,\mathcal{F}_\theta,\mathbb{T}^2\times\mathbb{C}^2)$, and its principal symbol is given by

$$\begin{aligned}&\sigma(K)(u,v,p_v,\tau)\\&=\sum_{(m,n)\in\mathbb{Z}^2}k_d(u+m,v+n,u-\tau,v-\theta\tau,p_v)|\omega_F(u,v,p_v)|^{1/2}|\omega_F(u-\tau,v-\theta\tau,p_v)|^{1/2},\\&\qquad (u,v,p_v,\tau)\in G_{\mathcal{F}_N}\cong\mathbb{T}^2\times\mathbb{R}\times\mathbb{R},\end{aligned}$$

where k_d is the (degree d) homogeneous component of k.

The orbits of the vector field e_H determine a foliation, $\mathcal{F}^{\pitchfork}$, transverse to $\mathcal{F}_\theta$. It is easy to see that, for any natural m, the space $\mathcal{H}^m$ coincides with the space of all $f\in L^2(\mathbb{T}^2)$ such that $e_H^j f\in L^2(\mathbb{T}^2)$ for $j=1,\dots,m$. Thus, for any $s\in\mathbb{R}$, the space $\mathcal{H}^s$ coincides with the anisotropic Sobolev space $H^{0,s}(\mathbb{T}^2,\mathcal{F}^{\pitchfork})$ [10].

Now Theorems 5 and 6 read as follows.

Theorem 11. *Let $D_{\mathcal{E}}$ be the transverse Dirac operator defined by* (11).

(1) *For any $K\in\Psi^{m,-\infty}(\mathbb{T}^2,\mathcal{F}_\theta,\mathbb{T}^2\times\mathbb{C}^2)$, there exists an operator $K(t)\in\Psi^{m,-\infty}(\mathbb{T}^2,\mathcal{F}_\theta,\mathbb{T}^2\times\mathbb{C}^2)$ such that $\Phi_t(K)-K(t)$, $t\in\mathbb{R}$, is a smooth family of operators of class $\mathcal{L}^1(\mathcal{H}^{-\infty},\mathcal{H}^\infty)$.*

(2) *If $\sigma\in S^m(G_{\mathcal{F}_N},\mathcal{L}(\pi^*\mathcal{E})\otimes|T\mathcal{G}_N|^{1/2})$ is the principal symbol of K:*

$$\sigma(u,v,p_v,\tau)=k(u,v,p_v,\tau)|\omega_F(u,v,p_v)|^{1/2}|\omega_F(u-\tau,v-\theta\tau,p_v)|^{1/2},$$

then the principal symbol $\sigma_t\in S^m(G_{\mathcal{F}_N},\mathcal{L}(\pi^*\mathcal{E})\otimes|T\mathcal{G}_N|^{1/2})$ of $K(t)$:

$$\sigma_t(u,v,p_v,\tau)=k(t,u,v,p_v,\tau)|\omega_F(u,v,p_v)|^{1/2}|\omega_F(u-\tau,v-\theta\tau,p_v)|^{1/2},$$

is the solution of the equation

$$\frac{\partial k}{\partial t}(t,u,v,p_v,\tau)=\big(\mathcal{H}+\tfrac12F(u,v)+\tfrac12F(u-\tau,v-\theta\tau)\big)k(t,u,v,p_v,\tau),$$

satisfying the condition $k(0,u,v,p_v,\tau)=k(u,v,p_v,\tau)$.

References

1. J. Block and E. Getzler, *Quantization of foliations*, Geometric Methods in Theoretical Physics, Vol. 1, 2 (New York, 1991), World Sci., River Edge, NJ, 1992, pp. 471–487.
2. A. Connes, *Sur la théorie non commutative de l'intégration*, Algèbres d'opérateurs (Les Plans-sur-Bex, 1978), Lecture Notes in Math., vol. 725, Springer, Berlin, 1979, pp. 19–143.
3. ———, *Noncommutative geometry*, Academic Press, San Diego, CA, 1994.
4. J. F. Glazebrook and F. W. Kamber, *On spectral flow of transversal Dirac operators and a theorem of Vafa - Witten*, Ann. Global Anal. Geom. **9** (1991), no. 1, 27–35.
5. ———, *Transversal Dirac families in Riemannian foliations*, Comm. Math. Phys. **140** (1991), no. 2, 217–240.
6. V. Guillemin and S. Sternberg, *Some problems in integral geometry and some related problems in microlocal analysis*, Amer. J. Math. **101** (1979), no. 4, 915–955.
7. ———, *Homogeneous quantization and multiplicities of group representations*, J. Funct. Anal. **47** (1982), no. 3, 344–380.
8. V. Guillemin and A. Uribe, *Circular symmetry and the trace formula*, Invent. Math. **96** (1989), no. 2, 385–423.
9. ———, *Reduction and the trace formula*, J. Differential Geom. **32** (1990), no. 2, 315–347.
10. Yu. A. Kordyukov, *Functional calculus for tangentially elliptic operators on foliated manifolds*, Analysis and Geometry in Foliated Manifolds (Santiago de Compostela, 1994), World Sci., River Edge, NJ, 1995, pp. 113–136.
11. ———, *Noncommutative spectral geometry of Riemannian foliations*, Manuscripta Math. **94** (1997), no. 1, 45–73.
12. ———, *Egorov's theorem for transversally elliptic operators on foliated manifolds and noncommutative geodesic flow*, Math. Phys. Anal. Geom. **8** (2005), no. 2, 97–119.
13. ———, *The Egorov theorem for transverse Dirac-type operators on foliated manifolds*, J. Geom. Phys. **57** (2007), no. 11, 2345–2364.
14. ———, *Noncommutative geometry of foliations*, J. K-Theory **2** (2008), no. 2, 219–327.
15. ———, *Vanishing theorem for transverse Dirac operators on Riemannian foliations*, Ann. Global Anal. Geom. **34** (2008), no. 2, 195–211.
16. P. Libermann and C.-M. Marle, *Symplectic geometry and analytical mechanics*, translated by B. E. Schwarzbach, Math. Appl., vol. 35, D. Reidel, Dordrecht, 1987.
17. A. Lichnerowicz, *Variété symplectique et dynamique associée à une sous-variété*, C. R. Acad. Sci. Paris Sér. A-B **280** (1975), A523–A527.
18. ———, *Les variétés de Poisson et leurs algèbres de Lie associées*, J. Differential Geometry **12** (1977), no. 2, 253–300.
19. J. Marsden and A. Weinstein, *Reduction of symplectic manifolds with symmetry*, Rep. Mathematical Phys. **5** (1974), no. 1, 121–130.
20. P. Molino, *Riemannian foliations*, translated by G. Cairns, with appendices by G. Cairns, Y. Carrière, É. Ghys, E. Salem, and V. Sergiescu, Progr. Math., vol. 73, Birkhäuser, Boston, MA, 1988.
21. B. L. Reinhart, *Differential geometry of foliations*, Ergeb. Math. Grenzgeb., vol. 99, Springer, Berlin, 1983.
22. T. Tate, *Quantum ergodicity at a finite energy level*, J. Math. Soc. Japan **51** (1999), no. 4, 867–885.
23. P. Xu, *Noncommutative Poisson algebras*, Amer. J. Math. **116** (1994), no. 1, 101–125.
24. S. Zelditch, *On a "quantum chaos" theorem of R. Schrader and M. Taylor*, J. Funct. Anal. **109** (1992), no. 1, 1–21.

INSTITUTE OF MATHEMATICS, RUSSIAN ACADEMY OF SCIENCES, 112 CHERNYSHEVSKY STR., UFA, 450077, RUSSIA

E-mail address: yurikor@matem.anrb.ru

Centre de Recherches Mathématiques
CRM Proceedings and Lecture Notes
Volume **52**, 2010

Eigenvalue Variations and Semiclassical Concentration

Luc Hillairet

1. Introduction

In this paper, we aim at explicating some relations between the behavior of analytic eigenvalue branches of a spectral problem and concentration properties of its eigenfunctions. In order to illustrate this we consider a semiclassical Schrödinger operator $-h^2\Delta+V$ on $\mathbb{R}^d$ (in which Δ is the nonpositive Euclidean Laplace operator). Viewing h as an analytic parameter, this problem enters the usual perturbation theory and the eigenvalues are organized into analytic eigenbranches $E_j(h)$. It is then natural to address the behavior of these eigenbranches when h goes to 0. This point of view is fundamentally different from the usual semiclassical perspective. Indeed, any analytic eigenbranch is expected to cross a typical energy window $[E_1, E_2]$ so that any semiclassical expansion in this window corresponds actually to different eigenbranches. Our point is that knowing how the eigenfunctions concentrate in the semiclassical regime yields some result on the behavior of eigenbranches. For instance, we have the following theorem (see Section 3).

Theorem 1. *Let $E_j(h)$ be an analytic eigenbranch of the family of operators $-h^2\Delta + V$ on $L^2(\mathbb{R}^d)$. The following then holds.*

(1) *The eigenbranch $E_j(h)$ converges to some limit E_0 when h tends to 0.*
(2) *The limit E_0 is necessarily a critical value of the potential V.*

We will also use the same kind of ideas to prove an integrated remainder estimate for an analytic variation of metrics on a smooth compact manifold X. (see Theorem 10 (in Section 4).

As we have already pointed out, the study of analytic eigenbranches and semiclassical analysis are two different worlds so that such a relation between them is a priori quite surprising. However, as it will become clear later, the formula that expresses the derivative of any analytic eigenbranch is actually a semiclassical quantity. This is the fact we will be exploiting here. This idea is also used in the two recent papers [8] and [5]. In [8] we deal with the limiting behavior of analytic eigenbranches in a singular setting. The main ingredient of the proof consists in controlling certain quadratic forms evaluated on the eigenfunctions of the spectral problem under consideration. We claim here that such a control may actually be

2010 *Mathematics Subject Classification.* Primary 35P20; Secondary 58J37.
This is the final form of the paper.

interpreted as a condition on the associated semiclassical measures. Such an interpretation is not necessarily helpful since rather little is known in general about semiclassical measures and the overall strategy of [8] relies more on direct estimates for solutions of one-dimensional ODEs. In [5], a part of the proof relies on the fact that an assumption on semiclassical measures allows to prove an integrated remainder estimate of the form we will consider here. Section 4 can actually be seen as a formalization of this argument of [5].

Acknowledgments. The ideas in this paper grew out from several discussions with Chris Judge and Andrew Hassell, I would like to thank both of them for their stimulating influence. I would also like to thank the CRM in Montréal and the organizers of the workshop "Spectrum and Dynamics" in April, 2008 and the MSRI in Berkeley where part of this work was done. Finally, I would like to thank Gilles Carron, Yves Colin de Verdière, Patrick Gérard and Andrew Hassell who made several comments on the first version of this paper that helped me improve the results and the exposition.

2. Quantization and semiclassical measures

In this section we will briefly recall some known facts about quantization and semiclassical measures. We will use both the classical and the semiclassical theory of pseudodifferential operators (See [4, 10] for background on these notions)

In the case of Schrödinger operators in $\mathbb{R}^d$, the principal symbol of a (h-) pseudodifferential operator is a function in $T^*(\mathbb{R}^d)$. When working on a smooth compact Riemannian manifold, we will denote by S^*X the quotient of $T^*X\backslash\{0\}$ by $(x,\xi) \sim (x, \lambda\xi)$, $\lambda > 0$. The subsets of S^*X are the conical subsets of $T^*X\backslash\{0\}$. The symbol of a pseudodifferential operator A is defined by testing A again fast oscillatory functions. It is then a function on $T^*M\backslash\{0\}$. The symbol of a classical pseudodifferential operator is conically invariant and thus defines a function on S^*X.

Remark 2. The principal symbol of a pseudodifferential operator may also be read off from the Schwartz kernel of A. Observe, however, that when performing this identification, one has to pay careful attention to the density that is used (especially when dealing with varying Riemannian metrics).

A crucial point is the ability of quantizing any reasonable function on S^*X. This is done via a so-called quantization procedure. There are several possible choices, but it is useful to choose a quantization that respects positivity (see [3, 7])[1]. Thus, for any function a on S^*X there exists a pseudodifferential operator $\mathrm{Op}^+(a)$ of order 0 whose principal symbol is a. If A is any other pseudodifferential operator with principal symbol a then $\mathrm{Op}^+(a) - A$ is a pseudodifferential operator of order -1. Moreover if, a is nonnegative, then $\mathrm{Op}^+(a)$ is also nonnegative, as an operator.

In the semiclassical setting, the homogeneity condition on a is dropped and we may quantize any smooth function on $T^*(X)$ with compact support.

Using this quantization, we define the so-called semiclassical measures. We present the construction in the semiclassical setting since in the classical one we will use a nonconcentration estimate rather than semiclassical measures.

---

[1] It is actually critical that such a quantization exists.

We start with a Schrödinger operator on $\mathbb{R}^d$, and we consider a sequence of normalized eigenfunctions u_h of the equation

$$-h^2\Delta u_h + V(x)u_h = E_h u_h,$$

such that the corresponding eigenvalues E_h converge to some noncritical energy E_0. To any compactly supported function a the map $a \mapsto \langle \mathrm{Op}^+(a)u_h, u_h\rangle$ defines a measure that converges weakly (up to extracting a subsequence). Any such limit is called a semiclassical measure.

Remark 3. One could in fact associate semiclassical measures to any bounded sequence in L^2. In the sequel we will consider only semiclassical measures that are associated with eigenfunctions of the Laplace operator.

Among the basic facts satisfied by any semiclassical measure, we mention only the two following properties (see [3, 7]).

(1) Any semiclassical measure is a probability measure supported on the energy surface $\Sigma_{E_0} := \{|\xi|^2 + V(x) = E_0\}$.
(2) Any semiclassical measure is invariant under the Hamiltonian flow associated with $|\xi|^2 + V(x)$ on Σ_{E_0}.

Remark 4. Under suitable hypotheses, we may extend the definition to smooth functions that are not necessarily compactly supported in $T^*\mathbb{R}^d$. In particular, for a sequence of eigenfunctions with energy E_h converging to E_0, the quantity $\int \|\nabla u_h\|^2\,\mathrm{d}x$ is uniformly bounded. This allows to prove that $\int |\xi|^2\,\mathrm{d}\mu < +\infty$ and then that

$$\int |\xi|^2\,\mathrm{d}\mu = \lim_{h\to 0} \int \|\nabla u_h\|^2\,\mathrm{d}x.$$

In the classical setting we can also define semiclassical measures. The main difference is that instead of a small parameter going to zero, we have to consider a sequence of eigenfunctions with eigenvalues growing to infinity and we recover a probability measure on S^*X. It can be pointed out that the estimate in Corollary 14 that we use in Section 4 is a kind of nonconcentration estimate. Indeed, such an estimate would give some information on any semiclassical measure when passing to the high energy limit. The assumptions on semiclassical measure in Section 3 and on geometric control in Section 4 are thus very similar in nature.

3. Schrödinger operators from the analytic viewpoint

In this section we consider the semiclassical Schrödinger equation in $X = \mathbb{R}^d$. The eigenvalue problem consists in looking for eigenfunctions u_h and eigenvalues E_h such that

$$-h^2\Delta u_h + V(x)u_h = E_h u_h, \tag{1}$$

in which Δ is the (nonpositive) Laplace operator associated with the Euclidean metric. Let us assume that V is smooth and confining (i.e., $\lim_{|x|\to\infty} V(x) = +\infty$) so that there is a complete orthonormal set of eigenfunctions and the corresponding eigenvalues grow to infinity.

This eigenvalue problem may be put into a generalized eigenvalue problem in the sense of Kato (cf. [9]). We change notations by letting $h = t$ thus emphasizing that we will now consider an analytic perturbation problem. An integration by parts leads to the following equivalent problem

$$q_t(u_t, v) = E_t \cdot n(u, v), \tag{2}$$

In which q_t is the quadratic form

$$q_t(u) = t^2 \int |\nabla u(x)|^2 \, \mathrm{d}x + \int V(x)|u(x)|^2 \, \mathrm{d}x$$

defined on $\mathcal{D} = H^1(X) \cap L^2\big((1 + V(x)^2)^{1/2} \, \mathrm{d}x\big)$ and n is the standard Riemannian quadratic form on $L^2(X)$. Observe that although q_t is real-analytic for $t \in \mathbb{R}$, $\mathcal{D}$ is the form domain of q_t only for $t \neq 0$. Consequently, usual analytic perturbation theory tells us that the eigenvalues organize into real-analytic eigenbranches for $t \in (0, 1]$. A standard question in this setting is to address the limiting behavior of these eigenbranches when t goes down to 0.

Theorem 5. *Let E_t be an eigenvalue branch of* (2), *then E_t converges to a limit E_0 when t goes to 0. Moreover, this limit E_0 is a critical value of V.*

Proof. Let E_t be an eigenvalue branch and u_t the corresponding normalized eigenvector branch. The derivative of E_t may be computed by differentiating (2) (see [9]), this yields

$$\dot{E}_t = 2t \int_X |\nabla u_t|^2(x) \, \mathrm{d}x. \tag{3}$$

This expression is obviously nonnegative so that E_t is a nondecreasing function of t. Since, for any t we have $E_t \geq \min_X V(x)$, E_t has a limit E_0 when t goes to 0.

The condition on concentration of semiclassical measures then enters in the following lemma.

Lemma 6. *Assume that for any semiclassical measure μ at the energy E_0,*

$$\int_{\Sigma_{E_0}} |\xi|^2 \, \mathrm{d}\mu(x) > 0. \tag{C}$$

Then, for any eigenbranch (E_t, u_t) such that $\lim E_t = E_0$,

$$\exists c > 0, \qquad \liminf_{t \to 0} t^2 \cdot \int_X |\nabla u_t(x)|^2 \, \mathrm{d}x \geq c. \tag{4}$$

Assuming this lemma, we finish the proof of the theorem. If we suppose, for a contradiction, that E_0 satisfies condition (C), then using (3), and (4) we get

$$\dot{E}_t \geq \frac{2c}{t},$$

and this is a contradiction since E_t has to converge when t goes down to 0. So condition (C) is not fulfilled and there exists a semiclassical measure at energy E_0 such that

$$\int_{\Sigma_{E_0}} |\xi^2| \, \mathrm{d}\mu = 0.$$

This implies that the support of μ is contained in the level set $\{\xi = 0\} \subset \Sigma_{E_0}$. Since $\mathrm{d}\mu$ has to be invariant by the Hamiltonian flow

$$\begin{cases} \dot{x} = 2\xi \\ \dot{\xi} = -\nabla V(x) \end{cases}$$

this forces E_0 to be a critical value of V. □

It remains to prove Lemma 6.

PROOF OF LEMMA 6. As it is standard in this kind of settings (see [6] or [2]), we actually work in the reverse direction, assuming that (4) is not true. We may thus find a sequence t_n going to 0 and corresponding eigenfunctions and eigenvalues E_{t_n} and u_{t_n}, such that

$$t_n^2 \int_X |\nabla u_{t_n}(x)|^2 \, \mathrm{d}x \leq \frac{1}{n}.$$

We now change notations again and let $t_n = h$ since the rest of the argument is of semiclassical nature (and, as usual in this setting h actually stands for h_n). We thus have

$$0 \leq h^2 \int_X |\nabla u_h(x)|^2 \, \mathrm{d}x \leq \varepsilon(h). \tag{5}$$

where $\varepsilon(h)$ is some function going to 0 with h. The sequence u_h is L^2 normalized so that we may extract subsequences and find associated semiclassical measures at energy E_0 since $\lim E_h = E_0$. Let $\mathrm{d}\mu$ be one of these semiclassical measures, by definition (see Remark 4 above)

$$\lim_{h\to 0} h^2 \cdot \int_X |\nabla u_h(x)|^2 \, \mathrm{d}x = \int_{\Sigma_{E_0}} |\xi^2| \, \mathrm{d}\mu, \tag{6}$$

where the limit is understood along the subsequence defining μ. Putting (5) and (6) together yields

$$\int_{\Sigma_{E_0}} |\xi^2| \, \mathrm{d}\mu = 0,$$

thus finishing the proof of the lemma. □

Remark 7. Since the energy surface is defined by $|\xi|^2 + V(x) = E_0$, we can replace everywhere

$$\int_{\Sigma_{E_0}} |\xi^2| \, \mathrm{d}\mu$$

by

$$\int_{\Sigma_{E_0}} (E_0 - V(x)) \, \mathrm{d}\mu.$$

In dimension 1 this result can be refined using that the spectrum of a one-dimensional Schrödinger operator is known to be simple. If we assume that the potential V has only one nondegenerate minimum, then the spectrum near the bottom of the energy is known (see [4]) and there is an infinite number of eigenvalues close to the minimum. Since the eigenvalue branches cannot cross, the nondegenerate minimum is then only possible limit.[2]

[2]I would like to thank San Vũ Ngọc and Frédéric Faure for pointing out that, in the one-dimensional case, the simplicity of the spectrum allows to improve the result.

4. Integrated remainder estimates

In this section, we consider a smooth compact manifold X of dimension $d \geq 2$ and a real-analytic family of Riemannian metrics $(g_\tau)_{\tau\in[-1,1]}$. We denote by Δ_τ the Laplace operator associated with g_τ and by $E_j(\tau)$ the associated analytic branches of eigenvalues. We will also denote by $\langle\cdot,\cdot\rangle_\tau$ the scalar product on $L^2(X, g_\tau)$.We now define the usual following counting functions:

$$\begin{aligned} N(\tau, E) &= \#\{j \mid E_j(\tau) \leq E\}, \\ R_M(\tau, E) &= \#\{j \mid E_j(\tau) \in (E-M, E+M]\} = N(\tau, E+M) - N(\tau, E-M). \end{aligned}$$

Remark 8. Observe that since (g_τ) is analytic in τ, the scalar products $\langle\cdot,\cdot\rangle_\tau$ define equivalent norms. The same is true at the level of H^1: the norms $\|\cdot\|_{H^1(X,g_\tau)}$ defined by:

$$\|u\|^2_{H^1(X,g_\tau)} = \int_M g_\tau(\nabla_\tau u, \nabla_\tau u)\, \mathrm{dvol}_{g_\tau} + \|u\|^2_\tau,$$

are all equivalent.

The following lemma expresses the variation of E_t using a family of pseudodifferential operators. It also serves as a definition of the family A_τ that will be used in the rest of the paper.

Lemma 9. *There exists a real-analytic family of symbols a_τ, and, for any τ_0, there exists a constant C such that, for any normalized eigenbranch $(E_n(\tau), u_n(\tau))$ we have*

$$\forall |\tau| \leq \tau_0, \qquad \left|\frac{\dot{E}_n(\tau)}{E_n(\tau)} - \langle \mathrm{Op}^+(a_\tau)u_n(\tau), u_n(\tau)\rangle_\tau\right| \leq C E_n(\tau)^{-1/2}.$$

We will denote by $A_\tau = \mathrm{Op}^+(a_\tau)$.

Proof. First, by differentiating the eigenvalue equation

$$E_n(\tau) = \langle \Delta_\tau u_n(\tau), u_n(\tau)\rangle_\tau,$$

we find a second-order differential operator $\dot{\Delta}_\tau$ such that

$$\dot{E}_n(\tau) = \langle \dot{\Delta}_\tau u_n(\tau), u_n(\tau)\rangle_\tau.$$

(Observe that since the branch is normalized and using the eigenvalue equation we have that $\langle \Delta\dot{u}, u\rangle = E\langle \dot{u}, u\rangle = 0$). Denote by a_τ the principal symbol of $\dot{\Delta}_\tau(\Delta_\tau+1)^{-1}$, and let $A_\tau = \mathrm{Op}^+(a_\tau)$. By definition $R_\tau = \dot{\Delta}_\tau(\Delta_\tau+1)^{-1} - A_\tau$ is an analytic family of pseudodifferential operators of order -1. In particular, R_τ is uniformly bounded from L^2 into H^1. This yields the bound

$$\left|\frac{\dot{E}_n(\tau)}{E_n(\tau)+1} - \langle \mathrm{Op}^+(a_\tau)u_n(\tau), u_n(\tau)\rangle_\tau\right| \leq C E_n(\tau)^{-1/2}.$$

We thus get

$$\left|\frac{\dot{E}_n(\tau)}{E_n(\tau)} - \langle (1+E_n^{-1})\, \mathrm{Op}^+(a_\tau)u_n(\tau), u_n(\tau)\rangle_\tau\right| \leq C E_n(\tau)^{-1/2}(1+E_n^{-1}),$$

from which the claimed bound follows since $|\langle \mathrm{Op}^+(a_\tau)u_n(\tau), u_n(\tau)\rangle_\tau|$ is uniformly bounded. □

The main result of this section is then the following theorem (the notion of geometric control is defined in Definition 11 in the following section).

Theorem 10. *Suppose that a_0 is nonnegative and that there exists $\varepsilon > 0$ such that the subset $\{(x,\xi) \in S^*X, a_0(x,\xi) > \varepsilon\}$ geometrically controls (S^*X, g_0), then for any fixed M, there exists τ_0, E_0 and K such that*

$$\forall E > E_0, \qquad \int_{-\tau_0}^{\tau_0} R_M(t,E)\,\mathrm{d}t \leq KE^{d/2-1}. \tag{7}$$

The universal remainder in Weyl's law gives that for a fixed t, $R_M(t,E) = O(E^{(d-1)/2})$ as E goes to infinity. The theorem thus says that averaging with respect to the perturbation greatly improves this estimate.

The rest of the section is devoted to the proof of this theorem. The central part of the argument is actually interesting by itself and consists in proving a somewhat explicit uniform control estimate.

4.1. A uniform control estimate. We begin by recalling the notion of geometric control observing that since the velocity is constant along a geodesic, the geodesic flow of a Riemannian metric is well-defined on S^*X.

Definition 11. We say that an open subset $\mathcal{U}$ of S^*X *geometrically controls* (S^*X, g) if for any $(x,\xi) \in S^*X$, there exists $T \in \mathbb{R}$ and $(x_0,\xi_0) \in \mathcal{U}$ such that $(x,\xi) = \Phi^T(x_0,\xi_0)$ where $\Phi^{\cdot}(\cdot,\cdot)$ denotes the geodesic flow of g.

It is well-known that geometric control implies a nonconcentration estimate for eigenfunctions. More precisely, if U is an open subset of X such that $\mathcal{U} = S^*U$ geometrically controls (S^*X, g) then, there exists some positive c and an energy E_0 such that

$$\int_U |u_n|^2\,\mathrm{dvol}_g \geq c\int_X |u_n|^2\,\mathrm{dvol}_g, \tag{8}$$

for any eigenfunction u_n of the Laplace operator associated with g such that $E_n \geq E_0$. (Actually the restriction on E for this estimate may be released using the principle of unique continuation for solutions of second order elliptic PDEs see Remark 13 below).

Such a control estimate can be proved by a contradiction argument relying on known properties of semiclassical measures ([2, 6] for the scheme of such an argument). Such an approach can probably be adapted to get a control estimate that is uniform in τ. We propose here a slightly different proof that also allows us to get some uniform control on c.

Proposition 12. *Let $\mathcal{U}$ be an open subset of S^*X that geometrically controls (S^*X, g_0). There exists $\tau_0 \in \mathbb{R}$, $E_0 \in \mathbb{R}$, a zeroth-order pseudodifferential operator Π on X and $c > 0$ such that:*

(1) *The symbol π of Π is supported in $\mathcal{U}$ and, $0 \leq \pi(x,\xi) \leq 1$ on S^*X.*
(2) *For any $|\tau| \leq \tau_0$ and any u eigenfunction of Δ_τ with energy greater than E_0 then the following nonconcentration estimate holds:*

$$\|\Pi u\|_\tau \geq c\|u\|_\tau. \tag{9}$$

Remark 13. Estimate (8) may be extended to any eigenfunction, without any restriction on the energy. Such an extension requires that one knows that an eigenfunction of a Laplace operator cannot vanish on an open subset (thus proving the control estimate for any finite number of eigenfunctions). The principle of unique continuation gives this property for eigenfunctions of the Laplace operator.

It is not clear to the author if this nonvanishing property holds in the cotangent bundle. It can also be observed that, strictly speaking, the formulation with the microlocal cutoff only has a meaning in the high energy limit since the microlocal cutoff is defined modulo smoothing operators.

PROOF. Let $(x,\xi)\in S^*X$, the geometric control assumption implies that there exists some T $(=T(x,\xi))$ and $(x_0,\xi_0)\in\mathcal{U}$ such that $\Phi_0^T(x_0,\xi_0)=(x,\xi)$. Using the continuity properties of solutions of an O.D.E. both with respect to initial conditions and with respect to parameters, we can find

- some $\tau_{\max}(x,\xi)$,
- two conical neighborhoods of (x,ξ): $\mathcal{V}=\mathcal{V}_{(x,\xi)}$ and $\mathcal{W}=\mathcal{W}_{(x,\xi)}$,
- a conical neighborhood $\widetilde{\mathcal{U}}=\widetilde{\mathcal{U}}_{(x,\xi)}$ of (x_0,ξ_0)

such that $\mathcal{V}$ is compactly included in $\mathcal{W}$ and $\Phi_\tau^{-T}(\mathcal{W})\subset\widetilde{\mathcal{U}}\Subset\mathcal{U}$ for any $|\tau|\leq\tau_{\max}(x,\xi)$. Since S^*X is compact, we can find a finite collection $\big((x_i,\xi_i)\big)_{1\leq i\leq N}$ such that $S^*X=\bigcup_{i=1}^N\mathcal{V}_{(x_i,\xi_i)}$. We now use the index i to denote any quantity formerly indexed by (or attached to) (x_i,ξ_i). Denote by τ_0 the minimum of all $\tau_{\max}(x_i,\xi_i)$. We can then find a smooth partition of unity π_i such that each π_i is identically 1 in $\mathcal{V}_i$ and identically 0 outside $\mathcal{W}_i$. Note that, by construction,

$$\forall|\tau|\leq\tau_0,\qquad \bigcup_{i=1}^N\Phi_\tau^{-T_i}(\mathcal{W}_i)\subset\bigcup_{i=1}^N\widetilde{\mathcal{U}}_i\Subset\mathcal{U}.$$

We denote by π_0 a smooth function which is identically 1 in $\bigcup_{i=1}^N\widetilde{\mathcal{U}}_i$ and 0 outside $\mathcal{U}$. Denote by $\Pi_i=\mathrm{Op}^+(\pi_i)$, any by $U_\tau(T)=\exp iT\sqrt{\Delta_\tau}$, we claim that for any $i>0$, and for any $|\tau|\leq\tau_0$, the operator $R_i=\Pi_iU_\tau(T_i)(\mathrm{Id}-\Pi_0)$ is smoothing. Indeed, its wave-front set is included

$$\{(y,\eta,x,\xi)\mid(y,\eta)\in\mathrm{supp}(\pi_i),(x,\xi)\in\mathrm{supp}(1-\pi_0),(y,\eta)=\Phi_\tau^{T_i}(x,\xi)\},$$

which is empty by construction. In particular, each $R_{i,\tau}$ is bounded from $H^{-1}(X)$ into $L^2(X)$. Moreover, a careful analysis of the construction of the Hadamard parametrix (as it is for instance presented in [1]) shows that there exists a uniform constant C such that:

$$\forall|\tau|\leq\tau_0,\forall w\in H^{-1}(X),\qquad \|R_iw\|_{L^2(X,g_0)}\leq C\|u\|_{H^{-1}(X,g_0)},\tag{10}$$

(see the appendix for a brief sketch of proof).

Since the $(\pi_i)_{i>1}$ are a partition of unity, $\sum_{i>1}\Pi_i=\mathrm{Id}$. We thus have

$$\|u\|_0^2=\sum_{i=1}^N\langle\Pi_iu,u\rangle_0.\tag{11}$$

Using that u is an eigenfunction of Δ_τ, we have

$$|\langle\Pi_iu,u\rangle_0|=|\langle\Pi_iU_\tau(T_i)u,u\rangle_0|\leq|\langle\Pi_iU_\tau(T_i)\Pi_0u,u\rangle_0|+|\langle R_iu,u\rangle_0|.$$

We use Cauchy – Schwarz inequality on both terms and the following facts on the norm of the different operators:

- Π_i is bounded from $L^2(X)$ to itself,
- $U_\tau(T)$ is an isometry from $L^2(X)$,
- and (10).

This yields:

$$|\langle \Pi_i u, u\rangle_0| \leq \left(c_i\|\Pi_0 u\|_0 + C\|u\|_{H^{-1}(X,g_0)}\right) \cdot \|u\|_0.$$

Using that the $H^{-1}(X, g_\tau)$ norms are all equivalent and that $\|u\|^2_{H^{-1}(X,g_\tau)} = (1+E)^{-1}\|u\|^2_\tau$ when u is an eigenfunction of Δ_τ, we get

$$|\langle \Pi_i u, u\rangle_0| \leq \left(c_i\|\Pi_0 u\| + C(1+E)^{-1/2}\|u\|_0\right) \cdot \|u\|_0,$$

for a constant C independent of $\tau \in [-\tau_0, \tau_0]$.

Summing all these inequalities and denoting by $\tilde{c}$ the maximum of the c_i we obtain:

$$\|u\|_0 \leq (N\tilde{c})\|\Pi_0 u\| + CN(1+E)^{-1/2}\|u\|_0.$$

This yields the estimate in the proposition with $\Pi = \Pi_0$ and $c^{-1} = 2\tilde{c}N$ as soon as $CN(1+E)^{-1/2}$ is less than $\frac{1}{2}$. □

This proposition has the following useful corollary.

Corollary 14. *Let a_τ be an analytically dependent family of nonnegative symbols of order 0 such that $\{a_0(x,\xi) > \varepsilon\}$ geometrically controls (S^*X, g_0) then there exist positive κ, τ_0 and E_0 such that, for any $|\tau| \leq \tau_0$ and any u eigenfunction of Δ_τ with energy greater than E_0 then*

$$\langle A_\tau u, u\rangle_0 \geq \kappa\|u\|^2_\tau,$$

where A_τ denotes $\mathrm{Op}^+(a_\tau)$.

Remark 15. In the following proof, we will use several times the fact that, for any symbols of order 0, $\mathrm{Op}^+(bc) - \mathrm{Op}^+(b) \circ \mathrm{Op}^+(c)$ is a pseudodifferential operator of order -1. We will denote by R any operator arising from this kind of operation. Observe that if b and c depend analytically on τ, then R also depends analytically on τ.

Proof. Let $\tau_0, \widetilde{E}_0$ and Π be given by applying Proposition 12 to $\mathcal{U} = \{a_0(x,\xi) > \varepsilon\}$. If needed, we restrict τ_0 so that, for any $|\tau| \leq \tau_0$, a_τ is bounded below by $\varepsilon/2$ on $\mathcal{U}$. Observe that

$$\begin{aligned}
\langle A_\tau u, u\rangle_0 &= \langle A_\tau \Pi^2 u, u\rangle_0 + \langle A_\tau(\mathrm{Id} - \Pi^2)u, u\rangle_0 \\
&= \langle \mathrm{Op}^+(a_\tau\pi^2)u, u\rangle_0 + \langle \mathrm{Op}^+(a_\tau(1-\pi^2))u, u\rangle_0 + \langle R_\tau u, u\rangle_0 \\
&\geq \langle \mathrm{Op}^+(a_\tau\pi^2)u, u\rangle_0 + \langle R_\tau u, u\rangle_0,
\end{aligned}$$

where the last inequality holds because $\pi(x,\xi) \leq 1$ and we have chosen a positive quantization. Since the support of π is included in $\mathcal{U}$ we have

$$\forall (x,\xi) \in S^*X, a_\tau(x,\xi)\pi^2(x,\xi) \geq \frac{\varepsilon}{2}\pi^2(x,\xi),$$

which is quantized into

$$\langle \mathrm{Op}^+(a_\tau\pi^2)u, u\rangle_0 \geq \frac{\varepsilon}{2}\langle \mathrm{Op}^+(\pi^2)u, u\rangle_0 \geq \frac{\varepsilon}{2}\|\Pi u\|_0^2 + \langle R_\tau u, u\rangle_0,$$

with another remainder operator R_τ. Putting these two inequalities together, we obtain

$$\langle Au, u\rangle_0 \geq \frac{\varepsilon}{2}\|\Pi u\|_0^2 + \langle R_\tau u, u\rangle_0. \tag{12}$$

Since R is of order -1 and the norms on H^{-1} associated with $(g_\tau)_{|\tau|\leq\tau_0}$ are uniformly equivalent, there exists some uniform C such that

$$|\langle R_\tau u, u\rangle_0| \leq CE^{-1/2}\|u\|_\tau.$$

Using Proposition 12, for $|\tau|\leq\tau_0$ and $E\geq E_0$, we have $\|\Pi u\|_0\geq c_1\|\Pi u\|_\tau\geq c_2\|u\|_\tau$ so that we get

$$\langle A_\tau u, u\rangle \geq \left(c\frac{\varepsilon}{2} - CE^{-1/2}\right)\cdot\|u\|_\tau^2. \tag{13}$$

The corollary then holds with $\kappa = c\varepsilon/4$ as soon as E is large enough so that $CE^{-1/2}\leq c\varepsilon/4$. □

4.2. Proof of Theorem 10. We consider normalized eigenbranches and we start from the formula for the derivative of the eigenfunctions: (in the following equation, everything except the constant C depends on τ)

$$\frac{\mathrm{d}}{\mathrm{d}\tau}\ln(E)\geq\langle Au, u\rangle - CE^{-1/2}.$$

Under the assumptions of the theorem, we may use Proposition 12 and its corollary. This provides us with $\kappa>0$, τ_0 and E_0 such that

$$\forall|\tau|\leq\tau_0, \forall E(\tau)\geq E_0, \qquad \frac{\mathrm{d}}{\mathrm{d}\tau}\ln(E)\geq\kappa. \tag{14}$$

The rest of the argument is merely reproduced from [5].

Denote by $R_M(E)=\int_{-\tau_0}^{\tau_0} R_M(t,E)\,\mathrm{d}t$. A straightforward manipulation of the integral yields:

$$R_M(E)=\sum_j \mathrm{Leb}\{\tau \mid E_j(\tau)\in(E-M,E+M]\}, \tag{15}$$

where Leb denotes the Lebesgue measure. We consider now E such that $E-M\geq E_0$. According to (14), any eigenbranch that crosses the energy $E-M$ has to do it in an increasing manner. This implies that any eigenbranch for which the set $I_j := \{t \mid E_j(\tau)\in(E-M,E+M]\}$ is not empty satisfies $E_j(-\tau_0)\leq E+M$ and $E_j(\tau_0)\geq E-M$. In particular, the sum in (15) has only a finite number of positive terms, and moreover, this number is bounded by $N_{-\tau_0}(E+M)$ which is bounded by $CE^{d/2}$ according to Weyl's law. Estimate (14) also implies that I_j is an interval whose extremities we denote by τ_j^- and τ_j^+.[3] Integrating (14) we obtain

$$\kappa(\tau_j^+-\tau_j^-)\leq\ln\left(\frac{E_j(\tau_j^+)}{E_j(\tau_j^-)}\right)\leq\ln\left(\frac{E+M}{E-M}\right)\leq\frac{C}{E},$$

as soon as E is large enough (independently of j). This bounds each term of the sum in (15), since the number of terms is also bounded explicitly using Weyl's law, we obtain

$$R_M(E)\leq C\cdot E^{d/2-1}.$$

Remark 16. The fact that a_0 is nonnegative implies in particularly that the derivative of $\mathrm{vol}_{g_\tau} X$ is increasing. In particular, the variation cannot be volume-preserving.

[3]We do not specify whether I_j is open or closed at each of its extremity since this fact depends on whether $\tau_j^\pm$ coincide or not with $\pm\tau_0$, and it is irrelevent when computing the Lebesgue measure.

Remark 17. Corollary 14 implies that for any semiclassical measure associated with Δ_τ, the following lower bound holds:

$$\int a \, d\mu \geq \kappa.$$

If we replace the control assumption in Theorem 10 by the following (weaker) assumption:

$$\forall \tau, \forall \mu \text{ semiclassical measure associated with} \Delta_\tau, \qquad \int a \, d\mu > 0,$$

then, following the same lines of proof, we would get a 'almost everywhere' result in the spirit of [5].

Remark 18. This result can probably be extended to manifolds with boundary with the precaution that, in this case we would have to control the possible escape of the mass to the boundary.

Examples. (1) A natural variation to look at is to simply take $g_\tau = \exp(\tau)g_0$. In this case the control assumption is trivially satisfied so that we get a uniform integrated remainder. However, this is not a new result since looking at a fixed energy interval and dilating the volume is equivalent to fixing the volume and dilating the energy interval so that the integrated remainder is actually a regularized remainder in Weyl's law. And it is known that integrating the remainder in Weyl's law with respect to the energy yields a much better estimate than the universal remainder term.

(2) For a more general conformal variation on a surface ($d = 2$) we have $g_\tau := \big(1 + \tau f(x)\big)^2 g_0$, the eigenvalue problem can be put in the form

$$\int_X \nabla_0 u \nabla_0 \phi \, \mathrm{dvol}_{g_0} = E \int_X u\phi \big(1 + \tau f(x)\big)^2 \, \mathrm{dvol}_{g_0}.$$

This gives $a_0(x, \xi) = 2f(x)$. Choosing f nonnegative and supported in a control region yields examples where the theorem applies.

(3) On the contrary, the variation considered in [5] on a Donnelly surface doesn't satisfy the control estimate because of the cylinder of periodic orbits and of periodic geodesic in the hyperbolic wings.

Appendix A. Uniform Hadamard parametrix

The aim of this appendix is to give a (rather sketchy) justification of the uniform estimate (10).

Let X be a smooth manifold without boundary and g_τ a real-analytic family of riemannian metrics.

Denote by $U_\tau(s)$ the propagator of the half-wave equation associated with Δ_τ:

$$U_\tau(s) = \exp(is\sqrt{\Delta_\tau}),$$

and denote by $k_\tau(s, x, y)$ its Schwartz kernel. This kernel $k_\tau(s, x, y)$ is obtained by restricting the Schwartz kernel of $\cos(s\sqrt{\Delta_\tau})$ to the region $\sigma > 0$ of the cotangent bundle $T^*(\mathbb{R} \times X \times Y)$

We start from the Hadamard parametrix construction as explained for instance in [1]. Concretely, for s less than some small ε and x, y such that $d_\tau(x, y) \leq 2\varepsilon$, we make the ansatz

$$\tilde{k}_N(\tau, s, x, y) = \sum_1^N u_k(\tau, x, y) T_{\alpha_d+k}\big(s^2 - d_\tau^2(x, y)\big),$$

in which the u_k are smooth functions of x, y and $T_{\alpha+k}(z)$ are oscillatory distributions of $\mathbb{R}$ with wave-front in $\{(z = 0, \zeta > 0)\}$. The coefficients u_k satisfy some transport equations from which it is straightforward that they also depend smoothly on τ. We fix a smooth cutoff function $\rho := \rho\big(d_\tau(x, y)\big)$ that cuts off in $d_\tau(x, y) \leq 2\varepsilon$ and we let $k_N = \tilde{k}_N \times \rho$. A computation then yields that k_N solves the half-wave equation up to a remainder term $r_N(\tau, s, x, y)$ which is $\mathcal{C}^{l(N)}$ where $l(N)$ tends to infinity with N. Applying Duhamel's principle, we get

$$k(\tau, s, x, y) - k_N(\tau, s, x, y) = \int_0^s k(\tau, s - s', x, w) r_N(\tau, s', w, y)\, \mathrm{d}s'\, \mathrm{d}w.$$

Here again, a detailed analysis shows that, for $s \leq \varepsilon$ and for any N, we finally have

$$k(\tau, s, x, y) = k_N(\tau, s, x, y) + R_N(\tau, s, x, y),$$

in which everything is smooth in τ, and R_N is $\mathcal{C}^{l(N)}$. Since k_N has a fixed order as a distribution of (s, x, y), applying it to a C^l function gives a $C^{l-\beta}$ with a loss β that is independent of N.

We now fix some T, and we use the semigroup properties of U to write

$$U_\tau(T) = U_\tau(\delta) \circ U_\tau(\delta) \circ \cdot \circ U_\tau(\delta),$$

with a fixed small δ. We now plug into this expression the decomposition $U_\tau = K_N + R_N$. Expanding everything gives a large sum in which each term is a composition of K_N factors with R_N factors. We address the regularity of each term. Each term has regularity $N - p\beta$ where p is the number of K_N factors. Thus any term except the one with only K_N factors may be made as regular as wanted by choosing N large enough. Moreover, everything is regular in τ.

It remains to address the term with only K_N factors. Each K_N is explicit and this composition is addressed using stationary phase. It is somewhat tedious in the general case because for instance of conjugate points. However it is much simpler when we have an initial and final microlocal cutoff that destroys everything in the wave-front set. Indeed a uniform nonstationary phase then applies that yields a uniform C^m bound for any m.

References

1. P. H. Bérard, *On the wave equation on a compact Riemannian manifold without conjugate points*, Math. Z. **155** (1977), no. 3, 249–276.
2. Nicolas Burq and Maciej Zworski, *Geometric control in the presence of a black box*, J. Amer. Math. Soc. **17** (2004), no. 2, 443–471 (electronic).
3. Y. Colin de Verdière, *Ergodicité et fonctions propres du laplacien*, Comm. Math. Phys. **102** (1985), no. 3, 497–502.
4. M. Dimassi and J. Sjöstrand, *Spectral asymptotics in the semi-classical limit*, London Math. Soc. Lecture Note Ser., vol. 268, Cambridge Univ. Press, Cambridge, 1999.
5. A. Hassell, *Ergodic billiards that are not quantum unique ergodic*, with an appendix by A. Hassell and L. Hillairet, Ann. of Math. (2) **171** (2010), no. 1, 605–618.
6. A. Hassell, L. Hillairet, and J. Marzuola, *Eigenfunction concentration for polygonal billiards*, Comm. Partial Differential Equations **34** (2009), no. 4-6, 475–485.
7. B. Helffer, A. Martinez, and D. Robert, *Ergodicité et limite semi-classique*, Comm. Math. Phys. **109** (1987), no. 2, 313–326.

8. L. Hillairet and C. Judge, *The eigenvalues of a domains with small slits*, Trans. Amer. Math. Soc., to appear.
9. T. Kato, *Perturbation theory for linear operators*, Classics Math., Springer, Berlin, 1995. Reprint of the 1980 edition.
10. M. E. Taylor, *Pseudodifferential operators*, Princeton Math. Ser., vol. 34, Princeton Univ. Press, Princeton, NJ, 1981.

Laboratoire de Mathématiques Jean Leray UMR CNRS 6629, Université de Nantes, 2 rue de la Houssinière, BP 92208, 44322 Nantes Cedex 3, France
E-mail address: Luc.Hillairet@math.univ-nantes.fr

Centre de Recherches Mathématiques
CRM Proceedings and Lecture Notes
Volume **52**, 2010

A Remainder Estimate for Weyl's Law on Liouville Tori

Hugues Lapointe

Abstract. The paper is concerned with the asymptotic distribution of Laplace eigenvalues on Liouville tori. Liouville metrics are the largest known class of integrable metrics on two-dimensional tori; they contain flat metrics and metrics of revolution as special cases. Using separation of variables, we reduce the eigenvalue counting problem to the problem of counting lattice points in certain planar domains. This allows us to improve the remainder estimate in Weyl's law on a large class of Liouville tori. For flat metrics, such an estimate has been known for more than a century due to classical results of W. Sierpiński and J. G. van der Corput. Our proof combines the method of Y. Colin de Verdière, who proved an analogous result for metrics of revolution on a sphere, with the techniques developed by P. Bleher, D. Kosygin, A. Minasov and Y. Sinai in their study of the almost periodic properties of the remainder in Weyl's law on Liouville tori.

1. Introduction and main results

1.1. Liouville tori. A Liouville torus $T = \mathbb{R}^2/(a_1\mathbb{Z} \oplus a_2\mathbb{Z})$ is a two-dimensional torus with the metric

$$ds^2 = \big(U_1(q_1) - U_2(q_2)\big)(dq_1^2 + dq_2^2), \tag{1.1.1}$$

where $U_1(q_1) > U_2(q_2) > 0$ are smooth periodic functions on $\mathbb{R}$, satisfying $U_i(q_i+a_i) = U_i(q_i)$ for all $q_i \in \mathbb{R}$, $i = 1, 2$. For simplicity, we assume that $a_1 = a_2 = 1$, but all the proofs work for arbitrary $a_i > 0$.

Definition 1.1.2. Let Ω be the set of pairs of functions (U_1, U_2), satisfying the following conditions for $i = 1, 2$:

(1) $U_i \in C^\infty(\mathbb{R})$.
(2) $U_i(q+1) = U_i(q)$ for all $q \in \mathbb{R}$.
(3) The function U_i has exactly one minimum and one maximum in $[0, 1)$, both nondegenerate.
(4) $U_1(q_1) > U_2(q_2)$ for any $q_1, q_2 \in \mathbb{R}$.

2010 *Mathematics Subject Classification.* Primary 58J50; Secondary 35P20.
Key words and phrases. Eigenvalue distribution, torus, Liouville metric, lattice counting.
Research was supported by the Canada Graduate Scholarships Program.
This is the final form of the paper.

If a Liouville torus T admits a finite group of translations G leaving the metric invariant, we may consider the quotient T/G. Such tori are called infra-Liouville and were considered in [17]. Our results hold for infra-Liouville tori under some additional assumptions, as shown in Section 5.

1.2. Weyl's law. The Laplacian on T is given by

$$\Delta = -\frac{1}{U_1(q_1) - U_2(q_2)}\left(\frac{\partial^2}{\partial q_1^2} + \frac{\partial^2}{\partial q_2^2}\right).$$

Its spectrum consists of an infinite sequence of eigenvalues

$$0 = \lambda_0 < \lambda_1 \leq \lambda_2 \leq \lambda_3 \leq \cdots \leq \lambda_n \leq \cdots \to \infty$$

of finite multiplicity. The corresponding eigenfunctions are smooth and form a basis in $L^2(T)$. The spectral counting function is defined by

$$N(\lambda) = \#(\{j \mid \sqrt{\lambda_j} \leq \lambda\}),$$

where each eigenvalue is counted with its multiplicity. According to Weyl's law (see [1, 12]), the following asymptotic formula holds on any compact surface:

$$R(\lambda) = N(\lambda) - \frac{\lambda^2}{4\pi}\operatorname{Area}(T) = O(\lambda).$$

The function $R(\lambda)$ is called the remainder term. The estimate $O(\lambda)$ is sharp and attained on a round sphere, since the spherical harmonics have high multiplicities. However, it can be improved under certain assumptions on the surface. For example, if the set of directions of periodic geodesics has Liouville measure zero in the unit cotangent bundle, it is shown in [7] that the remainder is of order $o(\lambda)$. This result has been extended to surfaces with boundary in [10]. Also, $R(\lambda) = O(\lambda/\log\lambda)$ for negatively curved surfaces (see [2]), and it is conjectured that for generic surfaces of negative curvature $R(\lambda) = O(\lambda^\epsilon)$ for any $\epsilon > 0$ (see [13]).

For a flat square torus, $R(\lambda)$ is the difference between the number of points of $\mathbb{Z}^2$ lying inside a circle of radius λ in $\mathbb{R}^2$ and the area of the disk bounded by this circle. Gauss's bound was $O(\lambda)$ and Sierpiński improved this estimate to $O(\lambda^{2/3})$ in [16]. This result can be obtained by smoothing the characteristic function of the domain and applying the Poisson summation formula (see [15]). A generalization of this method, applied to higher-dimensional cases, is found in [5]. It has been shown recently (see [8, 9]) that the remainder in the circle problem is of order $O(\lambda^{131/208}(\log\lambda)^{18627/8320})$. But we are still far from proving Hardy's conjecture that the bound $O(\lambda^{1/2+\epsilon})$ should hold for any $\epsilon > 0$, which would be optimal on the polynomial scale.

The upper bound on $R(\lambda)$ can also be improved for certain surfaces whose geodesic flow is completely integrable. Colin de Verdière showed in [6] that, for a generic convex sphere of revolution, the remainder is of order $O(\lambda^{2/3})$.

1.3. Main results. On a nondegenerate Liouville torus T, we estimate $R(\lambda)$ by representing the eigenvalues as lattice points in a planar domain. This correspondence, described in Section 2.3 and proved in [11], relies on the separation of variables and the asymptotic analysis of Sturm–Liouville problems. Certain techniques used to count points inside homothetic domains are then applied to bound $R(\lambda)$. The conditions required for a Liouville torus to qualify as nondegenerate will be explained in Section 2.2.

Our main results are as follows.

Theorem 1.3.1. *The spectral counting function of a nondegenerate Liouville torus admits the following bound on its remainder term:*

$$R(\lambda) = O(\lambda^{2/3}).$$

Theorem 1.3.2. *The set of nondegenerate metrics is dense in the set* Ω *(see Definition* 1.1.2) *in the Whitney* C^∞*-topology.*

In Sections 4 and 5, we show that analogous results also hold for nondegenerate tori of revolution and infra-Liouville tori.

Remark 1.3.3. Nondegenerate metrics actually form a set of first Baire category, hence they are not generic. This can be deduced from the results of Sections 3.4 and 3.5.

Remark 1.3.4. It is conjectured in [13] that the remainder term of any Liouville torus is of order $O(\lambda^{1/2+\epsilon})$ for $\epsilon > 0$ arbitrarily small. Note that in the particular case of a flat square torus, it is equivalent to Hardy's conjecture for the Gauss's circle problem.

2. Eigenvalues of a Liouville torus

In this section we review the known results regarding the geodesic flow and the eigenvalues of Liouville tori, obtained in [3, 4, 11].

2.1. Integrability of the geodesic flow. We study the geodesic flow on the cotangent bundle of T by introducing the Hamiltonian $H(p,q) : \mathbf{T}^*(T) \to \mathbb{R}$,

$$H(p,q) = \frac{1}{U_1(q_1) - U_2(q_2)}(p_1^2 + p_2^2).$$

The Hamiltonian system defined by $H(p,q)$ is integrable since it has the following additional first integral

$$S(p,q) = \frac{U_2(q_2)}{U_1(q_1) - U_2(q_2)}p_1^2 + \frac{U_1(q_1)}{U_1(q_1) - U_2(q_2)}p_2^2.$$

The Poisson bracket $\{H, S\}$ is identically zero so that the integrals H and S are in involution. On a fixed energy level set we can write $H(p,q) = L^2$ and $S(p,q) = cL^2$ for two constants L and c. We define

$$c_1 = \max_{0\le x\le 1} U_1(x) = U_1(M_1), \qquad c_2 = \min_{0\le x\le 1} U_1(x) = U_1(m_1),$$
$$c_3 = \max_{0\le x\le 1} U_2(x) = U_2(M_2), \qquad c_4 = \min_{0\le x\le 1} U_2(x) = U_2(m_2),$$

and assume that the second derivative of the U_i functions does not vanish at their respective critical points. Since

$$p_1^2 = (U_1(q_1) - c)L^2, \qquad p_2^2 = \big(c - U_2(q_2)\big)L^2$$

the action variables are given by

$$I_1(L,c) = \begin{cases} L\int_0^1 (U_1(q_1) - c)^{1/2}\,\mathrm{d}q_1 & \text{if } c_4 \le c \le c_2,\\ L\int_{U_1(q_1)\ge c} (U_1(q_1) - c)^{1/2}\,\mathrm{d}q_1 & \text{if } c_2 \le c \le c_1,\end{cases}$$

$$I_2(L,c) = \begin{cases} L\int_{U_2(q_2)\le c} \big(c - U_2(q_2)\big)^{1/2}\,\mathrm{d}q_2 & \text{if } c_4 \le c \le c_3,\\ L\int_0^1 \big(c - U_2(q_2)\big)^{1/2}\,\mathrm{d}q_2 & \text{if } c_3 \le c \le c_1.\end{cases}$$

We shall write $F_1(c) = I_1(1,c)$ and $F_2(c) = I_2(1,c)$. These functions define a curve $\gamma = \big(F_1(c), F_2(c)\big)$ in the plane for $c \in [c_4, c_1]$. Theorem 3.2 of [11] gives some information on the functions F_i, which we repeat here,

Theorem 2.1.1. *The functions F_i satisfy the following properties:*

(1) *$F_1(c)$ is a continuous function that is strictly decreasing. Moreover,*

$$F_1(c) \in C^\infty([c_4, c_2) \cup (c_2, c_1]) \quad \text{and} \quad F_1(c_1) = 0, \quad F_1'(c_1) = -\pi\big(-2U_1''(M_1)\big)^{-1/2}.$$

(2) *$F_2(c)$ is a continuous function that is strictly increasing. Moreover,*

$$F_2(c) \in C^\infty([c_4, c_3) \cup (c_3, c_1]) \quad \text{and} \quad F_2(c_4) = 0, \quad F_2'(c_4) = \pi\big(2U_2''(m_2)\big)^{-1/2}.$$

(3) *Close to their critical points, the derivatives of the F_i functions have the following asymptotics,*

$$\lim_{c\to c_2} \frac{\mathrm{d}F_1(c)}{\mathrm{d}c} \frac{1}{\log|c-c_2|} = \big(\tfrac{1}{2}U_1''(m_1)\big)^{-1/2},$$

$$\lim_{c\to c_3} \frac{\mathrm{d}F_2(c)}{\mathrm{d}c} \frac{1}{\log|c-c_3|} = -\big(-\tfrac{1}{2}U_2''(M_2)\big)^{-1/2}.$$

Asymptotics for higher derivatives of F_i can be found by differentiating the previous expressions.

(4) *For $c \in [c_4, c_1]$, the derivatives $\mathrm{d}F_i/\,\mathrm{d}c$ are different from zero.*

There exists a function $G(\alpha)$ such that γ is defined in polar coordinates by $\rho = G(\alpha), 0 \le \alpha \le \pi/2$, where

$$\begin{cases} G(\alpha) = \Big(F_1\big(c(\alpha)\big)^2 + F_2\big(c(\alpha)\big)^2\Big)^{1/2}, \\ \tan(\alpha) = F_2\big(c(\alpha)\big)/F_1\big(c(\alpha)\big). \end{cases}$$

We set

$$\alpha_0 = \alpha(c_4) = 0,$$

$$\alpha_1 = \alpha(c_3) = \arctan\frac{F_2(c_3)}{F_1(c_3)},$$

$$\alpha_2 = \alpha(c_2) = \arctan\frac{F_2(c_2)}{F_1(c_2)},$$

$$\alpha_3 = \alpha(c_1) = \frac{\pi}{2}.$$

The function $G(\alpha)$ is studied in Theorem 6.3 of [11]. We repeat here their results,

(1) $G(\alpha) \in C^1([0,\pi/2])$. The tangent to γ at the point with angular coordinates α_1 is vertical and the tangent to γ at the point with angular coordinates α_2 is horizontal.
(2) $G(\alpha) \in C^\infty([0,\alpha_1) \cup (\alpha_1,\alpha_2) \cup (\alpha_2,\pi/2])$
(3) We have $G^{(0)}(+0) = G(0) > 0$, $G^{(0)}(\pi/2-0) = G(\pi/2) > 0$ and $G^{(1)}(+0) \neq 0$, $G^{(1)}(\pi/2-0) \neq 0$.
(4) Close to the singularities at the angles α_1 and α_2, the second derivative of G has the following asymptotics,

$$\lim_{\alpha\to\alpha_i} \frac{\mathrm{d}^2G(\alpha)}{\mathrm{d}\alpha^2}(\alpha-\alpha_i)(\log|\alpha-\alpha_i|)^2 = \mathrm{const}(i), \qquad i = 1,2.$$

(5) We have the following inequalities

$$M \geq G(\alpha) \geq m > 0, \qquad \text{for } 0 \leq \alpha \leq \frac{\pi}{2},$$
$$\frac{\mathrm{d}G(\alpha)}{\mathrm{d}\alpha} \leq K, \qquad \text{for } 0 \leq \alpha \leq \frac{\pi}{2}.$$

Note that this implies the existence of a constant δ such that, for $\epsilon > 0$,

$$\operatorname{dist}(\gamma, (1+\epsilon)\gamma) \geq \delta\epsilon, \tag{2.1.2}$$

where $\operatorname{dist}(X, Y)$ is the Euclidean distance between two subsets X, Y of $\mathbb{R}^2$.

2.2. Nondegeneracy conditions. In what follows, we will require some quantities to be irrational and difficult to approximate using rational numbers (see [11, Section 11]).

Definition 2.2.1. A real number α is typical if there exists $\tau > 0$ such that

$$|k_1 + \alpha k_2| \geq \frac{\delta(\alpha)}{|k_2| \log(1+|k_2|)^\tau}, \qquad \forall k_1, k_2 \in \mathbb{Z}, k_2 \neq 0.$$

Given any $\tau > 1$, almost all real numbers satisfy the above inequality for some constant $\delta(\alpha)$. Indeed, for $k, n \in \mathbb{N}$, consider the set

$$S_{k,n} = \{x \in [0,1] | \exists l \in \mathbb{Z}, |l - xk| < (kn \log(1+k)^\tau)^{-1}\}.$$

The set of typical numbers associated to the exponent τ is the complement of $\bigcap_{n=1}^{+\infty} \bigcup_{k=1}^{+\infty} S_{k,n}$. The Lebesgue measure of $S_{k,n}$ is lower than $2(nk\log(1+k)^\tau)^{-1}$, and the measure of $\bigcup_{k=1}^{+\infty} S_{k,n}$ is lower than

$$\frac{1}{n} \sum_{k=1}^{+\infty} \frac{2}{k \log(1+k)^\tau} = \frac{C}{n}$$

for $\tau > 1$ and all n. Thus the set $\bigcap_{n=1}^{+\infty} \bigcup_{k=1}^{+\infty} S_{k,n}$ has measure zero and almost every real number is typical for the exponent $\tau > 1$. However, the set of typical numbers is also of first Baire category since it is the denumerable union of nowhere dense closed subsets.

Given a metric represented by (U_1, U_2), the curvature $\kappa(c)$ of γ at a point $(F_1(c), F_2(c))$ is given by $(F_2''F_1' - F_1''F_2')/((F_1')^2 + (F_2')^2)^{3/2}$. The only points where the curvature diverges are those corresponding to the singularities at c_2 and c_3. Note that on the interval $c \in (c_3, c_2)$,

$$8(F_2''F_1' - F_1''F_2')(c) = \int_0^1 \big(c - U_2(q_2)\big)^{-3/2} \mathrm{d}q_2 \int_0^1 (U_1(q_1) - c)^{-1/2} \mathrm{d}q_1 + \int_0^1 (U_1(q_1) - c)^{-3/2} \mathrm{d}q_1 \int_0^1 \big(c - U_2(q_2)\big)^{-1/2} \mathrm{d}q_2$$

so $\kappa(c)$ cannot vanish there.

Definition 2.2.2. The metric $\mathrm{d}s^2 = \big(U_1(q_1) - U_2(q_2)\big)(\mathrm{d}q_1^2 + \mathrm{d}q_2^2)$ on T is said to be nondegenerate if $(U_1, U_2) \in \Omega$ and the following conditions hold.

(1) The curvature $\kappa(c)$ has a finite number of zeros on $[c_4, c_3) \cup (c_2, c_1]$, each of first order.
(2) If $\kappa(\tilde{c}) = 0$, then $F_2'(\tilde{c})/F_1'(\tilde{c})$ is a typical number.

(3) The numbers $F_2'(c_i)/F_1'(c_i)$ are typical for $i = 1, 4$.

$$F_2'(c_1)/F_1'(c_1) = \frac{\int_0^1 \big(c_1 - U_2(q_2)\big)^{-1/2}\, dq_2}{-2\pi\big(-2U_1''(M_1)\big)^{-1/2}},$$

$$F_2'(c_4)/F_1'(c_4) = \frac{2\pi\big(2U_2''(m_2)\big)^{-1/2}}{\int_0^1 (U_1(q_1) - c_4)^{-1/2} dq_1}.$$

(4) The numbers $F_2(c_i)/F_1(c_i)$ are typical for $i = 2, 3$.

$$F_2(c_2)/F_1(c_2) = \frac{\int_0^1 \big(c_2 - U_2(q_2)\big)^{1/2}\, dq_2}{\int_0^1 (U_1(q_1) - c_2)^{1/2}\, dq_1},$$

$$F_2(c_3)/F_1(c_3) = \frac{\int_0^1 \big(c_3 - U_2(q_2)\big)^{1/2}\, dq_2}{\int_0^1 (U_1(q_1) - c_3)^{1/2}\, dq_1}.$$

Such conditions are also required in Theorem 3.1 of [4].

2.3. Eigenvalues on a Liouville torus. The Laplace operator on T has the form

$$\Delta = -\frac{1}{U_1(q_1) - U_2(q_2)}\left(\frac{\partial^2}{\partial q_1^2} + \frac{\partial^2}{\partial q_2^2}\right).$$

We associate to the first integral $S(p, q)$ another operator,

$$\widehat{S} = -\frac{U_2(q_2)}{U_1(q_1) - U_2(q_2)}\frac{\partial^2}{\partial q_1^2} - \frac{U_1(q_1)}{U_1(q_1) - U_2(q_2)}\frac{\partial^2}{\partial q_2^2}.$$

For a given pair of integers $m = (m_1, m_2)$, with $m_1 \geq 0$ and $m_2 \geq 0$, there is a pair of eigenvalues $(E_m, \widetilde{E}_m)$ of $(\Delta, \widehat{S})$. They correspond to solutions of the following periodic Sturm–Liouville problems, obtained after separation of variables,

$$\begin{cases} \Psi_1'' + (E_m U_1 - \widetilde{E}_m)\Psi_1 = 0, \\ \Psi_2'' + (\widetilde{E}_m - E_m U_2)\Psi_2 = 0. \end{cases} \tag{2.3.1}$$

More precisely, $\widetilde{E}_m$ is such that E_m is the m_1th eigenvalue of the first equation and the m_2th eigenvalue of the second equation. Note that given $\widetilde{E}$, the solutions E form an increasing sequence in the first case and a decreasing sequence in the second case of (2.3.1).

We set $E_m = \lambda^2$ and $c = \widetilde{E}_m/E_m$. Theorem 6.1 of [11] says that, for $|m|$ sufficiently large, $(E_m, \widetilde{E}_m)$ is the unique solution to the equation

$$\Phi_0(\lambda, c) + \Phi_1(\lambda, c) + \Phi_2(\lambda, c) = 2\pi\left(\left[\frac{m_1+1}{2}\right], \left[\frac{m_2+1}{2}\right]\right) \tag{2.3.2}$$

where $\Phi_0(\lambda, c) = \lambda\big(F_1(c), F_2(c)\big)$ and $|\Phi_2(\lambda, c)| \leq \text{Const}\,\lambda^{-2/3}\log\lambda$ uniformly for $c \in [c_4, c_1]$. The function $\Phi_1(\lambda, c)$ is of the form

$$\Phi_1(\lambda, c) = \big(\phi_1(\lambda, c), \phi_2(\lambda, c)\big)$$

and the following bounds apply, depending on the location of (m_1, m_2) in the plane,

$$\begin{cases} |\phi_1(\lambda,c) + (-1)^{m_1}\pi/2| \leq \text{Const}\,\lambda^{-2/3}\log\lambda, & \text{for } c_2 + \text{const}\,\lambda^{-2/3} \leq c \leq c_1, \\ |\phi_1(\lambda,c)| \leq \text{Const}\,\lambda^{-2/3}\log\lambda, & \text{for } c_4 \leq c \leq c_2 - \text{const}\,\lambda^{-2/3}, \\ |\phi_1(\lambda,c)| \leq \text{Const}, & \text{in other cases,} \end{cases}$$

$$\begin{cases} |\phi_2(\lambda,c)| \leq \text{Const}\,\lambda^{-2/3}\log\lambda, & \text{for } c_3 + \text{const}\,\lambda^{-2/3} \leq c \leq c_1, \\ |\phi_2(\lambda,c) + (-1)^{m_2}\frac{\pi}{2}| \leq \text{Const}\,\lambda^{-2/3}\log\lambda, & \text{for } c_4 \leq c \leq c_3 - \text{const}\,\lambda^{-2/3}, \\ |\phi_2(\lambda,c)| \leq \text{Const}, & \text{in other cases.} \end{cases}$$

Note that for $m_1, m_2 \geq 0$, each point of the form $(2\pi k_1, 2\pi k_2)$, $k_1, k_2 > 0$ can be written in four different ways as $2\pi([(m_1+1)/2], [(m_2+1)/2])$, the points lying on one axis in two, and the origin in a unique way.

The following domains will be used later

$$A_i = \{(\rho,\alpha) \mid \alpha_{i-1} \leq \alpha \leq \alpha_i, 0 \leq \rho \leq G(\alpha)\}, \qquad i = 1,2,3,$$
$$A = A_1 \cup A_2 \cup A_3.$$

We consider A_i and A as subsets of $\mathbb{R}^2$, via the mapping $(x,y) = (\rho\cos\alpha, \rho\sin\alpha)$.

Also, for $a = (a_1, a_2) \in \mathbb{R}^2$, we define the translated two-dimensional lattice

$$\Gamma_a = \{(2\pi k_1 + a_1, 2\pi k_2 + a_2) \mid (k_1,k_2) \in \mathbb{Z}^2\}$$

and, for D a subset of $\mathbb{R}^2$ with $rD = \{(x,y) \in \mathbb{R}^2 \mid (r^{-1}x, r^{-1}y) \in D\}$, let

$$N_a(D,r) = \sharp(\Gamma_a \cap r\mathring{D}) + \tfrac{1}{2}\sharp(\Gamma_a \cap r\partial D) + \sharp(\Gamma_{-a} \cap r\mathring{D}) + \tfrac{1}{2}\sharp(\Gamma_{-a} \cap r\partial D),$$

where $\mathring{D}$ is the interior of D and ∂D its boundary. The function $N_a(D,r)$ counts the number of points in rD of two opposite translates of $2\pi\mathbb{Z}^2$, giving a weight of $\frac{1}{2}$ to those lying on the boundary of rD.

3. Proofs of the main results

In Section 3.1, we explain the method used to count points in the domains rA_i, $i = 1,2,3$, and obtain several bounds on the Fourier transform of domains bounded by the curve γ. Additional results required for lattice counting are collected in Section 3.2. The proofs of Theorems 1.3.1 and 1.3.2 are contained in Sections 3.3 and 3.4, respectively. We show that nondegenerate metrics are not generic in Section 3.5.

3.1. Regularization of the counting function. We follow the approach used in Section 16 of [11]. Let ψ be a positive function such that $\psi \in C_0^\infty(\mathbb{R})$, $\operatorname{supp}\psi \subset (-1,1)$, $\int_{\mathbb{R}^2} \psi(\sqrt{x^2+y^2})\,dx\,dy = 1$ and $\psi \equiv 1$ in a neighborhood of 0. We will use the following cut-off function, $\Psi_\epsilon(x,y) = \epsilon^{-2}\psi(\epsilon^{-1}\sqrt{x^2+y^2})$. We write $\Psi(x,y)$ for $\Psi_1(x,y)$. Let χ_D be the characteristic function of the domain D,

$$\chi_D(x,y) = \begin{cases} 1 & \text{if } (x,y) \in D \setminus \partial D, \\ \frac{1}{2} & \text{if } (x,y) \in \partial D, \\ 0 & \text{else} \end{cases}$$

and $\widetilde{N}_a(D,r)$ be the regularized function,

$$\widetilde{N}_a(D,r) = \sum_{k \in 2\pi\mathbb{Z}^2 + a} (\Psi_{r^{-4/3}} * \chi_D)(r^{-1}k) + \sum_{k \in 2\pi\mathbb{Z}^2 - a} (\Psi_{r^{-4/3}} * \chi_D)(r^{-1}k)$$

which is an approximation to $N_a(D,r)$. Note that the points for which $\chi_D(r^{-1}k) \neq (\Psi_{r^{-4/3}} * \chi_D)(r^{-1}k)$ lie at distance of order $O(r^{-1/3})$ from $r\partial D$.

Using the Poisson summation formula we get

$$\widetilde{N}_a(D,r) = \frac{r^2}{2\pi^2} \sum_{k \in \mathbb{Z}^2} \cos(\langle a,k\rangle)\hat{\chi}_D(rk)\hat{\Psi}(r^{-1/3}k).$$

Since $\hat{\Psi}(0) = 1$ and $\hat{\chi}_D(0) = \mathrm{Area}(D)$, the term corresponding to $k = 0$ is $\mathrm{Area}(D)r^2/(2\pi^2)$. If $k \neq 0$, Stokes' formula gives,

$$\hat{\chi}_D(rk) = \int_D \mathrm{e}^{-\mathrm{i}r(xk_1+yk_2)}\,\mathrm{d}x\,\mathrm{d}y = \oint_{\partial D} \frac{1}{\mathrm{i}r}\frac{\mathrm{e}^{-\mathrm{i}r(xk_1+yk_2)}}{(k_1^2+k_2^2)}(k_2\,\mathrm{d}x - k_1\,\mathrm{d}y).$$

We have that $\hat{\Psi}(r^{-1/3}k)$ depends only on $r^{-1/3}|k|$ and is rapidly decreasing as its argument tends to infinity. However, the only bound needed is

$$|\hat{\Psi}(r^{-1/3}k)| \leq \frac{C}{(1+r^{-1/3}|k|)^\nu}$$

for a fixed $\nu > 2$. Under some restrictions on ∂D, we want to obtain

$$\text{(3.1.1)}\quad \frac{r}{2\pi^2}\sum_{k\neq 0} \cos(\langle a,k\rangle)\hat{\Psi}(r^{-1/3}k) \oint_{\partial D} \frac{\mathrm{e}^{-\mathrm{i}r(xk_1+yk_2)}}{(k_1^2+k_2^2)}(k_2\,\mathrm{d}x - k_1\,\mathrm{d}y) = O(r^{2/3})$$

so that

$$\widetilde{N}_a(D,r) = \frac{\mathrm{Area}(D)}{2\pi^2}r^2 + O(r^{2/3})$$

and

$$\widetilde{N}_a(D, r - \varkappa r^{-1/3}) \leq N_a(D,r) \leq \widetilde{N}_a(D, r + \varkappa r^{-1/3})$$

for $\varkappa$ sufficiently large and D star-shaped with respect to the origin, implies

$$N_a(D,r) = \frac{\mathrm{Area}(D)}{2\pi^2}r^2 + O(r^{2/3}).$$

This estimate is required in the proof of Theorem 1.3.1, where the region D considered is either A or A_i, with $i = 1,2,3$.

Note that the choice $\epsilon = r^{-4/3}$ is optimal with this method, since taking the exponent $-\frac{4}{3}+\delta$ instead would give at best a remainder of order

$$\frac{\mathrm{Area}(D)}{2\pi^2}\big((r+\varkappa r^{-1/3+\delta})^2 - (r-\varkappa r^{-1/3+\delta})^2\big) + O(r^{2/3-\delta/2})$$
$$= O\big(r^{2/3}(r^\delta + r^{-\delta/2})\big).$$

We separate ∂D in several pieces, and consider each of them separately. In particular, suppose ∂D contains some parts of γ. We reparametrize the curve γ from $\big(F_1(c), F_2(c)\big)$ to $\big(t, f(t)\big)$, with $t \in [0, F_1(c_4)]$. We use a partition to study f on distinct intervals $[t_j, t_{j+1}]$, where t_j is an increasing sequence with $t_0 = 0$ and $t_{m+1} = F_1(c_4)$,

$$[0, F_1(c_4)] = \bigcup_{j=0}^{m} [t_j, t_{j+1}]$$

and require that $f(t) \in C^0([t_j,t_{j+1}])$, $f(t) \in C^\infty\big((t_j,t_{j+1})\big)$ and $f''(t) \neq 0$ for $t \in (t_j, t_{j+1})$. The function $f(t)$ is singular at $F_1(c_2)$ and $F_1(c_3)$, since

$$f'(t) = \frac{F_2'\big(c(t)\big)}{F_1'\big(c(t)\big)} \quad\text{and}\quad f''(t) = \frac{(F_2''F_1' - F_1''F_2')\big(c(t)\big)}{F_1'\big(c(t)\big)^3}.$$

Here, the derivatives of the F_i functions are taken with respect to the c variable. We assume that the nondegeneracy conditions of Definition 2.2.2 are fulfilled by the metric. The points t_j will be the singular points of $f(t)$, the first order zeros of $f''(t)$, where $f'(t_j)$ is a typical number, and the ends of γ, 0 and $F_1(c_4)$.

Given a piece of γ, corresponding to the interval $[t_j, t_{j+1}]$, let

$$E = \{(k_1, k_2) \in \mathbb{Z}^2 \mid (k_1 + f'(s_k)k_2) = 0, s_k \in (t_j, t_{j+1})\}.$$

Since $f'(t)$ is monotone on (t_j, t_{j+1}), s_k is well defined for each $k \in E$ different from zero. We treat the cases of $k \notin E$ and $k \in E$ separately.

Theorem 3.1.2. *The following bound holds*

$$\sum_{k \notin E} \frac{r}{(1 + r^{-1/3}|k|)^{\nu}} \left| \int_{t_j}^{t_{j+1}} \frac{\mathrm{e}^{-\mathrm{i}r(tk_1 + f(t)k_2)}}{(k_1^2 + k_2^2)} (k_2 - k_1 f'(t))\, \mathrm{d}t \right| = O(r^{2/3}).$$

PROOF. For $k \notin E$, we can integrate by parts,

$$(3.1.3) \quad \int_{t_j}^{t_{j+1}} \frac{\mathrm{e}^{-\mathrm{i}r(tk_1+f(t)k_2)}}{(k_1^2 + k_2^2)} \big(k_2 - k_1 f'(t)\big)\, \mathrm{d}t$$
$$= \frac{1}{-\mathrm{i}r(k_1^2 + k_2^2)} \frac{\big(k_2 - k_1 f'(t)\big)}{\big(k_1 + k_2 f'(t)\big)} \mathrm{e}^{-\mathrm{i}r(tk_1+f(t)k_2)} \Big|_{t_j}^{t_{j+1}}$$
$$+ \frac{1}{-\mathrm{i}r} \int_{t_j}^{t_{j+1}} \frac{f''(t)}{\big(k_1 + k_2 f'(t)\big)^2} \mathrm{e}^{-\mathrm{i}r(tk_1+f(t)k_2)}\, \mathrm{d}t$$

if both $f'(t_j)$ and $f'(t_{j+1})$ are typical numbers. The contribution of the first term on the right-hand side of (3.1.3) will be bounded by

$$(3.1.4) \quad \frac{1}{r(k_1^2 + k_2^2)} \left(\left| \frac{k_2 - k_1 f'(t_{j+1})}{k_1 + k_2 f'(t_{j+1})} \right| + \left| \frac{k_2 - k_1 f'(t_j)}{k_1 + k_2 f'(t_j)} \right| \right).$$

Since $f''(t)$ is of constant sign on (t_j, t_{j+1}), we can integrate the absolute value of the integrand and bound the last term of (3.1.3) by

$$(3.1.5) \quad \frac{1}{r} \left| \frac{f'(t_j) - f'(t_{j+1})}{\big(k_1 + k_2 f'(t_j)\big)\big(k_1 + k_2 f'(t_{j+1})\big)} \right|.$$

Summing over $k \notin E$, after multiplying each term by the weight $r/(1 + r^{-1/3}|k|)^{\nu}$, we get a contribution of maximum order

$$\sum_{k \in \mathbb{Z}^2} \frac{1}{(1 + r^{-1/3}|k|)^{\nu}} + \sum_{n=1}^{+\infty} \frac{\log(1+n)^{\tau}}{(1 + r^{-1/3}n)^{\nu}} = O(r^{2/3}).$$

If t_j or t_{j+1} is equal to $F_1(c_2)$ or $F_1(c_3)$, we must be careful with the integration by parts because $f'\big(F_1(c_3)\big)$ diverges and $f'\big(F_1(c_2)\big) = 0$. We study only the case of $t_j = F_1(c_3)$ since the others are equivalent. The difficulty appears when handling the pairs (k_1, k_2) for which $k_2 = 0$. We know that the following asymptotic holds for t near enough t_j,

$$M \log(t - t_j) < f'(t) < m \log(t - t_j).$$

We deduce that for ϵ small enough,

$$\left| \int_{t_j}^{t_j+\epsilon} \frac{\mathrm{e}^{-\mathrm{i}rk_1 t}}{k_1} f'(t)\, \mathrm{d}t \right| \leq -C \frac{\epsilon \log(\epsilon)}{|k_1|}$$

and, using integration by parts, we also have

$$\left|\int_{t_j+\epsilon}^{t_{j+1}} \frac{\mathrm{e}^{-irk_1t}}{k_1} f'(t)\,\mathrm{d}t\right| \le -C\frac{\log(\epsilon)}{rk_1^2}.$$

We choose $\epsilon = (r|k_|)^{-1}$ so that the contribution of these terms is bounded by

$$\sum_{n=1}^{+\infty} \frac{C\log(rn)}{n^2(1+r^{-1/3}n)^\nu} = O\big(\log(r)\big).$$

□

Theorem 3.1.6. *The following bound holds*

$$\sum_{\substack{k\in E\\ k\neq 0}} \frac{r}{(1+r^{-1/3}|k|)^\nu}\left|\int_{t_j}^{t_{j+1}} \frac{\mathrm{e}^{-ir(tk_1+f(t)k_2)}}{(k_1^2+k_2^2)}\big(k_2-k_1f'(t)\big)\,\mathrm{d}t\right| = O(r^{2/3}).$$

PROOF. For $k\in E$, if $k_1+k_2f'(s_k)=0$, by definition $s_k\in(t_j,t_{j+1})$. In these cases we must use an approach similar to the stationary phase method to get an asymptotic as $r\to+\infty$. The stationary phase formula works for the integration of functions in $C_0^\infty(\mathbb{R})$, so that we cannot apply it directly here. Instead, we integrate separately on subintervals of (t_j,t_{j+1}) depending on r. The main term will correspond to the one given by the stationary phase but the error term will be easier to estimate. We integrate by parts on (t_j,s_k) and (s_k,t_{j+1}),

$$\begin{aligned}(3.1.7)\quad & -ir\int_{t_j}^{t_{j+1}} \frac{\mathrm{e}^{-ir(tk_1+f(t)k_2)}}{(k_1^2+k_2^2)}\big(k_2-k_1f'(t)\big)\,\mathrm{d}t\\ &= \lim_{\epsilon\to 0^+}\left(\int_{t_j}^{s_k-\epsilon}+\int_{s_k+\epsilon}^{t_{j+1}}\right)\mathrm{e}^{-ir(tk_1+f(t)k_2)}\frac{f''(t)}{\big(k_1+k_2f'(t)\big)^2}\,\mathrm{d}t\\ &\qquad + \frac{\mathrm{e}^{-ir(tk_1+f(t)k_2)}}{(k_1^2+k_2^2)}\frac{k_2-k_1f'(t)}{k_1+k_2f'(t)}\left(\Big|_{t_j}^{s_k-\epsilon}+\Big|_{s_k+\epsilon}^{t_{j+1}}\right).\end{aligned}$$

Note that

$$\frac{\mathrm{d}}{\mathrm{d}t}\left(\frac{k_2-k_1f'(t)}{k_1+k_2f'(t)}\right) = -\frac{(k_1^2+k_2^2)f''(t)}{\big(k_1+k_2f'(t)\big)^2}$$

so we can write (3.1.7) as

$$\begin{aligned}(3.1.8)\quad & \\ &\lim_{\epsilon\to 0^+}\left(\int_{t_j}^{s_k-\epsilon}+\int_{s_k+\epsilon}^{t_{j+1}}\right)\big(\mathrm{e}^{-ir(tk_1+f(t)k_2)}-\mathrm{e}^{-ir(s_kk_1+f(s_k)k_2)}\big)\frac{f''(t)}{\big(k_1+k_2f'(t)\big)^2}\,\mathrm{d}t\\ &\qquad + \frac{\big(\mathrm{e}^{-ir(tk_1+f(t)k_2)}-\mathrm{e}^{-ir(s_kk_1+f(s_k)k_2)}\big)}{(k_1^2+k_2^2)}\frac{k_2-k_1f'(t)}{k_1+k_2f'(t)}\left(\Big|_{t_j}^{s_k-\epsilon}+\Big|_{s_k+\epsilon}^{t_{j+1}}\right).\end{aligned}$$

Since $k_1+k_2f'(t)$ has only a first order zero at s_k,

$$\lim_{\epsilon\to 0}\frac{\mathrm{e}^{-ir((s_k+\epsilon)k_1+f(s_k+\epsilon)k_2)}-\mathrm{e}^{-ir(s_kk_1+f(s_k)k_2)}}{k_1+k_2f'(s_k+\epsilon)} = 0$$

and we can put $\epsilon=0$ in the second term of (3.1.8). Its contribution will then be bounded by (3.1.4) and we can proceed as in the case of $k\notin E$.

It now suffices to bound the first term of (3.1.8), which we write, after dividing by $\mathrm{e}^{-\mathrm{i}r(s_k k_1+f(s_k)k_2)}$, as

$$\int_{t_j}^{t_{j+1}} \frac{(\mathrm{e}^{-\mathrm{i}rk_2 f''(s_k)\frac{(t-s_k)^2}{2}h_k(t)}-1)}{(t-s_k)^2}\frac{f''(t)}{\big(k_2 g_k(t) f''(s_k)\big)^2}\,\mathrm{d}t$$

for $tk_1+f(t)k_2=(s_k k_1+f(s_k)k_2)+k_2 f''(s_k)(t-s_k)^2 h_k(t)/2$ and $k_1+k_2 f'(t)=k_2 g_k(t) f''(s_k)(t-s_k)$. Note that the integrand will be a continuous function, since it has a limit as $t\longrightarrow s_k$. We also have

$$h_k(t)=2\frac{f(t)-f(s_k)-f'(s_k)(t-s_k)}{f''(s_k)(t-s_k)^2}=2\frac{\int_{s_k}^{t}(t-u)f''(u)\,\mathrm{d}u}{f''(s_k)(t-s_k)^2}$$

and

$$g_k(t)=\frac{1}{f''(s_k)(t-s_k)}\int_{s_k}^{t} f''(u)\,\mathrm{d}u.$$

Our assumptions on $f''(t)$ at the points t_j and t_{j+1} imply the existence of H such that $0<1/H<g_k, h_k<H$ in the intervals $t\in[s_k-\alpha_k, s_k+\alpha_k]$, for $k\in E$ and

$$\alpha_k=\tfrac{1}{2}\min\{s_k-t_j, t_{j+1}\quad s_k\}. \tag{3.1.9}$$

A justification of this claim is contained in Lemma 3.1.17. We define a function B on $k\in E$,

$$B(k_1,k_2)=\alpha_k^{-2}|k_2 f''(s_k)|^{-1}. \tag{3.1.10}$$

It will be used to separate in two parts the sum in Theorem 3.1.6, depending on the value of r.

Using the estimate $|\mathrm{e}^z-1|\le \mathrm{e}^C|z|$ for $|z|\le C$, we deduce that for $r\ge B(k_1,k_2)$

$$\begin{aligned}
&\left|\int_{s_k-|rk_2 f''(s_k)|^{-1/2}}^{s_k+|rk_2 f''(s_k)|^{-1/2}} (\mathrm{e}^{-\mathrm{i}rk_2 f''(s_k)(t-s_k)^2 h_k(t)/2}-1)\frac{f''(t)}{(k_1+k_2 f'(t))^2}\,\mathrm{d}t\right| \\
&\quad\le \mathrm{e}^H H^3 \int_{s_k-|rk_2 f''(s_k)|^{-1/2}}^{s_k+|rk_2 f''(s_k)|^{-1/2}} \frac{r|k_2 f''(s_k)|(t-s_k)^2}{2}\frac{|f''(t)|\,\mathrm{d}t}{\big(k_2 f''(s_k)(t-s_k)\big)^2} \\
&\quad\le \mathrm{e}^H H^4 \frac{r^{1/2}}{|k_2^3 f''(s_k)|^{1/2}}
\end{aligned} \tag{3.1.11}$$

since $(k_2 f''(s_k)(t-s_k))^2\le H^2(k_1+k_2 f'(t))^2$ and

$$\begin{aligned}
&\int_{s_k-|rk_2 f''(s_k)|^{-1/2}}^{s_k+|rk_2 f''(s_k)|^{-1/2}} |f''(t)|\,\mathrm{d}t \\
&\qquad = |f'(s_k+|rk_2 f''(s_k)|^{-1/2})-f'(s_k-|rk_2 f''(s_k)|^{-1/2})| \\
&\qquad \le 2H|f''(s_k)|^{1/2}|rk_2|^{-1/2}.
\end{aligned}$$

Also

(3.1.12)

$$\left|\int_{t_j}^{s_k-|rk_2f''(s_k)|^{-1/2}} (\mathrm{e}^{-\mathrm{i}r(tk_1+f(t)k_2)} - \mathrm{e}^{-\mathrm{i}r(s_kk_1+f(s_k)k_2)})\frac{f''(t)}{\big(k_1+k_2f'(t)\big)^2}\,\mathrm{d}t\right|$$

$$+\left|\int_{s_k+|rk_2f''(s_k)|^{-1/2}}^{t_{j+1}} (\mathrm{e}^{-\mathrm{i}r(tk_1+f(t)k_2)} - \mathrm{e}^{-\mathrm{i}r(s_kk_1+f(s_k)k_2)})\frac{f''(t)}{\big(k_1+k_2f'(t)\big)^2}\,\mathrm{d}t\right|$$

$$\le \left|\frac{2}{k_2(k_1+k_2f'(t))}\left(\Big|_{t_j}^{s_k-|rk_2f''(s_k)|^{-1/2}} + \Big|_{s_k+|rk_2f''(s_k)|^{-1/2}}^{t_{j+1}}\right)\right|$$

$$\le 2\left|\frac{f'(t_j)-f'(t_{j+1})}{\big(k_1+k_2f'(t_j)\big)\big(k_1+k_2f'(t_{j+1})\big)}\right| + 4H\frac{r^{1/2}}{|k_2^3f''(s_k)|^{1/2}}.$$

We have obtained a bound for large enough r, on each individual term of the summation over $k \in E$. However, if we are to work with a fixed r, we must also have a bound for the terms such that $r < B(k_1,k_2)$. This is equivalent to $\alpha_k < |rk_2f''(s_k)|^{-1/2}$. In these cases

$$(3.1.13)\quad \left|\int_{s_k-\alpha_k}^{s_k+\alpha_k} (\mathrm{e}^{-\mathrm{i}rk_2f''(s_k)(t-s_k)^2h_k(t)dir} - 1)\frac{f''(t)}{\big(k_1+k_2f'(t)\big)^2}\,\mathrm{d}t\right|$$

$$\le \mathrm{e}^H H^4\frac{r\alpha_k}{|k_2|} \le \mathrm{e}^H H^4\frac{1}{k_2^2\alpha_k|f''(s_k)|}$$

and

(3.1.14)

$$\left|\left(\int_{t_j}^{s_k-\alpha_k} + \int_{s_k+\alpha_k}^{t_{j+1}}\right)(\mathrm{e}^{-\mathrm{i}r(tk_1+f(t)k_2)} - \mathrm{e}^{-\mathrm{i}r(s_kk_1+f(s_k)k_2)})\frac{f''(t)}{\big(k_1+k_2f'(t)\big)^2}\,\mathrm{d}t\right|$$

$$\le \left|\frac{2}{k_2\big(k_1+k_2f'(t)\big)}\left(\Big|_{t_j}^{s_k-\alpha_k} + \Big|_{s_k+\alpha_k}^{t_{j+1}}\right)\right|$$

$$\le 2\left|\frac{f'(t_j)-f'(t_{j+1})}{\big(k_1+k_2f'(t_j)\big)\big(k_1+k_2f'(t_{j+1})\big)}\right| + 4H\frac{1}{k_2^2\alpha_k|f''(s_k)|}.$$

Lemmas 3.1.15 and 3.1.16 complete the proof of Theorem 3.1.6. □

Lemma 3.1.15. *We have the following bound on the sum over $k \in E$ satisfying $r \ge B(k_1,k_2)$,*

$$\sum_{r\ge B(k_1,k_2)} \frac{r}{(1+r^{-1/3}|k|)^\nu}\left|\int_{t_j}^{t_{j+1}} \frac{\mathrm{e}^{-\mathrm{i}r(tk_1+f(t)k_2)}}{(k_1^2+k_2^2)}\big(k_2-k_1f'(t)\big)\,\mathrm{d}t\right| = O(r^{2/3}).$$

Proof. We know from (3.1.11) and (3.1.12) that we need only to bound

$$\sum_{r\ge\alpha_k^{-2}|k_2f''(s_k)|^{-1}} \frac{r^{1/2}}{(1+r^{-1/3}|k|)^\nu|k_2^3f''(s_k)|^{1/2}}$$

since the calculations made for the case $k \notin E$ can be applied to the sum over $\big|\big(f'(t_j)-f'(t_{j+1})\big)/\big(\big(k_1+k_2f'(t_j)\big)\big(k_1+k_2f'(t_{j+1})\big)\big)\big|$. For a given compact interval $[\beta_1,\beta_2]$ with $\beta_1 > 0$, the contribution of the pairs (k_1,k_2) such that $k_1 +$

$k_2 f'(s_k) = 0$ and $|f''(s_k)| \in [\beta_1, \beta_2]$ will be bounded, for some $C > 0$, by

$$Cr^{1/2} \sum_{k_2=1}^{+\infty} \frac{1}{|k_2|^{3/2}} \sum_{k_1=0}^{Ck_2} \frac{1}{(1 + r^{-1/3}|k|)^{\nu}} = O(r^{2/3}).$$

Now suppose $f''(t)$ vanishes at t_j. Since the curvature admits only zeros of first order, we will obtain for some ϵ

$$|f''(t)| \geq \epsilon |f'(t) - f'(t_j)|^{1/2}$$

if $0 \leq t - t_j \leq \epsilon$. Since $f'(t_j)$ is typical, we also have

$$|f'(s_k) - f'(t_j)| = \left| \frac{k_1}{k_2} + f'(t_j) \right| \geq \frac{\delta}{k_2^2 \log(1 + |k_2|)^{\tau}}.$$

The contribution of the terms such that $s_k - t_j \in [0, \epsilon]$ will be bounded by

$$\begin{aligned} r^{1/2} \sum_{k_2=1}^{+\infty} & \frac{C}{|k_2|^{3/2}(1 + r^{-1/3}k_2)^{\nu}} \sum_{n=0}^{\lfloor k_2 |f'(t_j+\epsilon) - f'(t_j)| \rfloor} \left(\frac{\delta}{k_2^2 \log(k_2+1)^{\tau}} + \frac{n}{k_2} \right)^{-1/4} \\ &\leq \frac{Cr^{1/2}}{\delta^{1/4}} \sum_{k_2=1}^{+\infty} \frac{\log(1+k_2)^{\tau/4}}{k_2(1 + r^{-1/3}k_2)^{\nu}} \\ &\qquad + Cr^{1/2} \sum_{k_2=1}^{+\infty} \frac{1}{|k_2|^{5/4}(1 + r^{-1/3}k_2)^{\nu}} \int_0^{k_2|f'(t_j+\epsilon) - f'(t_j)|} \frac{du}{u^{1/4}} = O(r^{2/3}). \end{aligned}$$

Note that

$$\sum_{k_2=1}^{+\infty} \frac{\log(1+k_2)^{\tau/4}}{k_2(1 + r^{-1/3}k_2)^{\nu}} \leq C \int_1^{+\infty} \frac{\log(1+u)^{\tau/4}\, du}{u(1 + r^{-1/3}u)^{\nu}}$$

and

$$\int_1^{r^{1/3}} \frac{\log(1+u)^{\tau/4}\, du}{u(1 + r^{-1/3}u)^{\nu}} = O(\log(r)^{1+\tau/4})$$

$$\int_{r^{1/3}}^{+\infty} \frac{\log(1+u)^{\tau/4}\, du}{u(1 + r^{-1/3}u)^{\nu}} = O(\log(r)^{\tau/4}).$$

We also have

$$\begin{aligned} \sum_{k_2=1}^{+\infty} \frac{1}{|k_2|^{5/4}(1 + r^{-1/3}k_2)^{\nu}} & \int_0^{k_2|f'(t_j+\epsilon) - f'(t_j)|} \frac{du}{u^{1/4}} \\ & \leq C \sum_{n=1}^{+\infty} \frac{1}{n^{1/2}(1 + r^{-1/3}n)^{\nu}} = O(r^{1/6}). \end{aligned}$$

The function $f''(t)$ diverges at $t = F_1(c_2)$ and $t = F_1(c_3)$. We will study the case of $F_1(c_3)$, as the other one is equivalent after swapping the axes. Suppose that the singular point is at t_j. When $t - t_j$ is small enough,

$$M \log(t - t_j) < f'(t) < m \log(t - t_j)$$

and

$$m(t - t_j)^{-1} < |f''(t)| < M(t - t_j)^{-1}$$

for some $0 < m < M$. We deduce

$$|f''(t)| > m\exp(-M^{-1}f'(t)).$$

The contribution of the terms for which s_k is near $F_1(c_3)$ will then be bounded by

$$r^{1/2}\sum_{k_2=1}^{\infty}\frac{1}{|k_2|^{3/2}}\sum_{k_1=0}^{+\infty}\frac{\exp(-k_1(2Mk_2)^{-1})}{m^{1/2}(1+r^{-1/3}|k|)^{\nu}} \le \frac{Cr^{1/2}}{m^{1/2}}\sum_{k_2=1}^{\infty}\frac{(1+2Mk_2)}{|k_2|^{3/2}(1+r^{-1/3}k_2)^{\nu}}$$
$$= O(r^{2/3}).$$

Similar bounds hold if we consider t_{j+1}, so the sum is of order $O(r^{2/3})$. □

Lemma 3.1.16. *We have the following bound on the sum over $k \in E$ satisfying $r < B(k_1, k_2)$,*

$$\sum_{r<B(k_1,k_2)}\frac{r}{(1+r^{-1/3}|k|)^{\nu}}\left|\int_{t_j}^{t_{j+1}}\frac{\mathrm{e}^{-\mathrm{i}r(tk_1+f(t)k_2)}}{(k_1^2+k_2^2)}(k_2-k_1f'(t))\,\mathrm{d}t\right| = O(r^{2/3}).$$

Proof. We use (3.1.13) and (3.1.14) so that it suffices to bound

$$\sum_{\alpha_k<|rk_2f''(s_k)|^{-1/2}}\frac{1}{(1+r^{-1/3}|k|)^{\nu}k_2^2\alpha_k|f''(s_k)|}.$$

If $m < |f''(t)| < M$ near t_j and $f'(t_j)$ is typical, then

$$2M\alpha_k > |f'(s_k)-f'(t_j)| = \left|\frac{k_1}{k_2}+f'(t_j)\right| \ge \frac{\delta}{k_2^2\log(1+|k_2|)^{\tau}}.$$

Suppose $f''(t)$ vanishes at t_j. Since the zeros of $f''(t)$ are of first order, $m(t-t_j) < |f''(t)| < M(t-t_j)$ near this point. Then $|f''(s_k)| > 2m\alpha_k$ and the fact that $f'(t)$ is typical at a zero of $f''(t)$ also implies

$$2M\alpha_k^2 > |f'(s_k)-f'(t_j)| = \left|\frac{k_1}{k_2}+f'(t_j)\right| \ge \frac{\delta}{k_2^2\log(1+|k_2|)^{\tau}}.$$

If $t_j = F_1(c_3)$, $f''(t)$ diverges like $(t-t_j)^{-1}$ so $\alpha_k|f''(s_k)|$ is bounded from below by a strictly positive constant when s_k approaches t_j. Similar bounds hold if s_k is near $F_1(c_2)$. We deduce

$$\sum_{\alpha_k<|rk_2f''(s_k)|^{-1/2}}\frac{1}{(1+r^{-1/3}|k|)^{\nu}k_2^2\alpha_k|f''(s_k)|}$$
$$\le \sum_{k_2=1}^{+\infty}\frac{C\log(1+k_2)^{\tau}}{(1+r^{-1/3}k_2)^{\nu}}+\sum_{k\neq 0}\frac{C}{(1+r^{-1/3}|k|)^{\nu}} = O(r^{2/3}).$$ □

Lemma 3.1.17. *There exists $H > 0$ such that the functions*

$$g_k(t) = \frac{1}{f''(s_k)(t-s_k)}\int_{s_k}^{t}f''(u)\,\mathrm{d}u$$

and

$$h_k(t) = 2\frac{\int_{s_k}^{t}(t-u)f''(u)\,\mathrm{d}u}{f''(s_k)(t-s_k)^2}$$

admit the bounds $0 < 1/H < g_k, h_k < H$ in the interval $t \in [s_k-\alpha_k, s_k+\alpha_k]$, for $k \in E$. We recall that $k_1 + k_2f'(s_k) = 0$ and $\alpha_k = \frac{1}{2}\min\{s_k-t_j, t_{j+1}-s_k\}$.

PROOF. The result is clear if we fix a compact interval $[T_j, T_{j+1}] \subset (t_j, t_{j+1})$ and consider the case of k such that $s_k \in [T_j, T_{j+1}]$. We are left to study the ends of the interval (t_j, t_{j+1}). Suppose that $f''(t)$ vanishes at t_j, we have $c(t-t_j) < |f''(t)| < C(t-t_j)$ for $(t-t_j)$ sufficiently small and, assuming $t \in [s_k - \alpha_k, s_k + \alpha_k]$,

$$\frac{c}{2C} < g_k(t) < \frac{3C}{2c}$$

with the same bounds for $h_k(t)$. Now suppose that $t_j = F_1(c_3)$, we know that in this case we have

$$c(t-t_j)^{-1} < |f''(t)| < C(t-t_j)^{-1}$$

and we deduce that

$$\frac{2c}{3C} < g_k(t) < \frac{2C}{c}$$

with the same bounds for $h_k(t)$. The same arguments can be applied for $t_j = F_1(c_2)$ or at t_{j+1}. This shows the existence of uniform bounds on g_k, h_k for $k \in E$. □

3.2. Lattice points near straight lines. Our previous calculations dealt with integrals taken over the curve γ. The boundaries of the domains considered in the proof of Theorem 1.3.1 also contain straight parts. We will need the following results to complete our study of $\widetilde{N}_a(r, D)$. Note that Ψ is the cut-off function introduced at the beginning of Section 3.1.

Theorem 3.2.1. *If ω is a line segment emanating from the origin, that has finite length, with rational or typical slope, then*

$$\sum_{k\neq 0} r\cos(\langle a,k\rangle)\hat{\Psi}(r^{-1/3}k)\int_\omega \frac{\mathrm{e}^{-\mathrm{i}r(xk_1+yk_2)}}{(k_1^2+k_2^2)}(k_2\,\mathrm{d}x - k_1\,\mathrm{d}y) = O(r^{2/3}).$$

PROOF. We first suppose that ω is rational and given by (pt, qt) for $t \in [0, l]$ and some $p, q \in \mathbb{Z}$. The summation over $k \neq 0$ is divided in two parts. The first part contains the terms such that $pk_1 + qk_2 = 0$. Since $\hat{\Psi}(r^{-1/3}k)$ depends only on $r^{-1/3}|k|$ and the terms of the sum are antisymmetric in k,

$$\begin{aligned}\sum_{\substack{pk_1+qk_2=0\\ k\neq 0}} & r\cos(\langle a,k\rangle)\hat{\Psi}(r^{-1/3}k)\int_0^l \frac{\mathrm{e}^{-\mathrm{i}r(pk_1+qk_2)t}}{(k_1^2+k_2^2)}(pk_2-qk_1)\,\mathrm{d}t \\ &= \sum_{\substack{pk_1+qk_2=0\\ k\neq 0}} rl\cos(\langle a,k\rangle)\hat{\Psi}(r^{-1/3}k)\frac{(pk_2-qk_1)}{(k_1^2+k_2^2)} = 0.\end{aligned}$$

The other part contains (k_1, k_2) such that $|pk_1 + qk_2| \geq 1$, and is bounded in absolute value by

$$\begin{aligned}\sum_{pk_1+qk_2\neq 0} & \frac{r}{(1+r^{-1/3}|k|)^\nu}\left|\int_0^l \frac{\mathrm{e}^{-\mathrm{i}r(pk_1+qk_2)t}}{(k_1^2+k_2^2)}(pk_2-qk_1)\,\mathrm{d}t\right| \\ &\leq \sum_{pk_1+qk_2\neq 0}\frac{2}{(1+r^{-1/3}|k|)^\nu|k|^2}\left|\frac{pk_2-qk_1}{pk_1+qk_2}\right| \\ &\leq C\int_1^{+\infty}\frac{du}{(1+r^{-1/3}u)^\nu} = O(r^{1/3}).\end{aligned}$$

If ω is represented by $(t, \alpha t)$, with $t \in [0, l]$ and α a typical number, we have

$$\sum_{|k_1+\alpha k_2|\geq 1} \frac{r}{(1+r^{-1/3}|k|)^\nu}\left|\int_0^l \frac{\mathrm{e}^{-ir(k_1+\alpha k_2)t}}{(k_1^2+k_2^2)}(k_2-\alpha k_1)\,\mathrm{d}t\right| \leq \sum_{|k_1+\alpha k_2|\geq 1} \frac{2}{(1+r^{-1/3}|k|)^\nu}\frac{|k_2-\alpha k_1|}{(k_1^2+k_2^2)} = O(r^{1/3}).$$

Since α is typical, we also have the following upper bound,

$$\begin{aligned}\sum_{\substack{|k_1+\alpha k_2|<1\\ k\neq 0}} \frac{r}{(1+r^{-1/3}|k|)^\nu}\left|\int_0^l \frac{\mathrm{e}^{-ir(k_1+\alpha k_2)t}}{(k_1^2+k_2^2)}(k_2-\alpha k_1)\,\mathrm{d}t\right| &\leq \sum_{\substack{|k_1+\alpha k_2|<1\\ k\neq 0}} \frac{2}{(1+r^{-1/3}|k|)^\nu|k|^2}\left|\frac{k_2-\alpha k_1}{k_1+\alpha k_2}\right| \\ &\leq \sum_{n=1}^{+\infty} \frac{C\log(1+n)^\tau}{(1+r^{-1/3}n)^\nu} = O(r^{1/3}\log(r)^\tau)\end{aligned}$$

for C sufficiently large. □

Lemma 3.2.2. *If ω is a ray of length $r > 1$, with rational or typical slope, then*

$$\#\{v \in \Gamma_a \mid 0 < \operatorname{dist}(v, \omega) \leq r^{-1/3}\} \leq Cr^{2/3}$$

for some $C > 0$ which depends on a and the direction of ω.

PROOF. If the slope is rational, ω is contained in $\bar{\omega} = \{(pt, qt) \mid t \in \mathbb{R}\}$ for some $(p, q) \in \mathbb{Z}^2$. In this case, there is a minimal distance between ω and the points of the lattice Γ_a not in $\bar{\omega}$. Thus the existence of the bound is clear.

Suppose the slope is typical and that ω is represented by $(t\cos\theta, t\sin\theta)$ with $\tan\theta$ a typical number and $t \in [0, r]$. Let B be the set of points in $\mathbb{R}^2$ which are at a distance of less than $2r^{-1/3}$ to ω, and χ_B the characteristic function of B. It suffices to bound

$$\sum_{k\in 2\pi\mathbb{Z}^2+a} (\Psi_{r^{-1/3}} * \chi_B)(k)$$

which, by the Poisson summation formula, is equal to

$$\frac{1}{4\pi^2}\sum_{k\in\mathbb{Z}^2} \exp(i\langle a, k\rangle)\hat{\Psi}(r^{-1/3}k)\hat{\chi}_B(k).$$

The term corresponding to $k = 0$ is the area of B, which is $4r^{2/3}(1 + \pi r^{-4/3})$. If $k \neq 0$,

$$\begin{aligned}
\mathrm{i}\hat{\chi}_B(k) &= \oint_{\partial B} \frac{\mathrm{e}^{-\mathrm{i}(xk_1+yk_2)}}{k_1^2+k_2^2}(k_2\,\mathrm{d}x - k_1\,\mathrm{d}y)\\
&= 2\sin\big(2r^{-1/3}(\cos\theta k_2 - \sin\theta k_1)\big)\int_0^r \frac{\mathrm{e}^{-\mathrm{i}t(\cos\theta k_1+\sin\theta k_2)}}{k_1^2+k_2^2}(\cos\theta k_2 - \sin\theta k_1)\,\mathrm{d}t\\
&\quad + 2r^{-1/3}\int_\theta^{\theta+\pi} \frac{\mathrm{e}^{-\mathrm{i}(-k_1\sin s+k_2\cos s)}}{k_1^2+k_2^2}(-k_2\cos s + k_1\sin s)\,\mathrm{d}s\\
&\quad + 2r^{-1/3}\mathrm{e}^{-\mathrm{i}r(k_1\cos\theta+k_2\sin\theta)}\int_\theta^{\theta+\pi} \frac{\mathrm{e}^{-\mathrm{i}(k_1\sin s-k_2\cos s)}}{k_1^2+k_2^2}(k_2\cos s - k_1\sin s)\,\mathrm{d}s.
\end{aligned}$$

The last two terms are of order $r^{-1/3}|k|^{-1}$, so their contribution in the sum is bounded by a constant

$$\sum_{k\neq 0}\frac{r^{-1/3}}{|k|(1+r^{-1/3}|k|)^\nu} < C.$$

We must then bound the integrals on the parts of ∂B parallel to ω. But

$$\left|\int_0^r \frac{\mathrm{e}^{-\mathrm{i}t(\cos\theta k_1+\sin\theta k_2)}}{k_1^2+k_2^2}(\cos\theta k_2 - \sin\theta k_1)\,\mathrm{d}t\right| \leq \frac{2}{k_1^2+k_2^2}\frac{|\cos\theta k_2 - \sin\theta k_1|}{|\cos\theta k_1 + \sin\theta k_2|}$$

and, since $\tan\theta$ is typical, their contribution is bounded by

$$\sum_{k\neq 0}\frac{1}{|k|(1+r^{-1/3}|k|)^\nu} + \sum_{n=1}^{+\infty}\frac{\log(1+n)^\tau}{(1+r^{-1/3}n)^\nu} = O(r^{1/3}\log(r)^\tau). \qquad \square$$

The following lemma estimates the variation of the total number of points from $\mathbb{Z}^2 + (0,\beta)$ and $\mathbb{Z}^2 - (0,\beta)$, contained in the region $\{(x,y) \mid 0 < x \leq r, 0 \leq y \leq \alpha x\}$ of the plane, in function of β. The points lying on the x axis are given a weight of $\frac{1}{2}$.

Lemma 3.2.3. *Let α be a typical number, and define the following function*

$$K(\alpha,r,\beta) = \begin{cases} r + 2\sum_{1\leq k\leq r}\lfloor \alpha k\rfloor & \text{if } \beta\in\mathbb{Z},\\ \sum_{1\leq k\leq r}(\lfloor \alpha k + 1 - \beta\rfloor + \lfloor \alpha k + \beta\rfloor) & \text{if } \beta\notin\mathbb{Z}\end{cases}$$

for $\beta\in\mathbb{R}$. Then for some $C_\alpha, \tau > 0$ and r sufficiently large we have

$$|K(\alpha,r,\beta_1) - K(\alpha,r,\beta_2)| < C_\alpha \log(r)^{1+\tau}$$

for any $\beta_1,\beta_2\in\mathbb{R}$.

PROOF. We consider the function $\sigma(t) = \lfloor t\rfloor - t + \frac{1}{2}$, so that it is sufficient to show

$$\left|\sum_{1\leq k\leq r}\sigma(\alpha k+\beta)\right| < C_\alpha\log(r)^{1+\tau}$$

for all $\beta\in\mathbb{R}$.

Suppose $\alpha \in \mathbb{R}$ is irrational, then we can write α in a unique way as a continued fraction

$$\alpha = a_0 + \cfrac{1}{a_1 + \cfrac{1}{a_2 + \cfrac{1}{a_3 + \cfrac{1}{\ddots}}}}$$

with $a_i \in \mathbb{Z}$ and $a_i > 0$ for $i > 0$. Also, any sequence $(a_0; a_1, a_2, a_3, \dots)$ respecting the previous conditions will represent an irrational number. The numbers $\{a_n\}$ are called the partial quotients of α. We define the convergents of α as

$$\frac{P_n}{Q_n} = a_0 + \cfrac{1}{a_1 + \cfrac{1}{a_2 + \cfrac{1}{\ddots + \cfrac{1}{a_n}}}}.$$

The denominators Q_n can also be defined by a recurrence relation, $Q_0 = 1$, $Q_1 = a_1$ and $Q_{n+1} = a_{n+1}Q_n + Q_{n-1}$ for $n > 0$.

Supposing r is an integer, let m be the largest integer such that $Q_m \leq r$ and consider

$$b_m = \lfloor r/Q_m \rfloor, \qquad r = b_m Q_m + N_{m-1}$$

with

$$b_j = \lfloor N_j/Q_j \rfloor, \qquad N_j = b_j Q_j + N_{j-1} \tag{3.2.4}$$

for $0 \leq j \leq m-1$. Note that $N_{j-1} < Q_j$ and $Q_0 = 1$, so $N_{-1} = 0$ and

$$r = \sum_{j=0}^{m} b_j Q_j.$$

Lemma 2.3.2 of [8] shows

$$\left| \sum_{1 \leq k \leq r} \sigma(\alpha k + b) \right| \leq 1 + m + \frac{5}{2} \sum_{j=0}^{m} b_j$$

so it suffices to find appropriate bounds on m and b_j using the fact that α is typical. We also know from Lemma 1.5.2 of [8] that

$$\left| \alpha - \frac{P_j}{Q_j} \right| < \frac{1}{Q_j Q_{j+1}} \leq \frac{1}{2^j}$$

but since for any $q > 0$

$$\left| \alpha - \frac{p}{q} \right| \geq \frac{\delta}{q^2 \log(1+q)^\tau}$$

we deduce

$$\delta 2^j < Q_j^2 \log(1 + Q_j)^\tau$$

and that for any fixed $\epsilon > 0$, and $C_\epsilon > 0$ large enough

$$(\delta 2^j)^{1/(2+\epsilon)} < Q_j < C_\epsilon Q_{j-1} \log(1 + Q_{j-1})^\tau, \tag{3.2.5}$$

$$m \leq \frac{1}{\log 2} \log\left(\frac{r^{2+\epsilon}}{\delta}\right). \tag{3.2.6}$$

Thus we are left to show that the sum $\sum_j b_j$ is of order $\log(r)^{1+\tau}$. Using (3.2.4) and (3.2.5) we deduce

$$\sum_j b_j \leq \frac{r}{Q_m} + \sum_{j=0}^{m-1} \frac{Q_{j+1}}{Q_j} \leq C_\epsilon \sum_{j=0}^{m} \log(1 + Q_j)^\tau \leq C \log(r) \log(1 + r)^\tau. \quad \square$$

3.3. Bound on the remainder term. Our results on lattice counting can be applied to the eigenvalue counting, using the correspondence established in Section 2.3, with a precision of order $O(r^{2/3})$. Note that the Riemannian volume of T and the Euclidean area of A are related in the following way

$$\frac{1}{4\pi} \operatorname{Area}(T) = \frac{1}{(2\pi)^2} \int_{H(p,q) \leq 1} \mathrm{d}p \, \mathrm{d}q = \frac{1}{\pi^2} \operatorname{Area}(A).$$

This is justified by the fact that for $c_3 < c < c_2$ the geodesic flow occur on 4 distinct tori in phase space, corresponding to the possible signs of $\dot{q}_1, \dot{q}_2$, on which I_1, I_2 take the same value. For $c_4 < c < c_3$ there are 2 tori, corresponding to the sign of $\dot{q}_1$, however the flow induced by I_2 has period 2 instead of 1. Thus the integration in the angle variables is doubled in this region. The case of $c_2 < c < c_1$ is similar if we consider $\dot{q}_2$ and I_1. The symplectic volume of the set satisfying $H(p,q) \leq 1$ in $\mathbf{T}^*(T)$ is then $4 \operatorname{Area}(A)$.

Proof of Theorem 1.3.1. Given r large enough, we can bound the difference between $R(r)$ and $2N_0(A, r)$ using equation (2.3.2). We know that to each pair (m_1, m_2), with $m_1, m_2 \geq 0$, correspond a pair (λ, c) which satisfies

$$\Phi_0(\lambda, c) + \Phi_1(\lambda, c) + \Phi_2(\lambda, c) = 2\pi\left(\left[\frac{m_1+1}{2}\right], \left[\frac{m_2+1}{2}\right]\right).$$

Since $\Phi_0(\lambda, c) = \lambda\big(F_1(c), F_2(c)\big)$, our first approximation is to count the number of pairs (m_1, m_2) such that

$$2\pi\left(\left[\frac{m_1+1}{2}\right], \left[\frac{m_2+1}{2}\right]\right) \in rA = r(A_1 \cup A_2 \cup A_3)$$

which is given by $2N_0(A, r)$. Theorems 3.1.2, 3.1.6 and 3.2.1 applied to the domain A show that

$$\widetilde{N}_0(A, r) - \frac{r^2}{2\pi^2} \operatorname{Area}(A) = O(r^{2/3}).$$

By (2.1.2), and since we give a weight of $\frac{1}{2}$ to the points of Γ_0 lying on the coordinate axes, we have

$$\widetilde{N}_0(A, r - \varkappa r^{-1/3}) \leq N_0(A, r) \leq \widetilde{N}_0(A, r + \varkappa r^{-1/3}) \tag{3.3.1}$$

for $\varkappa$ sufficiently large, so that

$$N_0(A, r) - \frac{r^2}{2\pi^2} \operatorname{Area}(A) = O(r^{2/3}).$$

We are then left to show that the corrections, brought by Φ_1 and Φ_2, to this approximation generate an error term bounded by $O(r^{2/3})$.

The first discrepancies considered are the points $2\pi([(m_1+1)/2],[(m_2+1)/2])$ for which the solution of (2.3.2) satisfies $|\lambda(m_1,m_2)-r|\leq$ Const, with $|c(m_1,m_2)-c_2|\leq \text{const}\,\lambda^{-2/3}$ or $|c(m_1,m_2)-c_3|\leq \text{const}\,\lambda^{-2/3}$. In such cases, Φ_1 and Φ_2 are bounded by a constant independent of (λ,c). The maximal number of such points is of order $r^{1/3}\log r$. Indeed, an interval of order $r^{-2/3}$ in the c variable, around c_2 or c_3, translates into an angular interval of order $r^{-2/3}\log r$. This is verified using the asymptotics (3) in Theorem 1, (also [11, Lemma 6.4]).

We then consider the points in rA_1. In this case, the function Φ_1 will induce a transformation from the lattice Γ_0 to $\Gamma_{(0,\pi/2)}$ and $\Gamma_{(0,-\pi/2)}$. Only points near the boundary γ, in $(r+\varkappa)A_1\setminus(r-\varkappa)A_1$ for some $\varkappa$ sufficiently large, might be affected by those corrections. Using Lemma 3.2.2, we can apply an inequality similar to (3.3.1) for A_1. Thus

$$N_a(A_1,r)-\frac{r^2}{2\pi^2}\operatorname{Area}(A_1)=O(r^{2/3})$$

for any $a\in\mathbb{R}^2$. However, we must only count the points leaving or entering rA_1 through $r\gamma$ during the transformation of the lattice. Lemma 3.2.3 shows that $\log(r)^{1+\tau}$ points pass through the line $(tF_1(c_3),tF_2(c_3))$, for $t\in\mathbb{R}$. This means that adding Φ_1 (which is equivalent to shifting the lattice in the region rA_1 only) yields a correction of order at most $O(r^{2/3})$. Since $|\Phi_2|<\text{Const}\,r^{-2/3}\log r$ around $r(\gamma\cap A_1)$, Φ_2 also changes the count by $O(r^{2/3})$ only.

A similar argument holds in rA_3, where we consider $\Gamma_{(\pi/2,0)}$ and $\Gamma_{(-\pi/2,0)}$, and rA_2, where we do not have to change the lattice. We deduce that, for large enough r, the errors occurring in our first approximation $2N_0(A,r)$ are all contained in a $O(r^{2/3})$ term and that

$$R(r)-\frac{r^2}{4\pi}\operatorname{Area}(T)=O(r^{2/3}).\qquad\square$$

3.4. Density of the nondegenerate metrics. Consider the set Ω of pairs of functions (U_1,U_2) satisfying the conditions described in Definition 1.1.2. We put on Ω the topology induced by the Whitney topology of $C^\infty(\mathbb{R}/\mathbb{Z})\times C^\infty(\mathbb{R}/\mathbb{Z})$. Note that Ω will be an open subset of $C^\infty(\mathbb{R}/\mathbb{Z})\times C^\infty(\mathbb{R}/\mathbb{Z})$.

Proof of Theorem 1.3.2. We first show that the set of metrics satisfying the condition (1) of Definition 2.2.2 is open and dense. It is open since the curvature $\kappa(c)$ and its derivatives are continuous functions of Ω. Any pair (U_1,U_2) can be approximated in Ω by analytic functions, for example using their partial Fourier series. The curvature $\kappa(c)$ of γ corresponding to a pair of analytic functions is also analytic. Since the curvature diverges at c_2 and c_3, its zeros are located in a compact subset of $[c_4,c_3)\cup(c_2,c_1]$. We deduce that $\kappa(c)$ has a finite number of zeros, each of finite order. If some zeros are of order greater than one, we must modify slightly (U_1,U_2) to make them first order. Remember that the curvature vanishes if and only if $(F_2''F_1'-F_1''F_2')(c)$ vanishes, and that their zeros are of the same order. Suppose $\kappa(\tilde c)=0$ and the order of vanishing is greater than one. It is sufficient to find $\widetilde U=(\widetilde U_1,\widetilde U_2)$ such that

$$\frac{\mathrm{d}}{\mathrm{d}\epsilon}\kappa_{U+\epsilon\widetilde U}(\tilde c)\bigg|_{\epsilon=0}\neq 0.$$

If $\tilde{c} \in [c_4, c_3)$, we might take $\widetilde{U}_2 = 0$ and

$$\widetilde{U}_1(q_1) = \frac{F_2''(\tilde{c})}{4} \frac{1}{(U_1(q_1) - \tilde{c})^{3/2}} - \frac{3F_2'(\tilde{c})}{8} \frac{1}{(U_1(q_1) - \tilde{c})^{5/2}}$$

so that

$$\frac{F_2''(\tilde{c})}{4} \int_0^1 \frac{\widetilde{U}_1(q_1)\,\mathrm{d}q_1}{(U_1(q_1) - \tilde{c})^{3/2}} - \frac{3F_2'(\tilde{c})}{8} \int_0^1 \frac{\widetilde{U}_1(q_1)\,\mathrm{d}q_1}{(U_1(q_1) - \tilde{c})^{5/2}} \neq 0.$$

If $\tilde{c} \in (c_2, c_1]$, we might take $\widetilde{U}_1 = 0$ and

$$\widetilde{U}_2(q_2) = -\frac{3F_1'(\tilde{c})}{8} \frac{1}{(\tilde{c} - U_2(q_2))^{5/2}} - \frac{F_1''(\tilde{c})}{4} \frac{1}{(\tilde{c} - U_2(q_2))^{3/2}}$$

so that

$$-\frac{3F_1'(\tilde{c})}{8} \int_0^1 \frac{\widetilde{U}_2(q_2)\,\mathrm{d}q_2}{(\tilde{c} - U_2(q_2))^{5/2}} - \frac{F_1''(\tilde{c})}{4} \int_0^1 \frac{\widetilde{U}_2(q_2)\,\mathrm{d}q_2}{(\tilde{c} - U_2(q_2))^{3/2}} \neq 0.$$

By taking ϵ small enough, the zeros of $\kappa_{U+\epsilon\widetilde{U}}$ in a neighborhood of $\tilde{c}$ will be of first order, where $U + \epsilon\widetilde{U} = (U_1 + \epsilon\widetilde{U}_1, U_2 + \epsilon\widetilde{U}_2)$. We deduce that the set of metrics for which (1) holds in Definition 2.2.2 is also dense.

The last part of the proof requires to find a perturbation $\widetilde{U}$ of (U_1, U_2) so that the derivatives in ϵ of the functions from conditions (2), (3) and (4), in Definition 2.2.2, do not vanish. Assuming (1) is satisfied, we can keep track of each zero $\tilde{c} = \tilde{c}(\epsilon)$ of $\kappa_{U+\epsilon\widetilde{U}}$. However, their variation does not influence the derivatives in ϵ of $F_2'(\tilde{c})/F_1'(\tilde{c})$. We might suppose that c_1 and c_4 are not zeros of $\kappa(c)$, and that $\widetilde{U}_i$ vanishes around the critical points of U_i, so the value of c_j remains constant, $j = 1, 2, 3, 4$. We assume additionally that the image of $\mathrm{supp}(\widetilde{U}_1)$ under U_1 is close enough to c_1 and does not contain any zero of κ. In the same way, the image of $\mathrm{supp}(\widetilde{U}_2)$ under U_2 must be near c_4 and not contain any zero of κ. In this setting, we require

$$F_1'(c) \int_{U_2(q_2) \leq c} \frac{\widetilde{U}_2(q_2)\,\mathrm{d}q_2}{(c - U_2(q_2))^{3/2}} - F_2'(c) \int_{U_1(q_1) \geq c} \frac{\widetilde{U}_1(q_1)\,\mathrm{d}q_1}{(U_1(q_1) - c)^{3/2}} \neq 0$$

at each zero for condition (2). Also

$$\int_0^1 \frac{\widetilde{U}_2(q_2)\,\mathrm{d}q_2}{(c_1 - U_2(q_2))^{3/2}} \neq 0 \quad \text{and} \quad \int_0^1 \frac{\widetilde{U}_1(q_1)\,\mathrm{d}q_1}{(U_1(q_1) - c_4)^{3/2}} \neq 0$$

for condition (3),

$$\int_0^1 \frac{\widetilde{U}_2(q_2)\,\mathrm{d}q_2}{(c_j - U_2(q_2))^{1/2}} \int_0^1 (U_1(q_1) - c_j)^{1/2}\,\mathrm{d}q_1 \\ + \int_0^1 (c_j - U_2(q_2))^{1/2}\,\mathrm{d}q_2 \int_0^1 \frac{\widetilde{U}_1(q_1)\,\mathrm{d}q_1}{(U_1(q_1) - c_j)^{1/2}} \neq 0$$

with $j = 2, 3$ for condition (4). We can obviously find such a pair $(\widetilde{U}_1, \widetilde{U}_2)$.

Then, for any zero $\tilde{c}$ of κ, the function $\big(F_2'(\tilde{c})/F_1'(\tilde{c})\big)(\epsilon)$ is a diffeomorphism between $(-\delta, \delta) \ni \epsilon$ and an interval in $\mathbb{R}$. Since κ has a finite number of zeros and almost all real numbers are typical, the set of ϵ for which $(U_1 + \epsilon\widetilde{U}_1, U_2 + \epsilon\widetilde{U}_2)$ meets condition (2) is also of full measure. The same argument can be applied in conditions (3) and (4). We deduce that there is an ϵ, as small as required, for which the metric given by $(U_1 + \epsilon\widetilde{U}_1, U_2 + \epsilon\widetilde{U}_2)$ is nondegenerate. □

3.5. Limitations of the method. We cannot show that the $O(\lambda^{2/3})$ bound holds for a set of second Baire category of metrics, using these methods. This would require that the following bound, assumed throughout Section 3.1, holds for α in a subset of second Baire category in $\mathbb{R}$,

$$\sum_{k=1}^{+\infty} \frac{1}{(1+r^{-1/3}k)^{\nu}} \frac{1}{k\|\alpha k\|} = O(r^{2/3}) \qquad \text{for some } \nu > 0$$

where $\|\alpha\|$ is the distance to the nearest integer from α. This is impossible since we can construct a denumerable intersection of open dense subsets of $\mathbb{R}$ in which it does not hold. Note that since

$$\sum_{k=1}^{+\infty} \frac{1}{(1+r^{-1/3}k)^{\nu}} \frac{1}{k\|\alpha k\|} \geq \frac{1}{2^{\nu}} \sum_{k=1}^{r^{1/3}} \frac{1}{k\|\alpha k\|}$$

it is sufficient to show

$$\sum_{k=1}^{N} \frac{1}{k\|\alpha k\|} \neq O(N^2) \tag{3.5.1}$$

for α in a subset of second Baire category. The construction uses the standard continued fractions expansion of the real numbers, explained in Lemma 3.2.3.

As shown in [14], we have $\|Q_n\alpha\| < 1/Q_{n+1}$ for any n, so that taking some of the quantities $1/(k\|\alpha k\|)$ in (3.5.1) to be large can be done by choosing sequences $Q_{n+1}/Q_n > a_{n+1}$ increasing sufficiently rapidly.

We put on $\mathbb{R} \setminus \mathbb{Q}$ the topology induced by $\mathbb{R}$. Given any finite sequence $\tilde{a} = (a_0; a_1, a_2, \ldots, a_n)$, the set of numbers having $\tilde{a}$ as their first partial quotients will form an open subset of $\mathbb{R} \setminus \mathbb{Q}$. Also, a dense subset of $\mathbb{R} \setminus \mathbb{Q}$ must contain a number having $\tilde{a}$ as its first partial quotients for any given $\tilde{a}$.

Since

$$\sum_{k=1}^{N} \frac{1}{k\|\alpha k\|} > \frac{1}{N\|\alpha N\|}$$

we will use sequences $(a_0; a_1, a_2, a_3, \ldots)$ such that for any $C > 0$,

$$\frac{Q_{n+1}}{Q_n} > a_{n+1} > CQ_n^2$$

for some n, so that the bound $O(N^2)$ does not hold. However, a_{n+1} has no dependence on Q_n and there are no restrictions to create such sequences.

We consider the following open dense subsets of $\mathbb{R} \setminus \mathbb{Q}$,

$$S(C) = \{(a_0; a_1, a_2, a_3, \ldots) \mid a_{n+1} > CQ_n^2 \text{ for some } n\}$$

and the denumerable intersection $S^* = \bigcap_{k=1}^{+\infty} S(k)$. Any element of S^*, which is of second Baire category, will satisfy (3.5.1).

Remark 3.5.2. It is mentioned in [11] that the main results of this paper hold for a set of Liouville metrics of second Baire's category, however no details of the proof of this statement are provided.

4. Tori of revolution

4.1. Statement of results. A torus of revolution $T = \mathbb{R}^2/(a_1\mathbb{Z} \oplus a_2\mathbb{Z})$ is a two-dimensional torus with the metric

$$\mathrm{d}s^2 = U_1(q_1)(\mathrm{d}q_1^2 + \mathrm{d}q_2^2), \tag{4.1.1}$$

where $U_1(q_1) > 0$ is a smooth periodic function on $\mathbb{R}$, satisfying $U_1(q_1+a_1) = U_1(q_1)$ for all $q_1 \in \mathbb{R}$. For simplicity, we assume that $a_1 = a_2 = 1$, but all the proofs work for arbitrary $a_i > 0$.

Definition 4.1.2. Let Ω_{rev} be the set of functions U_1, satisfying the following conditions:

(1) $U_1 \in C^\infty(\mathbb{R})$.
(2) $U_1(q_1 + 1) = U_1(q_1)$ for all $q \in \mathbb{R}$.
(3) The function U_1 has exactly one minimum and one maximum in $[0, 1)$, both nondegenerate.

The Hamiltonian $H(p, q) = L^2$ and the first integral $S(p, q) = cL^2$ are given by putting $U_2(q_2) = 0$ in Section 2.1. The eigenvalue estimates found in Section 2.3 still hold with $c_4 = c_3 = 0$. We keep the notations

$$c_1 = \max_{0\le x\le 1} U_1(x) = U_1(M_1), \qquad c_2 = \min_{0\le x\le 1} U_1(x) = U_1(m_1),$$

The curve $\gamma = \big(F_1(c), F_2(c)\big)$ for $c \in [0, c_1]$ is now defined by

$$F_1(c) = \int_{U_1(q_1)\ge c} (U_1(q_1) - c)^{1/2}\,\mathrm{d}q_1,$$
$$F_2(c) = c^{1/2},$$

and its curvature is denoted by $\kappa(c)$. One can show that for $\gamma = \big(t, f(t)\big)$, the function f satisfies the following asymptotics near $t = F_1(0)$,

$$m|t - F_1(0)|^{-1/2} < |f'(t)| < M|t - F_1(0)|^{-1/2},$$
$$m|t - F_1(0)|^{-3/2} < |f''(t)| < M|t - F_1(0)|^{-3/2}$$

for some $0 < m < M$.

The definition of a nondegenerate metric of revolution reads as follows:

Definition 4.1.3. The metric of revolution $ds^2 = U_1(q_1)(dq_1^2 + dq_2^2)$ on T is said to be nondegenerate if $U_1 \in \Omega_{\mathrm{rev}}$ and the following conditions hold:

(1) The curvature $\kappa(c)$ has a finite number of zeros on $(c_2, c_1]$, each of first order.
(2) If $\kappa(\tilde{c}) = 0$, then $F_2'(\tilde{c})/F_1'(\tilde{c})$ is a typical number (see Definition 2.2.1).
(3) The number $F_2'(c_1)/F_1'(c_1)$ is typical.

$$F_2'(c_1)/F_1'(c_1) = \frac{c_1^{-1/2}}{-2\pi(-2U_1''(M_1))^{-1/2}}.$$

(4) The number $F_2(c_2)/F_1(c_2)$ is typical.

$$F_2(c_2)/F_1(c_2) = \frac{c_2^{1/2}}{\int_0^1 (U_1(q_1) - c_2)^{1/2}\,\mathrm{d}q_1}.$$

Theorem 4.1.4. *The spectral counting function of a nondegenerate torus of revolution admits the following bound on its remainder term:*

$$R(\lambda) = O(\lambda^{2/3}).$$

Theorem 4.1.5. *The set of nondegenerate metrics of revolution is dense in Ω_{rev} in the Whitney C^∞-topology.*

All the proofs of Sections 3.1, 3.2 and 3.3 can be modified to treat the particular case of $U_2(q_2) = 0$. However the density property of the nondegenerate metrics of revolution has to be shown in another way.

4.2. Nondegenerate metrics of revolution are dense. The set of metrics satisfying the condition (1) of Definition 4.1.3 is open and dense. It is open since the curvature $\kappa(c)$ and its derivatives are continuous functions of Ω_{rev}. Any function U_1 can be approximated in Ω_{rev} by analytic functions, using its partial Fourier series. Suppose that $\kappa(c_1) \neq 0$, by slightly modifying the analytic approximation of U_1 if needed. We deduce that $\kappa(c)$ is analytic and has a finite number of zeros in (c_2, c_1), each of finite order. If some zeros are of order greater than one, we must perturb U_1 to make them first order. Remember that the curvature vanishes if and only if $(F_2''F_1' - F_1''F_2')(c)$ vanishes, and that their zeros are of the same order. Suppose $\kappa(\tilde{c}) = 0$ and the order of vanishing is greater than one. It is sufficient to find $\widetilde{U}_1$ such that

$$\frac{\mathrm{d}}{\mathrm{d}\epsilon}\kappa_{U_1+\epsilon\widetilde{U}_1}(\tilde{c})\bigg|_{\epsilon=0} \neq 0.$$

We have

$$\frac{\mathrm{d}}{\mathrm{d}\epsilon}(F_2''F_1' - F_1''F_2')(\tilde{c})\bigg|_{\epsilon=0}$$
$$= -\frac{\tilde{c}^{-3/2}}{16}\int_{U_1(q_1)\geq\tilde{c}} \frac{\widetilde{U}_1(q_1)\,\mathrm{d}q_1}{(U_1(q_1)-\tilde{c})^{3/2}} - \frac{3\tilde{c}^{-1/2}}{16}\int_{U_1(q_1)\geq\tilde{c}} \frac{\widetilde{U}_1(q_1)\,\mathrm{d}q_1}{(U_1(q_1)-\tilde{c})^{5/2}} < 0$$

if $\widetilde{U}_1$ takes nonnegative values and has support sufficiently close to M_1. Indeed, if the image of supp $(\widetilde{U}_1)$ under U_1 does not contain $\tilde{c}$, the previous integrands will vanish near the ends of the interval of integration. By taking ϵ small enough, the zeros of $\kappa_{U_1+\epsilon\widetilde{U}_1}$ in a neighborhood of $\tilde{c}$ will be of first order. We deduce that the set of metrics for which (1) holds in Definition 4.1.3 is also dense.

The last step of the proof requires to find a single perturbation $\widetilde{U}_1$ of U_1 so that the derivatives in ϵ of the functions in conditions (2), (3) and (4) do not vanish. We still suppose that the support of $\widetilde{U}_1$ is concentrated around M_1, and need

$$\int_{U_1(q_1)\geq\tilde{c}} \frac{\widetilde{U}_1(q_1)\,\mathrm{d}q_1}{(U_1(q_1)-\tilde{c})^{3/2}} \neq 0$$

at each zero $\tilde{c}$ of κ, for (2). In condition (3),

$$\frac{\mathrm{d}}{\mathrm{d}\epsilon}\Big(c_1(\epsilon)(U_1+\epsilon\widetilde{U}_1)''(M_1(\epsilon))\Big)$$

does not vanish if $\widetilde{U}_1(q_1)$ can be written as $(q_1 - M_1)^2\phi(q_1 - M_1)$ with ϕ smooth and $\phi(0) \neq 0$. Note that by definition $M_1(0) = M_1$, $c_1(0) = c_1$ and

$$c_1(\epsilon) = \max_{0\leq x\leq 1}(U_1+\epsilon\widetilde{U}_1)(x) = (U_1+\epsilon\widetilde{U}_1)\big(M_1(\epsilon)\big).$$

Finally, we need

$$\int_0^1 \frac{\widetilde{U}_1(q_1)\,\mathrm{d}q_1}{(U_1(q_1)-c_2)^{1/2}} \neq 0$$

for (4). We can obviously find $\widetilde{U}_1$ satisfying all these conditions.

As in the proof of Theorem 1.3.2, there is an ϵ, as small as required, for which $U_1+\epsilon\widetilde{U}_1$ satisfies simultaneously conditions (2), (3) and (4) of Definition 4.1.3.

5. Infra-Liouville tori

5.1. Infra-Liouville metrics. Assume that a Liouville torus T admits a finite group of translations G leaving the metric invariant, and consider T/G. If G is not of the form $G_1 \oplus G_2$, where G_i is generated by translations along q_i, the induced metric on T/G will not be Liouville. Such metrics have been studied in [17], and are called infra-Liouville. If T belongs to the conformal class of the square flat torus $\mathbb{R}^2/\mathbb{Z}^2$, then G is spanned by $\langle(r_1,r_2),(s_1,s_2)\rangle$ with $r_i, s_i \in \mathbb{Q}$, and there exists $(a_{i,j}) \in M_{2\times 2}(\mathbb{Z})$ such that

$$\begin{pmatrix} 1 & 0 \\ 0 & 1 \end{pmatrix} = \begin{pmatrix} r_1 & s_1 \\ r_2 & s_2 \end{pmatrix}\begin{pmatrix} a_{11} & a_{12} \\ a_{21} & a_{22} \end{pmatrix}.$$

The invariance of the metric implies

$$\begin{cases} U_i(q_i+r_i) - U(q_i) = v_i \in \mathbb{R}, \\ U_i(q_i+s_i) - U(q_i) = w_i \in \mathbb{R} \end{cases}$$

and, since $U_i(q_i+1) = U_i(q_i)$,

$$\begin{pmatrix} 0 & 0 \\ 0 & 0 \end{pmatrix} = \begin{pmatrix} v_1 & w_1 \\ v_2 & w_2 \end{pmatrix}\begin{pmatrix} a_{11} & a_{12} \\ a_{21} & a_{22} \end{pmatrix}.$$

We conclude that $v_i = 0$ and $w_i = 0$ for $i = 1, 2$.

5.2. Spectral properties of infra-Liouville tori. Let $1/n_i = \inf\{z > 0 \mid z = xr_i + ys_i \text{ and } x, y \in \mathbb{Z}\}$, so that $n_i \in \mathbb{N}$. We deduce $U_i(q_i + 1/n_i) = U_i(q_i)$, $i = 1, 2$. The eigenfunctions on T/G can be written as products $\Psi_1(q_1)\Psi_2(q_2)$ for which

$$\begin{cases} \Psi_1(q_1+1/n_1) = \mathrm{e}^{ii2\pi l_1/n_1}\Psi_1(q_1), \\ \Psi_2(q_2+1/n_2) = \mathrm{e}^{\mathrm{i}2\pi l_2/n_2}\Psi_2(q_2) \end{cases} \tag{5.2.1}$$

with $l_i \in \{0, \dots, n_i - 1\}$ and $(r_1 l_1 + r_2 l_2, s_1 l_1 + s_2 l_2) \in \mathbb{Z} \times \mathbb{Z}$.

If both U_i have only one nondegenerate maximum and one nondegenerate minimum on $[0, 1/n_i)$, the quantization rules found in Theorem 6.1 of [11] can be generalized to study solutions satisfying (5.2.1). We can also associate a set of metrics Ω_G to G and such pairs (U_1, U_2). A metric in Ω_G is nondegenerate if the rescaled pair $\big(U_1(q_1/n_1), U_2(q_2/n_2)\big)$, which belongs to Ω, is nondegenerate according to Definition 2.2.2.

Proposition 5.2.2. *The remainder term of a nondegenerate infra-Liouville torus is of order $O(\lambda^{2/3})$, and nondegenerate metrics are dense in Ω_G.*

Acknowledgements. This research was conducted under the supervision of Iosif Polterovich and supported by the NSERC Canada Graduate Scholarship. The problem was posed by Professor Polterovich and I would like to thank him for his assistance.

Note from the Editors. Hugues Lapointe tragically died shortly after submitting this article. It was supposed to be a part of his Ph.D. thesis that was almost ready for defense. Hugues had a big talent and was passionate about mathematics. He will be remembered by the mathematical community through several original contributions to geometric spectral theory that he made during his short but productive career.

References

1. V. G. Avakumović, *Über die Eigenfunktionen auf geschlossenen Riemannschen Mannigfaltigkeiten*, Math. Z. **65** (1956), 327–344.
2. P. H. Bérard, *On the wave equation on a compact Riemannian manifold without conjugate points*, Math. Z. **155** (1977), no. 3, 249–276.
3. P. M. Bleher, *Trace formula for quantum integrable systems, lattice-point problem, and small divisors*, Emerging Applications of Number Theory (Minneapolis, MN, 1996), IMA Vol. Math. Appl., vol. 109, Springer, New York, 1999, pp. 1–38.
4. P. M. Bleher, D. V. Kosygin, and Ya. G. Sinai, *Distribution of energy levels of quantum free particle on the Liouville surface and trace formulae*, Comm. Math. Phys. **170** (1995), no. 2, 375–403.
5. Y. Colin de Verdière, *Nombre de points entiers dans une famille homothétique de domaines de* **R**, Ann. Sci. École Norm. Sup. (4) **10** (1977), no. 4, 559–575.
6. ———, *Spectre conjoint d'opérateurs pseudo-différentiels qui commutent.* II: *Le cas intégrable*, Math. Z. **171** (1980), no. 1, 51–73 (French).
7. J. J. Duistermaat and V. W. Guillemin, *The spectrum of positive elliptic operators and periodic bicharacteristics*, Invent. Math. **29** (1975), no. 1, 39–79.
8. M. N. Huxley, *Area, lattice points, and exponential sums*, London Math. Soc. Monogr. (N.S.), vol. 13, Oxford Univ. Press, New York, 1996.
9. ———, *Exponential sums and lattice points.* III, Proc. London Math. Soc. (3) **87** (2003), no. 3, 591–609.
10. V. Ivrii, *The second term of the spectral asymptotics for a Laplace – Beltrami operator on manifolds with boundary*, Funktsional. Anal. i Prilozhen. **14** (1980), no. 2, 25–34 (Russian); English transl., Functional Anal. Appl. **14** (1980), no. 2, 98–106.
11. D. V. Kosygin, A. A. Minasov, and Ya. G. Sinai, *Statistical properties of the spectra of Laplace – Beltrami operators on Liouville surfaces*, Uspekhi Mat. Nauk **48** (1993), no. 4(292), 3–130 (Russian); English transl., Russian Math. Surveys **48** (1993), no. 4, 1–142.
12. B. M. Levitan, *On expansion in characteristic functions of the Laplace operator*, Doklady Akad. Nauk SSSR (N.S.) **90** (1953), 133–135 (Russian).
13. I. Polterovich, *Spectral asymptotics and dynamics on Riemannian manifolds*, Oberwolfach Rep. **4** (2007), no. 3, 2365–2367.
14. A. M. Rockett and P. Szüsz, *Continued fractions*, World Sci., River Edge, NJ, 1992.
15. W. Sierpiński, *O pewnem zagadnieniu z rachunku funckcyj asymptotycznych*, Prace mat.-fiz. **17** (1906), 77–118.
16. J. G. van der Corput, *Over roosterpunten in het platte vlak (de beteekenis van de methoden van Voronoi en Pfeiffer)*, Dissertation, Leiden, 1919.
17. Ya. B. Vorobets, *Asymptotics of the spectrum of the Laplace – Beltrami operator on tori with Liouville and infra-Liouville metrics*, Uspekhi Mat. Nauk **52** (1997), no. 2(314), 163–164 (Russian); English transl., Russian Math. Surveys **52** (1997), no. 2, 430–431.

Département de mathématiques et de statistique, Université de Montréal CP 6128 succ Centre-Ville, Montréal, QC H3C 3J7, Canada

E-mail address: `lapointe@dms.umontreal.ca`

Centre de Recherches Mathématiques
CRM Proceedings and Lecture Notes
Volume **52**, 2010

Embedded Eigenvalues for Cartan – Hadamard Manifolds

Harold Donnelly

1. Introduction

Let M be a simply connected complete manifold of negative curvature. If $p \in M$, then by a theorem of Cartan and Hadamard, the exponential map $\exp\colon T_pM \to M$ is a diffeomorphism. The simplest example is the hyperbolic space H^n of dimension n and constant negative curvature -1. For H^n, the Laplacian has purely continuous spectrum consisting of the half line $[(n-1)^2/4, \infty)$. More generally if $K \to -1$ at infinity, then the essential spectrum consists of the half-line $[(n-1)^2/4, \infty)$, see [5]. It is a more subtle question to determine whether or not there are any embedded eigenvalues $\lambda > (n-1)^2/4$.

A number of authors have formulated conditions which guarantee the absence of embedded eigenvalues. Pinsky [7] studied rotationally symmetric metrics. He showed that there are no embedded eigenvalues if $\int |K+1|\,\mathrm{d}r < \infty$, uniformly along all rays from p, $K \leq -1$ for sufficiently large r, and $K \to -1$ at infinity. Here r is the geodesic distance from a fixed basepoint p. The author removed the hypothesis of rotational symmetry, [3, 4]. In [3] decay conditions were required for the covariant derivatives of the curvature along geodesic spheres. This requirement was removed in [4], where it was assumed that $\int_0^\infty r^\beta |K+1| dr < \infty$ and $\lim_{r\to\infty} r^\beta |K+1| = 0$, uniformly in the spherical variable, with $\beta > 2$. More recently, Kumura [6] has given an improved result with a sharp decay rate.

The purpose of the present paper is to present examples which complement the above results. Rotationally symmetric metrics with global negative curvature are constructed where K decays slowly and oscillates so that embedded eigenvalues occur. The method is modeled on the analogous problem for the Schrödinger operator in Euclidean space. Here the basic idea is due to Wigner and von Neumann. There are examples of embedded eigenvalues for manifolds with a pole and negative curvature outside a compact set in the earlier works by the author [3] and Kumura [6]. However, some additional refinements are needed to define Cartan – Hadamard manifolds with embedded eigenvalues.

2010 *Mathematics Subject Classification.* 58J50, 58J53.

Research partially supported by NSFDMS 0504729.

This is the final form of the paper.

2. Embedded eigenvalues

Let M be a complete Riemannian manifold with a pole. This means that the exponential map $\exp\colon T_pM \to M$ is a diffeomorphism, for some basepoint $p \in M$. So there exists a global system of geodesic polar coordinates. We assume in addition that the full orthogonal group $\mathrm{O}(n)$, where $n = \dim M$, acts isometrically fixing the basepoint p. The metric may thus be written as a warped product $(\mathrm{d}r)^2 + g^2(r)(\mathrm{d}\theta)^2$, for $(r, \theta) \in \mathbb{R}^+ \times S^{n-1}$.

Consider the Laplacian Δ of M acting on functions that are rotationally symmetric, i.e., independent of θ. The natural Riemannian measure is $g^{n-1}(r)\,\mathrm{d}r\,\mathrm{d}\theta$ and Δ is self adjoint with respect to this measure. After renormalizing to Lebesgue measure $\mathrm{d}r$, we find that the action of Δ on the rotationally symmetric subspace is unitarily equivalent to a Schrödinger operator H acting on $L^2(\mathbb{R}^+)$. If $f = \log g$, then we have $H\psi = -\psi'' + \gamma(r)\psi$, where $\gamma(r) = \big((n-1)/2\big)f'' + \big((n-1)/2\big)^2 (f')^2$. Note that the radial sectional curvature of our metric is $K = -g''/g = -f'' - (f')^2$. Thus one has $\gamma = -(n-1)K/2 + \big[\big((n-1)/2\big)^2 - \big((n-1)/2\big)\big](f')^2$.

Our main result is

Theorem 1. *For each $n \geq 2$, there exist complete simply connected manifolds with sectional curvatures bounded between two negative constants, so that the associated Laplacians each have an embedded eigenvalue. More precisely, the essential spectra will be half-lines $[(n-1)^2/4, \infty)$ and the eigenvalues are larger than $(n-1)^2/4$.*

The remainder of this section is devoted to the proof of Theorem 1. We construct rotationally symmetric examples which meet the requirements. Our discussion above provides a partial reduction to the spectral theory of the Schrödinger operator. In general the potential function $\gamma(r)$ need not coincide with a multiple of the curvature. This is the main issue which must be addressed. It leads to separate discussions depending upon the dimension n of M.

The simplest case is when $n = 3$, so that $\gamma = -K$. We choose the radial sectional curvature $K(r)$ to satisfy $-K = c + V$, where $V(r)$ is the Wigner–von Neumann potential, [8]. The constant c is chosen sufficiently large to guarantee that K is globally negative. Recall that $V(r) \to 0$ as $r \to \infty$, so that $K(r) \to -c$. This implies that the essential spectrum is $[c, \infty)$ by the decomposition principle [5]. Moreover, $-\mathrm{d}^2/\mathrm{d}r^2 + V(r)$ has a positive eigenvalue 1, so that the Laplacian of our metric has an embedded eigenvalue $c+1$. The metric is obtained from K by solving the Jacobi equation $g'' = -Kg$, $g(0) = 0$, $g'(0) = 1$. In particular negative curvature assures that $g(r) \geq 1$. Using this fact and a standard formula for the curvature of warped products [2, p. 27], one checks that all sectional curvatures are negative. This also follows from the Gauss equations for the sectional curvatures of planes tangent to the geodesic spheres. Rescaling the metric gives essential spectrum $[1, \infty)$.

If $n \neq 3$, the basic idea of the proof is the same but there are certain technical difficulties. These difficulties stem from the fact that K is no longer a multiple of $\gamma(r)$. Moreover, note that the smoothness of our metric requires $\gamma(r) = \big((n-1)/2\big)\big((n-3)/2\big)r^{-2} + O(1)$, near $r = 0$. So the Wigner–von Neumann potential must be replaced by a potential which is singular near the origin. Let $V(r) = \big((n-1)/2\big)\big((n-3)/2\big)r^{-2}$ for $r < \epsilon$, and for $r > 1/\epsilon$ let $V(r)$ be equal to the Wigner–von Neumann potential. The associated Schrödinger operators have square integrable eigenfunctions $\psi(r)$ with eigenvalue 1. The existence of $\psi(r)$ is elementary for $r < \epsilon$, see [8] for $r > 1/\epsilon$. Extend $\psi(r)$ to be smooth and positive in the gap $\epsilon < r < 1/\epsilon$ and define $V(r) = 1 + \psi^{-1}\psi''$ there. This produces a Schrödinger operator with embedded eigenvalue one.

One wants to construct a metric with prescribed $\gamma(r)$. It is convenient to make some changes of dependent variable for the purpose of linearizing the problem and dealing with the singularity of $\gamma(r)$ at $r = 0$. Define $v = \big((n-1)/2\big)f$ so that $\gamma = v'' + (v')^2$. If $h = e^v$, then $h'' = \gamma h$. This equation is linear but $h = g^{(n-1)/2}$ will not be smooth at $r = 0$, when n is even. Define $w = g/r$, so that $w(0) = 1$. This gives $h = r^{(n-1)/2}u$, with $u = w^{(n-1)/2}$. So u is smooth at $r = 0$. Define $\beta(r)$ by $\beta(r) = \gamma(r) - \big((n-1)/2\big)\big((n-3)/2\big)r^{-2}$. Then u satisfies the degenerate elliptic equation $ru'' + (n-1)u' = \beta ru$. The roots of the indicial equation are 0 and $2-n$. So we get a unique smooth solution with $u(0) = 1$.

Suppose now that $n = 2$. We consider the behavior for small r of the solution for $ru'' + u' = \beta ru$, with constant $\beta > 0$. It is represented by the convergent power series $u(r) = \sum_{k=0}^{\infty}(\beta/4)^k r^{2k}/(k!)^2$. Now $g = ru^2$ and thus we may write $g'' = 2uu' + 2r(u')^2 + 2\beta ru^2$. So if $\beta > 0$ is constant on $(0, \epsilon)$ we deduce that $K < 0$ on $(0, \epsilon)$. Note that $g'' = -Kg$ so g will define a metric as long as $K < 0$. Moreover $\gamma = -K/2 - (f')^2/4$, so $\gamma > 0$ implies $K < 0$. Adding a large constant to $V(r)$ will make $\gamma(r) > 0$ for $r > \epsilon$. This means that we have a globally defined metric of negative curvature. So we take $\gamma(r) = c + V(r)$ to assure that the associated Laplacian will have an embedded eigenvalue. Here $V(r)$ is the Wigner–von Neumann potential modified as above to have the required singularity at $r = 0$. Rescaling gives the essential spectrum $[\frac{1}{4}, \infty)$. This proves Theorem 1 for $n = 2$. The method extends the earlier argument for $n = 3$.

If $n \geq 4$ some new difficulties arise because $\gamma > 0$ no longer implies $K < 0$, by inspection of the equation $\gamma = -\big((n-1)/2\big)K + \big[\big((n-1)/2\big)^2 - \big((n-1)/2\big)\big](f')^2$. We overcome this obstacle by employing the paper of Atkinson [1]. He puts the Wigner–von Neumann potential into a more general class of potentials whose associated Schrödinger operators have embedded eigenvalues. For sufficiently large $r > r_0$, we take the metric $g = e^f$, with $f(r) = a + \int_{r_0}^{r}[1 + bV(s+c)]\,ds$. The parameters a, b, c, and r_0 will be adjusted as required below, and V denotes the Wigner–von Neumann potential. Recall that $V(r) = -8\sin(2r)/r + 0(r^{-2})$ and $V'(r) = -16\cos(2r)/r + O(r^{-2})$, as $r \to \infty$. This determines the asymptotic behavior of the potential $\gamma(r)$ for the associated Schrödinger operator $H\psi = -\psi'' + \gamma(r)\psi$, where $\gamma(r) = \big((n-1)/2\big)f'' + \big((n-1)/2\big)^2(f')^2$ as above. In particular $\gamma(r) \to (n-1)^2/4$ as $r \to \infty$ and the potential approaches this constant value by oscillating and decaying slowly. More precisely $\gamma(r) - (n-1)^2/4$ is modeled upon the Wigner–von Neumann potential with a phase shift. The result of Atkinson [1, p. 369], then provides a $\lambda > \big((n-1)/2\big)^2$, so that $H\psi = -\lambda\psi$ has a square integrable solution on (r_0, ∞). In fact $\lambda = \big((n-1)/2\big)^2 + 1$, just like for the translate of the Wigner–von Neumann potential. One must choose b sufficiently large to invoke [1]. For this particular λ, the two linearly independent solutions of $H\psi = -\lambda\psi$ have the asymptotics $r^{\beta}\sin(r)$ and $r^{-\beta}\cos(r)$, for $\beta > \frac{1}{2}$, after translating r by a constant. For $r < r_0 - \epsilon$, we take $g(r) = \sinh r$, the standard hyperbolic metric with constant curvature -1. The Schrödinger operator associated to this metric has a square integrable solution ψ_0 on $(0, r_0)$. Note that the space of such square integrable solutions is one-dimensional because of the singularity of $\gamma(r)$ at $r = 0$. If we patch the two metrics along $(r_0 - \epsilon, r_0)$, $\psi_0(r)$ will extend to an eigenfunction of the form $c_1 r^\beta \sin(r) + c_2 r^{-\beta} \cos(r)$. Since both metrics have curvature close to -1, provided r_0 is sufficiently large, global negative curvature can be assured. If $c_1 = 0$, we have an embedded eigenvalue. If $c_1 \neq 0$, suppose $c_1 > 0$. By shifting the parameters r_0, a, and c in the definition of $f(r)$ we obtain the complementary case $c_1 < 0$. This is possible because $\psi_0(r)$ oscillates like $a_1 \sin r + a_2 \cos r$. So by the intermediate value theorem there is a choice of parameters which gives an embedded eigenvalue. The proof of Theorem 1 is complete. □

3. Point spectrum

We close with some remarks about the point spectrum. This includes all eigenvalues of the Laplacian, not just embedded eigenvalues. If M^n is a Cartan–Hadamard manifold with $K \to -1$ at infinity then Δ has essential spectrum, $[(n-1)^2/4, \infty)$, see [5]. Suppose now that $|K| < \frac{1}{2}$ on a large ball. The minimax principle guarantees finitely many eigenvalues less than $(n-1)^2/4$. If $\epsilon_n \to 0$ and $|K| < 1 - \epsilon_n$ on a sequence of sufficiently large annuli, then Δ has infinitely many eigenvalues less than $(n-1)^2/4$. Of course these eigenvalues must accumulate only at $(n-1)^2/4$, since this is the start of the essential spectrum. The existence of Cartan–Hadamard manifolds with infinite dimensional point spectrum and $c_1 < |K| < c_2$ answers a problem of Yau [9] in the negative.

References

1. F. V. Atkinson, *The asymptotic solution of second-order differential equations*, Ann. Mat. Pura Appl. (4) **37** (1954), 347–378.
2. R. L. Bishop and B. O'Neill, *Manifolds of negative curvature*, Trans. Amer. Math. Soc. **145** (1969), 1–49.
3. H. Donnelly, *Eigenvalues embedded in the continuum for negatively curved manifolds*, Michigan Math. J. **28** (1981), no. 1, 53–62.
4. ———, *Negative curvature and embedded eigenvalues*, Math. Z. **203** (1990), no. 2, 301–308.
5. ———, *On the essential spectrum of a complete Riemannian manifold*, Topology **20** (1981), no. 1, 1–14.
6. H. Kumura, *Geometry of an end and absence of eigenvalues in the essential spectrum*, Sūrikaisekikenkyūsho Kōkyūroku **1510** (2006), 141–150.
7. M. A. Pinsky, *Spectrum of the Laplacian on a manifold of negative curvature*. II, J. Differential Geom. **14** (1979), no. 4, 609–620.
8. M. Reed and B. Simon, *Methods of modern mathematical physics*. IV: *Analysis of operators*, Academic Press, New York, 1978.
9. S.-T. Yau, *Nonlinear analysis in geometry*, Enseign. Math. (2) **33** (1987), no. 1-2, 109–158.

Department of Mathematics, Purdue University, West Lafayette, IN 47907, USA
E-mail address: hgd@math.purdue.edu

Centre de Recherches Mathématiques
CRM Proceedings and Lecture Notes
Volume **52**, 2010

On Minimal Partitions: New Properties and Applications to the Disk

Bernard Helffer and Thomas Hoffmann-Ostenhof

ABSTRACT. In continuation of [15], we analyze the properties of spectral minimal partitions. We concentrate our analysis on 3-partitions in the case of the disk and discuss the conjecture that the minimal 3-partition for the disk is the Mercedes Star.

1. Introduction

In [15], we have analyzed the properties of spectral minimal partitions of a domain $\Omega \subset \mathbb{R}^2$ and clarified the links between this notion and the partitions obtained as the family of the nodal domains of an eigenfunction associated to an eigenvalue of the corresponding Dirichlet problem in Ω. We would like to propose some techniques for determining minimal partitions in concrete cases starting with the first non-trivial case $k = 3$. Although some of the arguments have a general character, we will concentrate our analysis on minimal 3-partitions in the case of the disk and discuss the conjecture that a minimal 3-partition for the disk should be the Mercedes Star.

1.1. Minimal partitions. Before presenting this conjecture more explicitly, let us first recall notations, definitions and results extracted essentially from [15].

Definition 1. We say that an open bounded domain $D \subset \mathbb{R}^2$ is *regular* if
- it satisfies an interior cone condition;
- ∂D is the union a finite number of simple arcs $\gamma_i(\overline{I_i})$, with $\gamma_i \in \mathcal{C}^{1,\alpha}(\overline{I}_i)$ (for some $\alpha > 0$) with no mutual nor self intersections, except possibly at the endpoints.

We consider the Dirichlet Laplacian $H = H(\Omega)$ on a bounded regular domain $\Omega \subset \mathbb{R}^2$. We are interested in the eigenvalue problem for $H(\Omega)$. Note that $H(\Omega)$ has discrete spectrum $\sigma\big(H(\Omega)\big)$, i.e., consisting of eigenvalues $\{\lambda_k\}_{k=1}^{\infty}$ with finite multiplicities which tend to infinity, so that

$$\lambda_1 < \lambda_2 \leq \lambda_3 \leq \cdots \leq \lambda_k \leq \dots \tag{1}$$

2010 *Mathematics Subject Classification.* 35B05.
Supported by Ministerium für Wissenschaft und Forschung der Republik Österreich.
This is the final form of the paper.

and such that the associated eigenfunctions u_k can be chosen to form an orthonormal basis for $L^2(\Omega)$. For any open domain D, we shall denote by $\lambda(D)$ the lowest eigenvalue of H with Dirichlet boundary condition

$$\lambda(D) = \lambda_1\big(H(D)\big). \tag{2}$$

Without loss we can assume that the u_k are real valued and by elliptic regularity $u_k \in C^\infty(\Omega) \cap C_0^0(\overline{\Omega})$. We know that u_1 can be chosen to be strictly positive in Ω, but the other eigenfunctions u_k must have non-empty zero sets. For any function $u \in C_0^0(\overline{\Omega})$, we define

$$N(u) = \overline{\{x \in \Omega \mid u(x) = 0\}} \tag{3}$$

and call the components of $\Omega \setminus N(u)$ the nodal domains of u. The number of nodal domains of such a function will be denoted by $\mu(u)$.

We now introduce the notion of partitions.

Definition 2. Let $1 \leq k \in \mathbb{N}$. We call a *k-partition* of Ω a family $\mathcal{D} = \{D_i\}_{i=1}^k$ of pairwise disjoint open regular domains such that

$$\bigcup_{i=1}^{k} D_i \subset \Omega. \tag{4}$$

It is called *strong* if

$$\mathrm{Int}\left(\overline{\bigcup_{i=1}^{k} D_i}\right) \setminus \partial\Omega = \Omega. \tag{5}$$

We denote by $\mathfrak{O}_k$ the set of such partitions.

We now introduce spectral minimal partitions:

Definition 3. Let $1 \leq k \in \mathbb{N}$. For $\mathcal{D} \in \mathfrak{O}_k$ we introduce

$$\Lambda(\mathcal{D}) = \max_i \lambda(D_i), \tag{6}$$

and

$$\mathfrak{L}_k(\Omega) = \inf_{\mathcal{D} \in \mathfrak{O}_k} \Lambda(\mathcal{D}) \tag{7}$$

We call a k-partition $\mathcal{D} \in \mathfrak{O}_k$ a spectral minimal k-partition if $\mathfrak{L}_k(\Omega) = \Lambda(\mathcal{D})$.

1.2. The case of the disk. If $k = 2$, the minimal value $\mathfrak{L}_2(\Omega)$ is the second eigenvalue and any minimal 2-partition is represented as the nodal partition associated to some second eigenfunction. For $k > 2$, the analysis of various domains like the rectangle shows that $\mathfrak{L}_k(\Omega)$ is not necessarily an eigenvalue.

We consider the unit disk $\mathbf{D} = \{|x| < 1\} \subset \mathbb{R}^2$ and study the corresponding $\mathfrak{L}_3(\mathbf{D})$ and some associated minimal partition $\mathcal{D} = (D_1, D_2, D_3)$. From now on, we choose always, possibly changing by sets of capacity 0, a regular representative which has been proved in [15] to exist.

Of course we do not know a priori that $\mathcal{D}$ is unique (up to rotations and reflections). But we have observed in [15] that $\mathcal{D}$ is not created by a nodal set. The argument was as follows. By analysis of the tables giving the zeros of Bessel functions [1], the first eigenvalue whose eigenfunction has three nodal domains is $\lambda_{15}(\mathbf{D}) > \lambda_4(\mathbf{D})$ and the eigenfunctions associated to $\lambda_4(\mathbf{D})$ have four nodal domains, so (as a consequence of Theorem 13 which will be recalled in Section 2) $\lambda_4(\mathbf{D}) = \mathfrak{L}_4(\mathbf{D})$. Hence

$$(8) \qquad \lambda_1(\mathbf{D}) < \mathfrak{L}_2(\mathbf{D}) = \lambda_2(\mathbf{D}) = \lambda_3(\mathbf{D}) < \mathfrak{L}_3(\mathbf{D}) < \lambda_4(\mathbf{D}) = \mathfrak{L}_4(\mathbf{D})\,.$$

Conjecture 4. *For the disk* $\mathbf{D}$, *there is a unique regular representative (up to rotation) of the minimal spectral* 3*-partition* $\mathcal{D}$ *associated to* $\mathfrak{L}_3(\mathbf{D})$. *It is given by three disjoint sectors with opening angle* $2\pi/3$, *i.e.,*

$$(9) \qquad D_1 = \{x = \rho e^{i\omega} \in \mathbf{D} \mid \omega \in]0, 2\pi/3[\}$$

and D_2, D_3 *are obtained by rotating* D_1 *by* $2\pi/3$, *respectively by* $4\pi/3$.

For short, we call this partition a Mercedes Star (MS). Unfortunately, we have not proved the conjecture yet. We note that this conjecture has been tested numerically without definite conclusion (see [4, 5]) but we will prove weaker versions of this conjecture developing along the way new tools for analyzing these questions for more general domains.

For example we have the following propositions.

Proposition 5. *If all the open sets of the minimal* 3*-partition of* $\mathbf{D}$ *are simply connected in the punctured disk* $\dot{\mathbf{D}} := \mathbf{D} \setminus \{0\}$, *then this partition is the Mercedes Star.*

This proposition will be proved in Section 3.

Proposition 6. *A minimal* 3*-partition of* $\mathbf{D}$ *cannot be invariant by inversion with respect to the origin.*

The proof of this proposition will be presented in Section 5.

Proposition 7. *If a minimal* 3*-partition* $\mathcal{D}$ *of* $\mathbf{D}$ *is such that* $(\bigcap_{i=1}^{3} \partial D_i) \cap \mathbf{D}$ *consists of just one point and is symmetric with respect to (say)* $\{x = 0\}$, *then it is the Mercedes Star.*

This proposition will be proved at the end of Section 7.

We will also prove the main conjecture assuming that we minimize over the 3-partitions which are invariant under rotation by $2\pi/3$.

1.3. Organization of the paper. The paper is organized as follows:

• Section 2 is devoted to a summary of the general results concerning spectral minimal partitions.

• Section 3 is devoted to the analysis of what can be obtained by going to the double covering.

• Section 4 is devoted to the discussion of the topological structure of minimal partitions. We first explain what can be done using the Euler formula. This is then applied to give a complete description of the possible minimal 3-partitions which cannot be colored by two colors. All these results are established for general simply connected domains in $\mathbb{R}^2$.

• Section 5 shows how one can exclude the existence of partitions which are symmetric by inversion.

• In Section 6, we discuss briefly the pair compatibility condition and explore an idea of G. M. Verzini which applies to the case of a disk.

• Section 7 is devoted to the proof of the main results in the case of a symmetry with respect to a given axis.

• Section 8 proves a weak form of the conjecture assuming the invariance by $2\pi/3$.

• We finally conclude in Section 9 by a presentation of other related problems which are discussed elsewhere or that we hope to treat in the near future.

Some of the results were announced in [11] or used in [4].

2. Previous results

We first give a brief summary of the subject.

2.1. Main definitions. Attached to a regular partition $\mathcal{D}$ we can associate its boundary set $N = N(\mathcal{D})$ which is the closed set in $\overline{\Omega}$ defined by

$$N(\mathcal{D}) = \overline{\bigcup_i (\partial D_i \cap \Omega)}. \tag{10}$$

This leads us to introduce the set $\mathcal{M}(\Omega)$ of the regular closed sets.

Definition 8. A closed set $N \subset \overline{\Omega}$ belongs to $\mathcal{M}(\Omega)$ if N meets the following requirements:

(i) There are finitely[1] many distinct $x_i \in \Omega \cap N$ and associated positive integers $\nu(x_i)$ with $\nu(x_i) \geq 3$ such that, in a sufficiently small neighborhood of each of the x_i, N is the union of $\nu(x_i)$ smooth arcs (non-self-crossing) with one end at x_i (and each pair defining at x_i a positive angle in $(0, 2\pi)$) and such that in the complement of these points in Ω, N is locally diffeomorphic to a regular curve. We denote the set of these critical points of N by $X(N)$.

(ii) $\partial\Omega \cap N$ consists of a (possibly empty) finite set of points z_i, such that at each z_i, $\rho(z_i)$ arcs hit the boundary with $\rho(z_i) \geq 1$. Moreover for each $z_i \in \partial\Omega$, N is near z_i the union of ρ_i distinct smooth arcs which hit at z_i the (two arcs constituting the) boundary with strictly positive distinct angles.

We denote the set of these critical points of $N \cap \partial\Omega$ by $Y(N)$.

We actually will need later[2] a more general definition.

We consider for $N \in \mathcal{M}(\Omega)$ the decomposition of $X(N)$

$$X(N) = X_{\text{even}}(N) \cup X_{\text{odd}}(N), \tag{11}$$

where

$$X_{\text{even}}(N) = \{x \in X(N) \mid \nu(x) = \text{even}\},$$

and

$$X_{\text{odd}}(N) = \{x \in X(N) \mid \nu(x) = \text{odd}\}.$$

Definition 9. We will say that a closed regular set has the *equal angle meeting property* if the arcs meet with equal angles at each critical $x_i \in N \cap \Omega$ point of N and also with equal angles at the $z_i \in N \cap \partial\Omega$. For the boundary points z_i we mean that the two arcs in the boundary are included.

---

[1]The set of the x_i's could be empty

[2]When going to the covering $\dot{\Omega}^{\mathcal{R}}$ of $\Omega \setminus \{0\}$, we simply need to explain what is going on around $\{0\}$ and this is achieved by counting the number of half-lines meeting (with equal angle) at $\{0\}$.

Let us also recall the relations between graphs and partitions. If $\mathcal{D}$ is a strong partition, we say $D_i, D_j \subset \mathcal{D}$ are *neighbors* if

$$\text{Int}(\overline{D_i \cup D_j}) \setminus \partial\Omega \text{ is connected} \tag{12}$$

and denote this by $D_i \sim D_j$.

We can associate to $\mathcal{D}$ a graph $G(\mathcal{D})$ by associating to each $D_i \in \mathcal{D}$ a vertex v_i and to each pair of neighbours $D_i \sim D_j$ an edge $e_{i,j}$. Then, we will say that the graph $G(\mathcal{D})$ is *bipartite* if its vertices can be colored by two colors (two neighbours having different colors). In this case, we say that the partition is *admissible.* We recall that a collection of nodal domains of an eigenfunction is always admissible.

2.2. Main former results. It has been proved by Conti – Terracini – Verzini [8–10] that

Theorem 10. *For any k there exists a minimal regular strong k-partition.*

Analogous (somewhat weaker in our particular case) results have been obtained by (or referred in) Bucur – Buttazzo [6], Henrot [18] and Caffarelli – Lin [7]. It has been proved in [15] the

Theorem 11. *Any minimal spectral k-partition admits a representative which is regular and strong.*

A natural question is whether a minimal partition is the nodal partition induced by an eigenfunction. The next theorem [15] gives a simple criterion for a partition to be associated to a nodal set:

Theorem 12. *Assume that there is a minimal spectral k-partition (we choose then a regular strong representative) of Ω such that the associated graph is bipartite. Then this partition*[3] *is associated to the nodal set of an eigenfunction of $H(\Omega)$ corresponding to an eigenvalue equal to $\mathfrak{L}_k(\Omega)$.*

This theorem was already obtained in [14] by adding a strong a priori regularity and the assumption that Ω is simply connected. Any subpartition of cardinality 2 corresponds indeed to a second eigenvalue (as we will recall in property P4 at the end of Section 2) and the criterion of pair compatibility of [14] can be applied.

A natural question is now to determine how general is the situation described in the previous theorem. The surprise was that this only occurs in the Courant-sharp situation explained below.

Suppose that u is an eigenfunction of the Dirichlet Laplacian in Ω attached to the eigenvalue λ_k:

$$H(\Omega)u = \lambda_k u,$$

Courant's theorem says that the number of nodal domains $\mu(u)$ satisfies $\mu(u) \leq k$. Then we say, as in [2], that u is *Courant-sharp* if $\mu(u) = k$. For any integer $k \geq 1$, we denote by L_k the smallest eigenvalue such that its eigenspace contains an eigenfunction with k nodal domains. In general we have (by the variational principle)

$$\lambda_k(\Omega) \leq \mathfrak{L}_k(\Omega) \leq L_k(\Omega). \tag{13}$$

The last result [15] gives the full picture of the equality cases:

[3] From now on, we call such a partition a nodal partition.

Theorem 13. *Suppose* $\Omega \subset \mathbb{R}^2$ *is regular. If* $\mathfrak{L}_k(\Omega) = L_k(\Omega)$ *or* $\lambda_k(\Omega) = \mathfrak{L}_k(\Omega)$, *then*

$$\lambda_k(\Omega) = \mathfrak{L}_k(\Omega) = L_k(\Omega).$$

In addition any minimal k-partition admits a representative which is the nodal partition of some eigenfunction u associated to $\lambda_k(\Omega)$.

2.3. Useful additional results. We first note the following proposition:

Proposition 14. *Suppose that* $\Omega_2 \subset \Omega_1$. *Then*

$$\lambda_k(\Omega_1) \leq \lambda_k(\Omega_2), \tag{14}$$

and

$$\mathfrak{L}_k(\Omega_1) \leq \mathfrak{L}_k(\Omega_2). \tag{15}$$

PROOF. Coming back to the definition of $\mathfrak{L}_k$, this is immediate using the min-max characterization of the ground state energy and the property that a k-partition of Ω_2 is a k-partition of Ω_1. □

There are other rather immediate consequences of the proofs of [14] or [15] which are useful to mention for the regular representative of a minimal k-partition $\mathcal{D}$ and its associate $N(\mathcal{D})$:

P1. The number $\nu(x)$ of half-lines of $N(\mathcal{D})$ meeting at a singular point x of $N(\mathcal{D})$ satisfies

$$\nu(x) \geq 3. \tag{16}$$

P2. If all the $\nu(x)$ are even (for any singular point x), then, by classical graph theory, the graph $G(\mathcal{D})$ is bipartite. Hence, by Theorem 12, $\mathcal{D}$ is the nodal domain of an eigenfunction of $H(\Omega)$ associated with $\mathfrak{L}_k$.

P3. All the $\lambda(D_j)$ are equal.

P4. For any (i,j) s.t. $D_i \sim D_j$, the family (D_i, D_j) is a nodal 2-partition of D_{ij}. This property is called the pair compatibility condition in [14]. Note that in addition $\mathfrak{L}_k(\Omega) = \lambda_2(D_{ij})$.

We finally recall necessary conditions for spectral partitions which were only implicit in [15]. We call a domain D **nice** in Ω, if it satisfies

$$\mathrm{Int}(\overline{D}) \setminus \partial\Omega = D. \tag{17}$$

Proposition 15. *Suppose* Ω *is regular and that* $\mathcal{D} = (D_1, \ldots, D_\mu)$ *is a regular representative of a spectral minimal partition of* $H(\Omega)$, *then all the* D_i*'s are nice in* Ω.

PROOF. The proof is indeed a byproduct of the proof (see [15, Proposition 4.1]) that any minimal partition has a regular representative.

Let us give an alternative proof (assuming that the D_i's are regular). Suppose by contradiction that there is a $D_i \in \mathcal{D}$ that does not satisfy (17), then we can find a point, say $x_0 \in \partial D_i$ such that for a sufficiently small $\epsilon > 0$, $\{|x - x_0| < \epsilon\} \subset \mathrm{Int}(\overline{D}_i)$. We can assume in addition that x_0 is a regular point of the boundary. So we can remove this part of the boundary of D_i without changing the number of domains in $\mathcal{D}$ and this leads to a new minimal partition with D_i replaced by D_i^{new} and $\lambda(D_i^{\mathrm{new}}) < \lambda(D_i)$. This is in contradiction with property P3. □

3. Going to the double covering

Given some $\Omega \subset \mathbb{R}^2$ such that $0 \in \Omega$. We consider the double covering[4] $\dot{\Omega}^{\mathcal{R}}$ of the punctured $\dot{\Omega} := \Omega \setminus \{0\}$ and denote by $\pi^{\mathcal{R}}$ the projection of $\dot{\Omega}^{\mathcal{R}}$ onto $\dot{\Omega}$. Let $\mathcal{D}$ be a minimal k-partition of Ω and let us assume that all its open sets are simply connected in $\dot{\Omega}$. One can then associate to this k-partition a $(2k)$-partition of $\dot{\Omega}^{\mathcal{R}}$, by considering for each i the two components $D_i^{\pm}$ of $(\pi^{\mathcal{R}})^{-1}(D_i)$. This partition has the symmetry property $\sigma(D_i^+) = D_i^-$ relatively to the deck transformation σ which associates to each point x of $\dot{\Omega}^{\mathcal{R}}$ the other point $\sigma(x)$ of $\dot{\Omega}^{\mathcal{R}}$ which has the same projection on $\dot{\Omega}$.

It is then clear that, under this assumption,

$$\mathfrak{L}_k(\Omega) \geq \mathfrak{L}_{2k}(\dot{\Omega}^{\mathcal{R}}). \tag{18}$$

A particular situation occurs when $\mathfrak{L}_{2k}(\dot{\Omega}^{\mathcal{R}})$ is attained by a $(2k)$-partition which has the σ-symmetry that is, which satisfies for k disjoint pairs (D_i^+, D_i^-) $(i = 1, \ldots, k)$

$$\sigma(D_i^{\pm}) = D_i^{\mp}. \tag{19}$$

One gets indeed in this case the reverse inequality

$$\mathfrak{L}_k(\Omega) \leq \mathfrak{L}_{2k}(\dot{\Omega}^{\mathcal{R}}). \tag{20}$$

Proof of Proposition 5. Let us consider the particular case of the disk $\mathbf{D}$ and of $k = 3$. As seen from inspection of the tables in [1] (see the discussion in [15] or in Appendix B) , one can verify that the 6th eigenvalue of the Laplacian on $\dot{\mathbf{D}}^{\mathcal{R}}$ is Courant sharp and corresponds to a minimal 6-partition. The corresponding eigenfunction is symmetric with respect to σ. Hence, the partition has the covering symmetry, its projection gives the Mercedes Star and we obtain

$$\mathfrak{L}_3(\mathbf{D}) \geq \mathfrak{L}_6(\dot{\mathbf{D}}^{\mathcal{R}}) = \Lambda(\mathrm{MS}). \tag{21}$$

The proof of Proposition 5 is then immediate. □

Remark 16. The argument does not work for k odd, $k \geq 5$. By using a notion of Courant-sharp for symmetric partitions with respect to σ, one can probably discuss the case $k = 5$. We plan to present this kind of argument in [16].

4. On the topological structure of minimal partitions

4.1. Around Euler's formula. We explore now some consequences of Euler's formula. We first recall

Proposition 17. *For G a connected planar graph with n vertices, e edges, the number f of components of $\mathbb{R}^2 \setminus G$ satisfies*

$$f = e - n + 2. \tag{22}$$

Note that in our applications this graph will not be the graph of the partition as introduced after Definition 9. The vertices will be the singular points of the boundary of the partition and the edges will be the arcs joining these singular points or the closed arcs.

Here comes a relation which can be derived from (22). We refer for the proof to [19], where a similar result was obtained if N corresponds to an (abstract) nodal set.

[4] with its natural flat metrics

Proposition 18. *Let Ω an open set in $\mathbb{R}^2$ with piecewise $C^{1,+}$ (i.e., $C^{1,\alpha}$ for some $\alpha > 0$) boundary and let $N \in \mathcal{M}(\Omega)$ such that the associate $\mathcal{D}$ consists of μ domains $D_1, \ldots, D_\mu$. Let b_0 be the number of components of $\partial\Omega$ and b_1 be the number of components of $N \cup \partial\Omega$. Denote by $\nu(x_i)$ and $\rho(z_i)$ the numbers of arcs associated to the $x_i \in X(N)$, respectively $z_i \in Y(N)$. Then*

$$\mu = b_1 - b_0 + \sum_{x_i \in X(N)} \left(\frac{\nu(x_i)}{2} - 1\right) + \frac{1}{2} \sum_{z_i \in Y(N)} \rho(z_i) + 1. \tag{23}$$

4.2. The case of non-admissible minimal 3-partitions. We now give the application of these results in the case of a minimal 3-partition, which is not admissible.[5] Using Theorem 13, all the minimal 3-partitions are non-admissible if

$$\lambda_3(\Omega) < \mathfrak{L}_3(\Omega). \tag{24}$$

This will be in particular the case of the disk.

Proposition 19. *Consider a non-admissible spectral minimal 3-partition $\mathcal{D} = (D_1, D_2, D_3)$ of Ω associated to $\mathfrak{L}_3(\Omega)$. Let us assume that $\partial\Omega$ has one component. Let $N = N(\mathcal{D})$ and let $X(N)$, respectively $Y(N)$ be defined as in (11).*

Then there are three possibilities.

(a) *$X(N)$ consists of one point x with $\nu(x) = 3$ and $Y(N)$ consists of either three distinct y_1, y_2, y_3 points with $\rho(y_1) = \rho(y_2) = \rho(y_3) = 1$, two distinct points y_1, y_2 with $\rho(y_1) = 2$, $\rho(y_2) = 1$ or one point y with $\rho(y) = 3$.*

(b) *$X(N)$ consists of two distinct points x_1, x_2 so that $\nu(x_1) = \nu(x_2) = 3$. $Y(N)$ consists either of two points y_1, y_2 such that $\rho(y_1) + \rho(y_2) = 2$ or of one point y with $\rho(y) = 2$.*

(c) *$X(N)$ consists again of two distinct points x_1, x_2 with $\nu(x_1) = \nu(x_2) = 3$, but $Y(N) = \varnothing$.*

The proof of Proposition 19 relies essentially on Proposition 18 but we need some preliminary results.

Preliminaries. Our assumption (24) implies that $G(\mathcal{D})$ is not bipartite hence just a triangle and N is not a nodal partition. Then we note that in (23): $b_0 = 1$, $\mu = 3$ and $b_1 \geq 1$. Hence this formula reads:

$$3 - b_1 = \sum_{x_i} \left(\frac{\nu(x_i)}{2} - 1\right) + \frac{1}{2} \sum_{y_j} \rho(y_j), \tag{25}$$

This implies that

$$2 \geq \sum_{x_i} \left(\frac{\nu(x_i)}{2} - 1\right) + \frac{1}{2} \sum_{y_j} \rho(y_j), \tag{26}$$

which immediately implies, using (16), that

$$4 \geq \#X(N) + \#Y(N), \tag{27}$$

where $\#X(N)$ and $\#Y(N)$ denote respectively the cardinality of $X(N)$ and $Y(N)$.

When $Y(N) = \varnothing$, we observe in addition that $b_1 \geq 2$, so the same proof gives

$$1 \geq \sum_{x_i} \left(\frac{\nu(x_i)}{2} - 1\right), \qquad \text{if } Y(N) = \varnothing, \tag{28}$$

[5]We recall that a strong regular partition is admissible if its corresponding graph is bipartite

and, using again (16),

$$2 \geq \#X(N), \qquad \text{if } Y(N) = \varnothing. \tag{29}$$

We now observe :

$$\#Y(N) \neq 1. \tag{30}$$

Indeed, this would imply that there is only one domain $D \in \mathcal{D}$ such that $\partial\Omega \subset \partial D$. Hence the arc would be contained in D, so D would not be nice.

Consequently, we get from either (29) or (27)-(30) that

$$\#X(N) \leq 2. \tag{31}$$

Next we show, under the assumptions of the proposition,

Lemma 20. *$X(N)$ consists*

- *either of one point x_0 with $\nu(x_0) = 3$,*
- *or of two distinct points x_0, x_1 with $\nu(x_i) = 3$ for $i = 0, 1$.*

Proof of Lemma 20. N being not nodal (hence with non-bipartite graph), $X(N)$ cannot be empty. Moreover (26) and (16) show that any $x_c \in X_{\text{odd}}$ satisfies $3 \leq \nu(x_c) \leq 5$, so $\nu(x_c) = 3$ or $\nu(x_c) = 5$.

If $\nu(x_c) = 5$, (26) shows that either Y is empty (but this is excluded using (28)) or $Y(N)$ consists of a single point y with $\rho(y) = 1$ and (30) applies. So we have proven that any $x_c \in X_{\text{odd}}$ should satisfy

$$\nu(x_c) = 3. \tag{32}$$

Using (31), we get immediately the lemma. □

Proof of Proposition 19. Let us first analyze further the two cases appearing in the previous lemma.

In the second case, (25) leads to

$$\sum_{y_j} \rho(y_j) = 2(2 - b_1), \tag{33}$$

which gives immediately the cases (b) or (c) of the proposition. Note that we have the additional information that $b_1 = 1$ in case (b) and that $b_1 = 2$ in case (c).

In the first case, we have necessarily $b_1 = 1$ and (25) gives in addition that $\sum \rho(y_i) = 3$. So we are effectively in case (a). □

5. On minimal partitions invariant with respect to inversion at the origin

If I is the inversion

$$(x, y) \mapsto I(x, y) = (-x, -y), \tag{34}$$

and if Ω is invariant by I, then the associated map I_* defined, for $u \in L^2(\Omega)$, by

$$(I_* u)(x, y) = (u \circ I)(x, y) = u(-x, -y), \tag{35}$$

commutes with the Laplacian. More generally, we will say that a k-partition is invariant by I, if there is a permutation σ_I of $\{1, \ldots, k\}$, such that for any i in $\{1, \ldots, k\}$,

$$I(D_i) = D_{\sigma_I(i)}. \tag{36}$$

PROOF OF PROPOSITION 6. Suppose for contradiction that $\mathcal{D}$ is invariant with respect to I.

Step 1. First we observe that this excludes case (a) in Proposition 19. The cardinality of $Y(N)$ (with $N = N(\mathcal{D})$) has to be even due to the I-symmetry. In the case when $\#Y(N) = 2$, we also obtain a contradiction by observing that by the I-symmetry, we should have $\rho(y_1) = \rho(y_2)$. Hence we have to consider the cases (b) and (c).

Step 2. In these two cases, we can label the D_i's such that $\sigma_I(1) = 3$ and $\sigma_I(2) = 2$. So D_2 is invariant by I and I exchanges D_1 and D_3. One can choose the corresponding positive ground states as previously but with the additional property that

$$I_* \varphi_i = \varphi_{\sigma_I(i)}, \qquad \forall i. \tag{37}$$

Consider

$$\widehat{\Omega} = \mathrm{Int}(D_1 \cup \overline{D_2} \cup D_3). \tag{38}$$

Then $\lambda_3(\widehat{\Omega}) = \mathfrak{L}_3$ by assumption. Note that some associated eigenvector can be defined in $\widehat{\Omega}$ by

$$u := \varphi_1 - \varphi_2 + \varphi_3, \tag{39}$$

which satisfies

$$I_* u = u. \tag{40}$$

If $\widetilde{\Omega}$ is an I-invariant open set, let $\lambda_2^s(\widetilde{\Omega})$ be the second eigenvalue if we consider $H(\widetilde{\Omega})$ restricted to the subspace of the functions which are symmetric with respect to I.

The previous computation shows that

$$\mathfrak{L}_3(\Omega) \geq \lambda_2^s(\widehat{\Omega}), \tag{41}$$

and by domain monotonicity we get

$$\lambda_2^s(\widehat{\Omega}) \geq \lambda_2^s(\Omega). \tag{42}$$

The case of the disk. We now assume that $\Omega = \mathbf{D}$. An easy computation gives:

$$\lambda_2^s(\mathbf{D}) = \lambda_4(\mathbf{D}).$$

So on one side we have obtained that

$$\lambda_2^s(\mathbf{D}) = \lambda_4(\mathbf{D}) \leq \mathfrak{L}_3(\mathbf{D}), \tag{43}$$

but on the other hand $\mathfrak{L}_3(\mathbf{D}) < \lambda_4(\mathbf{D})$ (as already observed in (8)) and we have hence a contradiction. □

Remark 21. The argument works in the same way for any I-invariant domain Ω such that $\lambda_3(\Omega) < \mathfrak{L}_3(\Omega) < \lambda_2^s(\Omega)$.

This is easy to verify for the square or more generally for rectangles $]-a/2, a/2[\times]-b/2, b/2[$ such that $1 \leq b/a < \sqrt{\frac{8}{3}}$ (see in [15] or in [3]).

6. Eigenfunctions extensions associated to a minimal partition

Let us start from a minimal k-partition $\mathcal{D}$ of Ω. Of course, when the associated graph is not bipartite, we cannot find an eigenfunction which has this partition as nodal domain. But we can nevertheless construct non-trivial solutions of $-\Delta\Psi = \mathfrak{L}_k\Psi$ in suitable subdomains.

6.1. Pair compatibility condition. We recall[6] from [14] or [15] that, if we take any pair of neighbours (D_i, D_j) in a minimal k-partition ($k \geq 2$), we can find a constant $d_{i,j} \neq 0$ so that $\phi_{i,j} = \varphi_i + d_{i,j}\varphi_j$ satisfies

$$-\Delta\phi_{i,j} = \mathfrak{L}_k\phi_{i,j}, \text{ in } D_{i,j}. \tag{44}$$

Actually, as done in [15], we can choose the functions φ_i in $W_0^1(\Omega)$ such that their restrictions to D_i are positive ground states for $H_0(D_i)$ and such that $d_{i,j} = -1$, and we will keep this convention so we have

$$\phi_{i,j} = \varphi_i - \varphi_j. \tag{45}$$

Moreover this choice holds for any pair in the partition.

If the associated graph is bipartite, we can extend this construction to k open sets $(D_1, D_2, \ldots, D_k)$.

6.2. Invariance by rotation. The important observation is that the Laplacian commutes with the operator $\partial/\partial\omega = x\partial_y - y\partial_x$. In particular, if Φ is a local solution of $-\Delta\Phi = \lambda\Phi$, then

$$-\Delta(\partial_\omega\Phi) = \lambda(\partial_\omega\Phi). \tag{46}$$

Very roughly, this will give us a way to construct new eigenfunctions of the Laplacian in suitable domains. We have to be careful in the application of this technique if we apply this to a boundary problem. One should indeed verify that the new object $\partial_\omega\Phi$ or a suitable extension of it belongs to the form domain of the operator. Particular attention is needed at cracks or corners. So this trick cannot be usually extended to higher derivatives.

7. On the minimal partitions symmetric with respect to one axis of reflection

In this section we collect some partial results. In particular we assume that the 3-partition $\mathcal{D}_3$ satisfies some symmetry condition and then show that this implies or excludes one of the topological types described in Proposition 19. Suppose that associated to $\mathfrak{L}_3(\mathbf{D})$ for the unit disk $\mathbf{D}$ there is (up to rotation and reflection) a minimal partition which is now invariant by the map

$$(x, y) \mapsto T(x, y) = (x, -y). \tag{47}$$

We observe that what we called the *Mercedes Star* has (after a rotation by $\pi/2$) this symmetry.

We would like to analyze a weak form of the main conjecture.

Conjecture 22. *Suppose that there exists some minimal 3-partition for the disk which is symmetric with respect to T. Then, possibly after a rotation, this partition corresponds to the* Mercedes Star.

[6] This is the property of pair compatibility discussed in P4.

While we do not have a complete proof of this conjecture we can show special cases and this we shall do below.

First remember that Proposition 19 tells us that there are just three topologically different candidates for a minimal 3-partition.

We have of course

$$\mathfrak{L}_3(\mathbf{D}) \leq \Lambda(\mathcal{D}_{\mathrm{MS}}). \tag{48}$$

Proposition 7 will be an immediate consequence of Proposition 19 and of the following statement.

Lemma 23. *Suppose we are considering a minimal* 3*-partition* $\mathcal{D}$ *which is* T*-symmetric and has the properties of case* (a) *in Proposition* 19*. Then* $(0,0)$ *is the critical point of the partition.*

PROOF. First we observe that $X = X_{\mathrm{odd}}$ consists of one point $z_0 = (\alpha_0, 0)$ and we pick the x-axis as the symmetry axis.

We also know that $\mathfrak{L}_3(\mathbf{D})$ is the ground state energy of $H(D_j)$ and the second eigenvalue of the Dirichlet problem in $D_{i,j}$.

For any $\alpha \in [-1,+1]$, let us introduce the segment $l(\alpha)$ defined by

$$l(\alpha) := \{x \in [-1,\alpha], y = 0\}.$$

We can assume w.l.o.g.[7] that the segment $l(\alpha_0)$ belongs to N, so that $\alpha_0 \in]-1,1[$.

The corresponding domains D_1, D_2, D_3 satisfy

$$TD_1 = D_2, \qquad TD_2 = D_1, \qquad TD_3 = D_3,$$

where $T(x,y) = (x,-y)$. We label the partition such that $D_1 \subset \{y > 0\}$.

Another consequence of [15] (see also [4] for an extensive use of this property) is that

$$\mathfrak{L}_3(\mathbf{D}) = \lambda_3(D(\alpha_0)), \tag{49}$$

with $\mathbf{D}(\alpha) = \mathbf{D} \setminus l(\alpha)$.

We can indeed show that $\mathcal{D}$ is a minimal 3-partition of $\mathbf{D}(\alpha_0)$ and that $\mathfrak{L}_3(\mathbf{D})$ is an eigenvalue of $H(\mathbf{D}(\alpha_0))$ corresponding to an eigenfunction with $\mathcal{D}$ as nodal partition. We can then use the results recalled in Section 2.

We now consider different cases depending on α_0.

Case a1: $\alpha_0 \geq 0$.

In this case, we can indeed apply Proposition 5.

Case a2: $\alpha_0 < 0$.

Let us consider $V = \mathbf{D} \setminus \{\{y=0\} \times [\alpha_0, 1[\}$. We now use the ideas developed in Section 6. Observing that φ_3 is symmetric with respect to $\{y=0\}$ and that $\partial_\omega \varphi_3$ vanishes at 0, we can consider the symmetric function Ψ on V which is defined on $V^+ = V \cap \{y \geq 0\}$ by $\partial_\omega \varphi_1 - \partial_\omega \varphi_3$. Using the pair compatibility condition for D_{13}, D_{12} and D_{23} and the local analysis[8] (we refer to [15, Section 3] and references therein) at $z_0 = (\alpha_0, 0)$ of $\varphi_1 - \varphi_2$, $\varphi_1 - \varphi_3$, we obtain that Ψ is in the form domain of the Dirichlet Laplacian on V and an eigenfunction attached to $\mathfrak{L}_3(\mathbf{D})$.

Let us look at the nodal set $N(\Psi)$ of Ψ. If we consider the situation in D_1, we first observe that ∂D_1 consists of three parts:

- the segment $[-1, \alpha_0]$ on the x-axis,

---

[7]Possibly after a rotation by π.

[8]We have a behavior in $c\rho^{3/2}\cos 3\theta$ in polar coordinates (ρ,θ) centered at $z_0 = (\alpha_0, 0)$

- an arc on the upper half-circle starting from $(-1,0)$ and ending at a point $\hat{z}_0$, and
- an arc γ connecting $\hat{z}_0$ with z_0 that hits the circle at $\hat{z}_0$ transversally.

The first claim is that there exists a point $z_1 := N(\partial_\omega \varphi_1) \cap \partial \mathbf{D} \cap \{y > 0\}$ and an arc γ_1 in $\mathbf{D}$ starting from z_1. If indeed there is no such path, there would be a circle (centered at 0 with radius between $|\alpha_0|$ and 1) crossing the first and the third part of ∂D_1 and avoiding $N(\alpha_0) \cap D_1$. We will then find on this circle an arc on which φ_1 would be strictly positive, zero at the two ends and with $\partial_\omega \varphi_1$ not zero (contradiction).

The second claim is that one can extend γ_1 as a piecewise smooth path which ends either on $\{y = 0, x \geq \alpha_0\}$ or on the upper half-circle and cannot cross (applying Hopf's boundary point lemma to φ_1) the segment $l(\alpha_0)$.

Hence the nodal set $N(\Psi)$ of Ψ has empty intersection with $l(\alpha_0)$ and contains $\gamma_1 \cup T\gamma_1$, with the additional property that $V \setminus N(\Psi)$ has (at least) three components.

Coming back to $\mathbf{D}$, we obtain a partition $\mathcal{D}$, with $\Lambda(\mathcal{D}) = \mathfrak{L}_3(\mathbf{D})$, whose boundary is

$$N(\alpha_0) = N(\Psi) \cup \{[\alpha_0, 1] \times \{y = 0\}\}.$$

We get a priori the following three possibilities.

- There are l components (with $l > 3$), but this contradicts

$$\mathfrak{L}_3(\mathbf{D}) \leq \Lambda(\text{Mercedes Star}) < \mathfrak{L}_4(\mathbf{D}) \leq \mathfrak{L}_l(\mathbf{D}), \tag{50}$$

  which was observed in [15].
- There are three components and γ_1 ends at another point z_1 of the half-circle. In this case, we can observe that we arrive at a symmetric bipartite 3-partition which cannot be a minimal 3-partition.
- There are three components and γ_1 ends at $(\alpha_1, 0)$ for some $\alpha_1 \in [\alpha_0, 1[$, but, up to a inversion at the origin, we are in Case a1 which was already treated.

This achieves the proof of Lemma 23. □

Toward the exclusion of the other types. Next we consider the cases (b) an (c). There are two possibilities to place the two points in X_{odd} once we have fixed the x-axis as the reflection line. The two possibilities are:

(i) $X_{\text{odd}} = \{(x_0, y_0), (x_0, -y_0)\}$.
(ii) $X_{\text{odd}} = \{(x_1, 0), (x_2, 0)\}$.

These two possibilities correspond to the cases when the associated permutaion σ_T (defined by $T(D_i) = D_{\sigma_T(i)}$) is either the identity (case (i)) or exchanging 1 and 2 (case (ii)).

Lemma 24. *For the cases* (b) *and* (c), *a 3-partition satisfying* (i) *cannot be minimal.*

Proof. Consider as before $\mathbf{D}_+ = \{(x,y) \in \mathbf{D} \mid y > 0\}$. Define also H_+ as $-\Delta$ on $\mathbf{D}_+$ with Dirichlet boundary condition on $\partial \mathbf{D}_+ \cap \{y > 0\}$ and Neumann on $\partial \mathbf{D}_+ \cap \{y = 0\}$. We can associate to H_+ also minimal spectral partitions. Looking up the tables of eigenvalues of the disk and the corresponding eigenfunctions as recalled in Appendix B reveals that

$$\lambda_1(\mathbf{D}) = \lambda_1(\mathbf{D}_+) < \lambda_2(\mathbf{D}) = \lambda_2(\mathbf{D}_+) < \mathfrak{L}_3(\mathbf{D}_+) = \lambda_4(\mathbf{D}) = \lambda_3(\mathbf{D}_+) > \mathfrak{L}_3(\mathbf{D}).$$

But by inspection cases (b) and (c) lead to three domains in $\mathbf{D}_+$ and we have hence a contradiction. □

Another result is

Lemma 25. *The origin cannot belong to* $N(\mathcal{D})$ *for a minimal* 3*-partition* $\mathcal{D}$ *of type*(b).

PROOF. If $0 \in N(\mathcal{D})$ then the double covering $\dot{\mathbf{D}}^{\mathcal{R}}$ will have a nodal pattern with six domains. This is excluded by the proof of Proposition 5 and the second part of Theorem 13. □

8. On minimal partitions invariant under rotation by $2\pi/3$

One can also consider the minimization over 3-partitions generated by one open set and its rotations by $2\pi/3$. Then the proof of [15] goes through.[9] The analysis of the different types of partitions leads to the fact that we have necessarily a 3-partition of type (a) with 0 as the unique singular point. We can then go to the double covering of the punctured disk and conclude.

Proposition 26. *The* 3*-partition which is minimal inside the class of the* 3*-partitions invariant under rotation by* $2\pi/3$ *is the Mercedes Star* (*up to rotation*).

A more direct proof is to see by going to the quotient that this problem is equivalent to looking for the minimization of $\lambda_1(\omega)$ over the simply connected sets ω in $\dot{\mathbf{D}}$ and to use the results of [13].

9. Conclusion

We have presented general properties of minimal partitions and explained how they can be used for analyzing more deeply minimal 3-partitions. These considerations appear to be useful in the analysis of various conjectures. This has been explored numerically in [4]. Here we have emphasized on the Mercedes Star problem. We analyze elsewhere the case of the rectangle [3], of the sphere [16] and hope to discuss the same kind of problems in larger dimension [17]. We hope also to develop the Courant theory for domains with symmetries and also on the double covering of the corresponding punctured domains, in connection with Aharonov–Bohm Hamiltonians.

Appendix A. Zeros of Bessel functions

Let $j_{l,k}$ the kth zero of the Bessel function corresponding to the integer $l \in \mathbb{N}$.

Here is a list of approximate values after the classical handbook of [1, p. 409] we keep only the values which are less than 10.

(51)

$k \backslash l$	0	1	2	3	4	5	6
1	2.40	3.83	5.14	6.38	7.59	8.77	9.93
2	5.52	7.02	8.42	9.76	·	·	·
3	8.65	·	·	·	·	·	·

This leads to the following ordering of the zeros:

(52) $j_{0,1} < j_{1,1} < j_{2,1} < j_{0,2} < j_{3,1} < j_{1,2} < j_{4,1} < j_{2,2} < j_{0,3} < j_{5,1} < j_{3,2} < j_{6,1}.$

---

[9]We thank S. Terracini and G. Verzini for discussions on this problem.

It is also useful to have the half-integer results (see in [1, p. 467])

(53)

$k \backslash l$	$\frac{1}{2}$	$\frac{3}{2}$	$\frac{5}{2}$	$\frac{7}{2}$	$\frac{9}{2}$	$\frac{11}{2}$
1	π	4.49	5.76	6.99	8.18	9.36
2	2π	7.73	9.09	·	·	·
3	3π	·	·	·	·	·

Appendix B. On the lowest eigenvalues of the Dirichlet Laplacian on the disk and its double covering

In this appendix, we recall for completion the detailed analysis of the disk which was presented in [15]. Although the spectrum is explicitly computable, we are mainly interested in the ordering of the eigenvalues corresponding to different angular momenta.

Consider the Dirichlet realization H_0 in the unit disk $\mathbf{D} \subset \mathbb{R}^2$. We have in polar coordinates:

$$-\Delta = -\frac{\partial^2}{\partial r^2} - \frac{1}{r}\frac{\partial}{\partial r} - \frac{1}{r^2}\frac{\partial^2}{\partial \theta^2}$$

and the Dirichlet boundary conditions require that any eigenfunction u satisfies $u(r,\theta) = 0$ for $r = 1$. We analyze for any $l \in \mathbb{N}$ the eigenvalues $\lambda_{l,j}$ of

$$\left(-\frac{\mathrm{d}^2}{\mathrm{d}r^2} - \frac{1}{r}\frac{\mathrm{d}}{\mathrm{d}r} + \frac{l^2}{r^2}\right) f_{l,j} = \lambda_{l,j} f_{l,j}, \qquad \text{in } (0,1).$$

We observe that the operator is self-adjoint for the scalar product in $L^2\big((0,1), r\,\mathrm{d}r\big)$.

The corresponding eigenfunctions of the eigenvalue problem take the form

$$u(r,\theta) = f_{l,j}(r)(a\cos l\theta + b\sin l\theta), \qquad \text{with } a^2 + b^2 > 0, \tag{54}$$

where the $f_{l,j}(r)$ are suitable Bessel functions satisfying for $l = 0$, $f'_{0,j}(0) = 0$ and $f_{0,j}(1) = 0$ and for $l > 0$, $f_{l,j}(0) = f_{l,j}(1) = 0$. For the corresponding $\lambda_{l,j}$'s, we find the following ordering.

$$\lambda_1 = \lambda_{0,1} < \lambda_2 = \lambda_3 = \lambda_{1,1} < \lambda_4 = \lambda_5 = \lambda_{2,1} < \lambda_6 = \lambda_{0,2}. \tag{55}$$

We recall that the zeros of the Bessel functions are related to the eigenvalues by the relation

$$\lambda_{l,k} = (j_{l,k})^2. \tag{56}$$

We hence have from (54)

$$\begin{aligned} &\mu(u_1) = 1, \\ &\mu(u) = 2, \qquad \text{for any eigenfunction } u \text{ associated to } \lambda_2 = \lambda_3, \\ &\mu(u) = 4, \qquad \text{for any eigenfunction } u \text{ associated to } \lambda_4 = \lambda_5, \\ &\mu(u_6) = 2. \end{aligned} \tag{57}$$

Hence

$$L_1 = \lambda_1, \qquad L_2 = \lambda_2, \qquad L_4 = \lambda_4 \tag{58}$$

and one can also verify that $L_3 = \lambda_{15}$. This implies

$$L_3 > \lambda_6 > L_4 > \lambda_3, \tag{59}$$

and

$$L_3 > \mathfrak{L}_3. \tag{60}$$

More generally, it is proven in [15]:

Proposition 27. *Except the cases $k = 1, 2$ and 4, minimal k-partitions never correspond to nodal domains.*

The analysis of the Laplacian on the double covering of the punctured disk $\dot{\mathbf{D}}^{\mathcal{R}}$ is obtained by taking $\theta \in]-2\pi, 2\pi]$ which leads to consider l's which are also half-integers.

Using the tables of zeros of Bessel functions, we get the following ordering:

$$\begin{aligned}\lambda_1^{\mathcal{R}} = \lambda_{0,1} < \lambda_2^{\mathcal{R}} = \lambda_3^{\mathcal{R}} = \lambda_{\frac{1}{2},1} < \lambda_4^{\mathcal{R}} = \lambda_5^{\mathcal{R}} = \lambda_{1,1} \\ < \lambda_6^{\mathcal{R}} = \lambda_7^{\mathcal{R}} = \lambda_{\frac{3}{2},1} < \lambda_8^{\mathcal{R}} = \lambda_9^{\mathcal{R}} = \lambda_{2,1}.\end{aligned} \tag{61}$$

For the nodal domains, we obtain:

$$\begin{aligned} &\mu(u_1) = 1, \\ &\mu(u) = 2, \quad \text{for any eigenfunction } u \text{ associated to } \lambda_2^{\mathcal{R}} = \lambda_3^{\mathcal{R}}, \\ &\mu(u) = 4, \quad \text{for any eigenfunction } u \text{ associated to } \lambda_4^{\mathcal{R}} = \lambda_5^{\mathcal{R}}, \\ &\mu(u) = 6, \quad \text{for any eigenfunction } u \text{ associated to } \lambda_6^{\mathcal{R}} = \lambda_7^{\mathcal{R}}. \end{aligned} \tag{62}$$

In particular, we have

Proposition 28. *In the case $k = 6$, any minimal 6-partition is nodal.*

PROOF. We observe indeed from (62), that an eigenfunction corresponding to the sixth eigenvalue has six nodal domains, hence is Courant sharp. We can then apply Theorem 13 which is valid on $\dot{\mathbf{D}}^{\mathcal{R}}$. □

Remark 29. It could be interesting to determine when $\mathfrak{L}_k$ is an eigenvalue of the Laplacian on the double covering. Again, one can show that this cannot be the case for k large.

Acknowledgements. The authors would like to thank Gian Maria Verzini who mentioned that for the disk we can look also at the derivative with respect to the angle since the Laplacian commutes with this operation. This was indeed the starting point for our investigations. The authors thank also Susanna Terracini for preliminary discussions about the problem.

Discussions with (and numerical computations of) Virginie Bonnaillie-Noël and Gregory Vial were very helpful. We send the reader to the website [5] for various numerical simulations.

References

1. M. Abramowitz and I. A. Stegun, *Handbook of mathematical functions with formulas, graphs, and mathematical tables*, National Bureau of Standards Applied Mathematics Series, vol. 55, U.S. Government Printing Office, Washington, DC, 1964.
2. A. Ancona, B. Helffer, and T. Hoffmann-Ostenhof, *Nodal domain theorems à la Courant*, Doc. Math. **9** (2004), 283–299 (electronic).
3. V. Bonnaillie-Noël, B. Helffer, and T. Hoffmann-Ostenhof, *Aharonov–Bohm Hamiltonians, isospectrality and minimal partitions*, J. Phys. A **42** (2009), no. 18, 185203.
4. V. Bonnaillie-Noël, B. Helffer, and G. Vial, *Numerical simulations for nodal domains and spectral minimal partitions*, ESAIM Control Optim. Calc. Var. **16** (2010), no. 1, 221-246.
5. V. Bonnaillie-Noël and G. Vial, numerical simulations, available at `http://w3.bretagne.ens-cachan.fr/math/simulations/MinimalPartitions/`.

6. D. Bucur and G. Buttazzo, *Variational methods in shape optimization problems*, Progr. Nonlinear Differential Equations Appl., vol. 65, Birkhäuser, Boston, MA, 2005.
7. L. A. Cafferelli and F. H. Lin, *An optimal partition problem for eigenvalues*, J. Sci. Comput. **31** (2007), no. 1-2, 5–18.
8. M. Conti, S. Terracini, and G. Verzini, *An optimal partition problem related to nonlinear eigenvalues*, J. Funct. Anal. **198** (2003), no. 1, 160–196.
9. ———, *A variational problem for the spatial segregation of reaction-diffusion systems*, Indiana Univ. Math. J. **54** (2005), no. 3, 779–815.
10. ———, *On a class of optimal partition problems related to the Fučík spectrum and to the monotonicity formulae*, Calc. Var. Partial Differential Equations **22** (2005), no. 1, 45–72.
11. B. Helffer, *Domaines nodaux et partitions spectrales minimales (d'après B. Helffer, T. Hoffmann-Ostenhof et S. Terracini)*, Séminaire : Équations aux Dérivées Partielles. 2006–2007, Sémin. Équ. Dériv. Partielles, École Polytech., Palaiseau, 2007, Exp. No. VIII.
12. B. Helffer, M. Hoffmann-Ostenhof, T. Hoffmann-Ostenhof, and M. P. Owen, *Nodal sets for groundstates of Schrödinger operators with zero magnetic field in non-simply connected domains*, Comm. Math. Phys. **202** (1999), no. 3, 629–649.
13. ———, *Nodal sets, multiplicity and superconductivity in non-simply connected domains*, Connectivity and Superconductivity, Lecture Notes Phys. Monogr., vol. 62, Springer, Berlin, 2000, pp. 63–86.
14. B. Helffer and T. Hoffmann-Ostenhof, *Converse spectral problems for nodal domains*, Mosc. Math. J. **7** (2007), no. 1, 67–84.
15. B. Helffer, T. Hoffmann-Ostenhof, and S. Terracini, *Nodal domains and spectral minimal partitions*, Ann. Inst. H. Poincaré Anal. Non Linéaire **26** (2009), no. 1, 101–138.
16. ———, *On spectral minimal partitions: the case of the sphere*, Around the Research of Vladimir Maz′ya III, Int. Math. Ser. (N. Y.), vol. 13, Springer, New York, 2010, pp. 153–178.
17. ———, *On spectral minimal partitions: the 3D-case*, in progress.
18. A. Henrot, *Extremum problems for eigenvalues of elliptic operators*, Front. Math., Birkhäuser, Basel, 2006.
19. T. Hoffmann-Ostenhof, P. W. Michor, and N. Nadirashvili, *Bounds on the multiplicity of eigenvalues for fixed membranes*, Geom. Funct. Anal. **9** (1999), no. 6, 1169–1188.

Département de Mathématiques, Bat. 425, Université Paris-Sud, 91405 Orsay Cedex, France

E-mail address: `Bernard.Helffer@math.u-psud.fr`

Institut für Theoretische Chemie, Universität Wien, Währinger Strasse 17, 1090 Wien, Austria & International Erwin Schrödinger Institute for Mathematical Physics, Boltzmanngasse 9, 1090 Wien, Austria

E-mail address: `thoffman@esi.ac.at`

Centre de Recherches Mathématiques
CRM Proceedings and Lecture Notes
Volume **52**, 2010

Asymptotic Vertex Growth for Graphs

Mark Pollicott

1. Introduction

In this note, we shall describe natural invariants for weighted graphs and characterize the minima of these quantities over weightings which are suitably normalized.

Let G be a connected finite graph with vertex set V. If we assume for convenience that there is at most one edge e joining a pair of vertices $(i,j) \in V \times V$ then we can identify the directed edges with a subset of $V \times V$. However, since we will be primarily interested in undirected graphs we can also use E to denote directed edges provided $(i,j) \in E$ if and only if $(j,i) \in E$. Let $|E|$ denote the number of (undirected) edges and $|V|$ denote the number of vertices. Ordinarily, we would think of G as having edges of unit length. However, additional structure can be added in the form of an edge weighting $w\colon E \to \mathbb{R}^+$ (which satisfies $w(i,j) = w(j,i)$, where $(i,j),(j,i) \in E$). Let us consider two particular weightings:

(1) w_u corresponds to the unweighted graph by setting $w_u(e) = 1$ for every edge $e \in E$; and
(2) w_c is defined by

$$w_c(i,j) = |E|\left(\frac{\log(k_i k_j)}{\sum_{(i,j)\in E}\log(k_i k_j)}\right)$$

where $k_i + 1$ is degree of the vertex i (i.e., the number of edges $(i,j) \in E$ which have the same vertex $i \in V$).

Definition 1 (cf. [2, p. 5]**).** We define the volume of G with edge weighting w by $\mathrm{Vol}(G,w) = \sum_{e\in E} w(e)$.

From the definitions we see that $\mathrm{Vol}(G,w_u) = |E|$ and $\mathrm{Vol}(G,w_c) = |E|$. More generally, we denote by $W = \{w : \mathrm{Vol}(G,w) = |E|\}$ the space of all weightings of the same volume. This finite-dimensional space W is easily identified with a simplex in $(\mathbb{R}^+)^{|E|}$. The universal cover $\widetilde{G}$ for the graph G is a tree. For any reference point $x_0 \in \widetilde{G}$ we can denote by $\mathrm{Vert}_{\widetilde{G}}(x_0,T)$ the number of vertices in $\widetilde{G}$ whose distance from x_0 (using the lifted edge weighting) is at most $T > 0$.

2010 *Mathematics Subject Classification.* 37D35, 37B40, 05C22.
This is the final form of the paper.

Definition 2. We define the asymptotic vertex growth (or entropy) by

$$h(w) = \limsup_{T \to \infty} \frac{1}{T} \log \mathrm{Vert}_{\widetilde{G}}(x_0, T). \tag{1}$$

It is easy to see that this definition is independent of the particular choice of x_0 since given any other point $x_1 \in \widetilde{G}$ we can estimate

$$\mathrm{Vert}_{\widetilde{G}}(x_1, T) \leq \mathrm{Vert}_{\widetilde{G}}\big(x_0, T + d_{\widetilde{G}}(x_0, x_1)\big).$$

Moreover, by subadditivity the limit supremum is actually a limit.

Our first result is the following.

Theorem 1. *Let G be a connected graph containing more than one closed loop. The function $h\colon W \to \mathbb{R}$ is convex, nonzero, C^ω (i.e., real analytic) and the critical point occurs at w_c.*

A graph G which has the same number $k+1$ of edges connected to each vertex is called a $(k+1)$-regular graph (i.e., $k_i = k$ for each $i \in V$). For such graphs it is easy to see that $w_u = w_c$ and that $h(w_c) = \log k$. We have the following immediate corollary.

Corollary 1. *The unweighted graph is the critical point for $h(w)$ if G is a regular graph.*

The definition (1) is motivated by the classical characterization of topological entropy for negatively curved manifolds [5]. In that context, Besson – Courtois – Gallot showed that the critical point for the entropy over Riemannian metrics of normalized volume typically occurs at locally symmetric metrics [1]. Fortunately, in the context of graphs we only require simple matrix analysis and elementary real analysis.

There is a less well known generalization due to Ruelle of the topological pressure of a function using the universal cover of a Riemannian manifold [11]. In the present context the appropriate analogy would be to consider a weighting of the vertices by a function $g\colon V \to \mathbb{R}$. We denote by $\widetilde{G}\colon \widetilde{V} \to \mathbb{R}$ the natural lift to $\widetilde{g}$ and by $\mathrm{Vert}^g_{\widetilde{G}}(x_0, T)$ the sum of the function $\widetilde{g}$ over the vertices of $\widetilde{G}$ whose distance from x_0 (using the lifted edge weighting) is at most $T > 0$.

Definition 3. We define the asymptotic weighted vertex growth (or pressure) for g by

$$P(g, w) = \limsup_{T \to \infty} \frac{1}{T} \log \mathrm{Vert}^g_{\widetilde{G}}(x_0, T). \tag{2}$$

Again the definition is independent of the choice of x_0 and the limit supremum is actually a limit.

Henceforth, we will want to impose the condition that whenever $i, j \in V$ are joined by an edge than we have that

$$g(i) + g(j) > -\log(k_i k_j). \tag{3}$$

Condition (3) is necessary for the resulting edge weights to be positive.

Let us then consider the weighting

$$w_c^g(i) = |E| \left(\frac{\log(k_i k_j) + g(i) + g(j)}{\sum_{(i,j) \in E} \log(k_i k_j) + g(i) + g(j)} \right).$$

Our second result is the following.

Theorem 2. *Let G be a connected graph containing more than one closed loop. Let $g\colon V \to \mathbb{R}$ then the function $P(g,\cdot)\colon W \to \mathbb{R}$ is convex, nonzero, C^ω and if* (3) *holds then a critical point occurs at w_c^g.*

It is easy to see that without some hypothesis on g the conclusion of Theorem 2 will be false (see Example 3, in Section 6).

Corollary 2. *The unweighted graph is the critical point for $P(w,g)$ if $g(i) = 1 - \log k_i$, for $i = 1,\dots,|V|$.*

Remark 1. One might reasonably ask if there is a geometric analogue for negatively curved manifolds V of the above theorem. Since the pressure of a fixed function $g\colon V \to \mathbb{R}$ varies smoothly as the Riemannian metric ρ changes, one could look for critical points using the formula for the derivative of the pressure. One could speculate that the geodesic flow $\phi^{(\rho)}$ for the critical metric(s) ρ would be one that satisfies

$$\int \left(\Xi g + \rho^{(1)}(v,v)\right) \mathrm{d}\mu_g^{(\phi)} = 0$$

where $\rho^{(1)}$ is any first order volume preserving perturbation of the metric, Ξ is the corresponding Morse correspondence and $\mu_g^{(\phi)}$ is the Gibbs measure associated to $\phi^{(\rho)}$ and g.

Theorem 1 was originally presented by the author in a seminar at MSRI in the Spring of 1992, during the program on "Lie Groups and Ergodic Theory with Applications to Number Theory and Geometry." A special case of Theorem 1 also appeared in [9] for regular graphs. Theorem 1 has also been proved independently by S. Lim using a somewhat different approach [6]. See also [5].

2. Matrices and derivatives of $h\colon W \to \mathbb{R}$

We can associate to the graph G a directed graph $\mathcal{G}$ whose vertices are the directed edges e associated to G. Every undirected edge gives rise to the two directed edges. If $e = (i,j) \in V \times V$ then we denote $e(0) = i$ and $e(1) = j$. We define a square matrix A by $A(e,f) = 1$ if the final vertex $e(1)$ of the directed edge e of $\mathcal{G}$ coincides with the initial vertex $f(0)$ of the directed edge f of $\mathcal{G}$, and they are not simply the same edge reversed. We set $A(e,f) = 0$ otherwise. For simplicity, we shall assume in the sequel that A is aperiodic (i.e., $\exists n \geq 1$, $\forall (e,f)$ we have $A^n(e,f) > 0$). To accommodate the weighting $w \in W$ we can associate, for each $t \in \mathbb{R}$, a weighted $|E| \times |E|$ matrix A_{tw} defined by

$$A_{tw}(e,f) = \begin{cases} \mathrm{e}^{-tw(e)} & \text{if } A(e,f) = 1, \\ 0 & \text{if } A(e,f) = 0. \end{cases}$$

If we shall assume that G contains more than one closed path then the maximal eigenvalue $\lambda(A_{tw})$ of A_{tw} is strictly bigger than 1. The following is a standard result on aperiodic matrices [3].

Lemma 1 (Perron – Frobenius theorem). *The maximal eigenvalue $\lambda(A_{tw})$ is simple and the associated right eigenvector $v = (v_1,\dots,v_{|V|})^T$ has all of its entries strictly positive.*

It follows from perturbation theory for matrices that the map $\mathbb{R} \times W \ni (h,w) \mapsto \lambda(A_{hw}) \in \mathbb{R}$ is analytic. The next lemma gives an implicit characterization of $h(w)$ in terms of this eigenvalue.

Lemma 2. *$h = h(w)$ is the unique value such that the maximal eigenvalue of A_{hw} satisfies $\lambda(A_{hw}) = 1$.*

PROOF. We can consider the Stieltjes integral

$$\eta_w(t) = \int e^{-tT} d\,\mathrm{Vert}_{\widetilde{G}}(x_0, T) = \frac{1}{2} \sum_{e \in E(x_0)} \sum_{f} \left(\sum_{n=0}^{\infty} (A_{tw})^n(e, f) \right)$$
$$= \frac{1}{2} \frac{\sum_{e \in E(x_0)} \sum_{f \in E} \mathrm{adj}[A_{tw}](e, f)}{\det(I - A_{tw})},$$

where $E(x_0) \subset V \times V$ are the directed edges in G which start at the vertex corresponding to x_0. (The factor of $\frac{1}{2}$ appears because every undirected edge corresponds to two directed edges.) In particular, by the definition of $h(w)$ we see that this is the abscissa of convergence of the first expression in (1). Therefore, h is precisely the smallest t such that $\det(I - A_{tw}) = 0$ or, equivalently, $\lambda(A_{hw}) = 1$. □

By Lemma 2, it follows from the implicit function theorem applied to $(t, w) \mapsto \lambda(A_{tw}) \in \mathbb{R}$ that the entropy $h(w)$ has a C^ω dependence on $w \in W$.

Lemma 3. *The map $h \colon W \to \mathbb{R}$ is strictly convex (i.e., if $w_1, w_2 \in W$ and $0 < \alpha < 1$ then $h(\alpha w_1 + (1-\alpha) w_2) \leq \alpha h(w_1) + (1-\alpha) h(w_2)$). In particular, there is at most one critical point for $h \colon W \to \mathbb{R}$ which will be a minimum.*

PROOF. We can write

$$\eta_w(t) = \frac{1}{2} \sum_{e \in E(x_0)} \sum_{f} \left(\sum_{n=0}^{\infty} \sum_{\substack{\underline{e} = e_1 \cdots e_n \\ e_1 = e, e_n = f \\ A(e_r, e_{r+1}) = 1}} e^{-tw(\underline{i})} \right)$$

where the inner most summation is over all allowed paths $\underline{e}$ through n consecutive edges $e_1, \ldots, e_n$ in G and $w(\underline{e}) = w(e_1, e_2) + \cdots + w(e_{n-1}, e_n)$. From the proof of Lemma 2, the value $h(w)$, is the abscissa of convergence of the series. In particular, for any $\epsilon > 0$ we have that $\eta([h(w_1) + \epsilon] w_1) < \infty$ and $\eta([h(w_2) + \epsilon] w_2) < \infty$. By the convexity of $x \to e^{-x}$ we see that for each $\underline{e}$

$$e^{-\alpha[h(w_1)+\epsilon] w_1(\underline{e}) - (1-\alpha)[h(w_2)+\epsilon] w_2(\underline{e})} \leq \alpha e^{-[h(w_1)+\epsilon] w_1(\underline{e})} + (1-\alpha) e^{-[h(w_2)+\epsilon] w_2(\underline{e})}.$$

Summing over $\underline{e}$ we see that

$$\eta(\alpha[h(w_1) + \epsilon] w_1 + (1-\alpha)[h(w_2) + \epsilon] w_2) \leq \alpha \eta([h(w_1) + \epsilon] w_1) + (1-\alpha) \eta([h(w_2) + \epsilon] w_2) < \infty, \tag{4}$$

for all $\epsilon > 0$. We can write

$$\frac{\alpha h(w_1) w_1 + (1-\alpha) h(w_2) w_2}{\alpha h(w_1) + (1-\alpha) h(w_2)} = \beta w_1 + (1-\beta) w_2$$

where $\beta := \alpha / \big(\alpha h(w_1) + (1-\alpha) h(w_2)\big)$. We can then deduce from (4) that

$$h(\beta w_1 + (1-\beta) w_2) \leq \alpha h(w_1) + (1-\alpha) h(w_2). \tag{5}$$

Assume for definiteness that $h(w_2) \geq h(w_1)$, the other case being similar, then $\beta \leq \alpha$ and it follows from (5) that

$$h(\beta w_1 + (1-\beta) w_2) \leq \alpha h(w_1) + (1-\alpha) h(w_2) \leq \beta h(w_1) + (1-\beta) h(w_2).$$

This completes the proof. □

To describe the derivative we can associate to A_{hw} the stochastic matrix P defined by

$$P(e,f) = \frac{e^{-hw(e,f)}v_e}{v_f}A_{hw}(e,f) \tag{6}$$

where v is the maximal eigenvector, i.e., $A_{hw}v = \lambda(A_{hw})v$. Let $p = (p_1, \ldots, p_n)$ be the left eigenvector such that $pP = p$.

Lemma 4. *We can estimate the derivatives:*
(1) $D_1\lambda(A_{hw}) = \lambda(A_{hw})\mu(w)$; *and*
(2) $D_2\lambda(A_{hw}) = \big(\lambda(A_{hw})\mu(e_1), \ldots, \lambda(A_{h,w})\mu(e_{|E|})\big)$,
where we denote $\mu(w) = \sum_{e\in E} p_e w(e)$ *and* $\mu(e) = p_e$.

PROOF. We shall prove only part (1), the proof of part (2) being similar. Starting from the identity $A_{tw}v_{t,w} = \lambda(A_{tw})v_{tw}$ we can use analytic perturbation theory and differentiate in the t variable to write

$$\sum_{e:A(e,f)=1} e^{-tw(e)}\big(v_{tw}(e)w(e,f) + v'_{tw}(e)\big) = \lambda'(A_{tw})v_{tw}(f) + \lambda(A_{tw})v'_{tw}(f), \tag{7}$$

for each directed edge f. If we set $t = h$ then we observe that $u = (u_1, \ldots, u_n)$, where $u_i = p_i/v_i$, is the corresponding left eigenvalue for A_{hw}, i.e., $uA_{hw} = \lambda(A_{hw})u$. We can choose the normalization:

$$\sum_e u(e)v_{hw}(e) = 1. \tag{8}$$

Observe that

$$\sum_f u(f)\Big(\sum_{e:A(e,f)=1} e^{-hw(e)}v'_{hw}(e)\Big) = \sum_f v'_{hw}(e)\Big(\sum_{e:A(e,f)=1} u(f)e^{-hw(e)}\Big) \tag{9}$$
$$= \lambda(A_{hw})\sum_e v'_{hw}(e)u(e).$$

Furthermore,

$$\sum_f u(f)\Big(\sum_{e:A(e,f)=1} e^{-hw(e)}v_{hw}(e)w(e)\Big) = \sum_f p_f\Big(\sum_{e:A(e,f)=1} P(e,f)w(e)\Big) \tag{10}$$
$$= \sum_e \mu(e)w(e).$$

We can now weight each of the terms in (7) by $u(f)$ and sum over the directed edges f. Using (9), the terms corresponding to the first and last expressions in (7) cancel. Applying (8) and (10) to the remaining terms results in

$$\lambda'(A_{hw}) = \sum_e \mu(e)w(e).$$

This completes the proof. □

The following lemma will be useful in Section 4.

Lemma 5. *We can express the derivative for* $h\colon W \to \mathbb{R}$ *as*

$$Dh(w) = \big(\mu(e_1), \ldots, \mu(e_{|E|})\big).$$

PROOF. We can apply the implicit function theorem to $\lambda(A_{tw}) = 1$ to write

$$Dh(w) = \frac{D_2\lambda(A_{hw})}{D_1\lambda(A_{hw})}.$$

The result then follows immediately from Lemma 4. □

3. Matrices and derivatives of $P(g, \cdot)\colon W \to \mathbb{R}$

In this section we modify the results of the preceding section to deal with weighted matrices.

We can associate to each function $g(i)$ on directed edges, and each $t \in \mathbb{R}$, the matrix A^g_{tw} defined by

$$A^g_{tw}(e, f) = \begin{cases} e^{g(e)-tw(e)} & \text{if } A(e, f) \neq 0, \\ 0 & \text{if } A(e, f) = 0. \end{cases}$$

It again follows from perturbation theory that the map $\mathbb{R} \times W \ni (t, w) \mapsto \lambda(A^g_{tw}) \in \mathbb{R}$ is analytic.

The next lemma gives an implicit characterization of $P(g, w)$.

Lemma 6. *$P = P(g, w)$ is the unique value such that the maximal eigenvalue $\lambda(A^g_{Pw})$ of A^g_{Pw} satisfies $\lambda(A^g_{Pw}) = 1$.*

PROOF. The proof is analogous to that of Lemma 2. □

By Lemma 6, it follows from the implicit function theorem applied to $(t, w) \mapsto \lambda(A^g_{Pw}) \in \mathbb{R}$ that the entropy $P(g, w)$ has a real analytic dependence on $w \in W$.

Lemma 7. *The map $P(g, \cdot)\colon W \to R$ is strictly convex (i.e., if $w_1, w_2 \in W$ and $0 < \alpha < 1$ then $P(g, w_1 + (1-\alpha)w_2) \leq \alpha P(g, w_1) + (1-\alpha)P(g, w_2)$). Moreover, the critical point of $P(g, \cdot)\colon W \to \mathbb{R}$ is a minimum.*

PROOF. We can write

$$\eta^g(tw) = \frac{1}{2}\sum_f \left(\sum_{n=0}^{\infty} \sum_{\substack{\underline{e}=e_1,e_n \\ e_1=e,\ldots,e_n=f \\ A(e_r,e_{r+1})=1}} e^{g(e)-tw(e)} \right)$$

where the summation is over all allowed paths $\underline{e}$ through n vertices and $g(\underline{e}) = g(e_1, e_2) + \cdots + g(e_{n-1}, e_n)$. From the definition, the value $P(g, w)$ is the abscissa of convergence of this series. The rest of the proof of the lemma is analogous to that of Lemma 3. □

We can associate to A^g_{Pw} the stochastic matrix P^g defined by

$$P^g(e, f) = \frac{e^{g(e)-Pw(e)}v_e}{v_f} \tag{11}$$

where $A^g_{Pw}v = \lambda(A^g_{Pw})v$. Let $p = (p_1, \ldots, p_n)$ be the left eigenvector such that $pP^g = p$.

Lemma 8. *We can estimate the derivatives*

(1) *$D_1\lambda(A^g_{Pw}) = \lambda(A_{Pw})\mu^g(w)$ and*
(2) *$D_2\lambda(A^g_{Pw}) = (\lambda(A^g_{Pw})\mu^g(e_1), \ldots, \lambda(A^g_{Pw})\mu^g(e_{|E|}))$*

where we denote $\mu^g(w) = \sum_e p_e w(e)$ and $\mu^g(e) = p_e$.

PROOF. The proof of this lemma is analogous to that of Lemma 4. □

The following lemma will be useful in Section 5.

Lemma 9. *We can express the derivative for $P(g,\cdot)\colon W \to \mathbb{R}$ as*

$$DP(w) = \big(\mu^g(e_1), \ldots, \mu(e^g_{|E|})\big).$$

PROOF. We can apply the implicit function theorem to $\lambda(A^g_{tw}) = 1$ to write $DP(g,w) = D_2\lambda(A^g_{Pw})/D_1\lambda(A^g_{Pw})$. The result then follows immediately from Lemma 8. □

4. Proof of Theorem 1

We want to find the critical point w_c for $h\colon W \to R$ subject to the restriction that $\mathrm{Vol}(G,w) = |E|$. By the Lagrange multiplier theorem this occurs where

$$Dh(w_c) = \beta D\big(\mathrm{Vol}(G,w_c)\big) \tag{12}$$

for some nonzero $\beta \in \mathbb{R}$. We immediately see that $D\big(\mathrm{Vol}(G,w)\big) = (1,1,\ldots,1)$, for any $w \in W$. Therefore, by comparing (12) with Lemma 5 we deduce that the matrix P is doubly stochastic, i.e.,

$$\sum_{f:A(e,f)=1} P(e,f) = 1.$$

In particular, since $P(e,f)$ depends only on e we easily see that $P(e,f) = 1/k_j$, where $j = f(0) = e(1)$. By (6) we see that $\mathrm{e}^{-hw(e)} = (1/k_j)v_f/v_e$ for any f such that $A(e,f) = 1$. In particular, $v(j) := v_f$ depends only on j. Thus

$$hw(e) = \log k_j + \log\big(v(i)/v(j)\big). \tag{13}$$

We can use (13) with edge $\bar{e} = (j,i)$ (i.e., the same edge with different orientations) and the fact that $w(e) = w(\bar{e})$ to write

$$hw(e) = \log k_j + \log\big(v(j)/v(i)\big). \tag{14}$$

Comparing (13) and (14) gives that $\log(k_i/k_j) = 2\log\big(v(j)/v(i)\big)$. Finally, Substituting this identity back into (13) gives that

$$hw(e) = \tfrac{1}{2}\log(k_i k_j)$$

up to normalization. This completes the proof of Theorem 1. □

5. Proof of Theorem 2

This is a modification of the proof of Theorem 1. We want to find the critical point w^g_c for $P(g,\cdot)\colon W \to \mathbb{R}$ subject to the restriction that $\mathrm{Vol}(G,w) = |E|$. By the Lagrange multiplier theorem this occurs where

$$DP(g,w^g_c) = \beta D\big(\mathrm{Vol}(G,w^g_c)\big) \tag{15}$$

for some non-zero β. We recall that $D(\mathrm{Vol}(G,w)) = (1,1,\ldots,1)$, for any $w \in W$.

We deduce again that the matrix P^g is doubly stochastic and that $P(e,f) = 1/k_j$. By (11) we see that $\mathrm{e}^{g(e)-Pw^g(e)} = (1/k_j)v_f/v_e$ for any f such that $A(e,f) = 1$. In particular, $v(j) := v_f$ depends only on $j = f(0)$ $(= e(1))$. Thus

$$g(i) - Pw^g(e) = \log k_j + \log\big(v(i)/v(j)\big). \tag{16}$$

We can use (16) with edge $\bar{e} = (j, i)$ (i.e., the same edge with different orientations) and the fact that $w^g(e) = w^g(\bar{e})$ to write

$$g(j) - Pw^g(e) = \log k_j + \log\big(v(j)/v(i)\big). \tag{17}$$

Comparing (16) and (17) gives that $\log(k_i/k_j) = 2\log\big(v(j)/v(i)\big)$. Finally, substituting this identity back into (16) gives that

$$w^g(e) = \tfrac{1}{2}\big(\log(k_i k_j) + g(i) + g(j)\big)$$

up to normalization. This completes the proof of Theorem 2.

6. Examples

In this section we illustrate our results by considering a number of standard examples.

The following lemma gives explicit expressions for $h(w_c)$ and $P(g, w_c)$.

Lemma 10. *We have that:*

$$h(w_c) = \frac{1}{|E|} \sum_{(i,j)\in E} \tfrac{1}{2}\log(k_i k_j) \tag{1}$$

and

$$P(g, w_c^g) = \frac{1}{|E|} \sum_{(i,j)\in E} \tfrac{1}{2}\big(\log(k_i k_j) + g(i) + g(j)\big), \tag{2}$$

providing (3) *holds.*

This lemma is easily deduce from the proofs in Sections 4 and 5 (or by elementary thermodynamic manipulations).

Example 1 (Regular graphs). There are a number of interesting examples of graphs which are $(k+1)$-regular and for which $w_c = w_u$ (by Corollary 1):

(1) The full graph in which every pair of vertices are connected by a pair of edges [2, p. 6]. In this case $h(w_c) = \log(|V| - 2)$.
(2) The n-cube with a vertex set which can be indexed by the vertices $v \in \{0,1\}^n$ of the n-dimensional hypercube, and edges corresponding to the edges of the cube [2, p. 6]. In this case, $k = n$ and $h(w_c) = \log(n-1)$.

Other examples of regular graphs are Platonic graphs, Paley graphs, Margulis graphs and Ramanujan graphs cf. [2, Section 6.3].

Example 2 (Bipartite graphs (cf. [2, p. 96])). For $n, m \geq 1$ we let the bipartite graph $K_{n,m}$ have vertices $V = \{1 \leq i \leq n+m\}$ with edges:

$$(i,j) \in E \iff \begin{cases} 1 \leq i \leq m \text{ and } m+1 \leq j \leq m+n, \text{ or} \\ 1 \leq j \leq m \text{ and } m+1 \leq i \leq m+n. \end{cases}$$

If $m \neq n$ then the graph is not regular. However, for every edge $(i,j) \in E$ we have that either $k_i + 1 = n$ and $k_j + 1 = m$ or $k_i + 1 = m$ and $k_j + 1 = n$. Therefore, for any $(i,j) \in E$ we see by Theorem 1 that the critical weighting is

$$w_c = |E|\left(\frac{\log\big((n-1)(m-1)\big)}{|E|\log\big((n-1)(m-1)\big)}\right) = 1.$$

Thus, we have a simple example of non-regular graphs for which we still have $w_c = w_u$. We can easily compute that $h(w_c) = \frac{1}{2}\log\big((n-1)(m-1)\big)$.

Example 3. Let G be a tetrahedron with vertices $V = \{v_1, \ldots, v_4\}$. We define $g_\epsilon : V \to \mathbb{R}$ by

$$g_\epsilon(i) = \begin{cases} -\log 4 + \epsilon & \text{if } i = 1, \\ 0 & \text{if } i \in \{2,3,4\} \end{cases}$$

for $\epsilon \geq 0$ and let $w_c^{g_\epsilon}$ be the critical point $w_c^{g_\epsilon}$ for $\epsilon > 0$. As $\epsilon \to 0$ we have that the lengths of the edges with vertex 1 shrink to zero.

Example 4 (The Grötzsch graph). This is a standard example of a graph with 11 vertices and 20 undirected edges. Amongst the edges (i, j) the value $k_i k_j = 2.3 = 6$ occurs 10 times; the value $k_i k_j = 2.4 = 8$ occurs 5 times and $k_i k_j = 3.3 = 9$ occurs 5 times. We can compute the edge weightings for w_c to be

$$w_c(i,j) = \begin{cases} 0.91181\ldots & \text{for 10 edges,} \\ 1.05822\ldots & \text{for 5 edges,} \\ 1.11815\ldots & \text{for 5 edges.} \end{cases}$$

We can compute $h(w_c) = 0.98252\ldots$.

References

1. G. Besson, G. Courtois, and S. Gallot, *Minimal entropy and Mostow's rigidity theorems*, Ergodic Theory Dynam. Systems **16** (1996), no. 4, 623–649.
2. F. R. K. Chung, *Spectral graph theory*, CBMS Regional Conf. Ser. in Math., vol. 92, Conf. Board Math. Sci., Washington, DC, 1997.
3. F. R. Gantmacher, *The theory of matrices*. Vol. 1, Chelsea, New York, 1977.
4. D. Jakobson and I. Rivin, *Extremal metrics on graphs*. I, Forum Math. **14** (2002), no. 1, 147–163.
5. I. Kapovich and T. Nagnibeda, *The Patterson – Sullivan embedding and minimal volume entropy for outer space*, Geom. Funct. Anal. **17** (2007), no. 4, 1201–1236.
6. S. Lim, *Minimal volume entropy for graphs*, Trans. Amer. Math. Soc. **360** (2008), no. 10, 5089–5100.
7. A. Manning, *Topological entropy for geodesic flows*, Ann. of Math. (2) **110** (1979), no. 3, 567–573.
8. W. Parry and M. Pollicott, *Zeta functions and the periodic orbit structure of hyperbolic dynamics*, Astérisque **187-188** (1990).
9. I. Rivin, *Growth in free groups (and other stories)*, available at `arXiv:math/9911076`.
10. D. Ruelle, *Thermodynamic formalism*, with a foreword by G. Gallavotti and G.-C. Rota, Encyclopedia Math. Appl., vol. 5, Addison-Wesley, Reading, MA, 1978.
11. ———, *The pressure of the geodesic flow on a negatively curved manifold*, Bol. Soc. Brasil. Mat. **12** (1981), no. 1, 95–99.

Department of Mathematics, Warwick University, Coventry CV4 7AL, UK

E-mail address: `masdbl@warwick.ac.uk`

Centre de Recherches Mathématiques
CRM Proceedings and Lecture Notes
Volume **52**, 2010

Nearest λ_q-Multiple Fractions

Dieter Mayer and Tobias Mühlenbruch

ABSTRACT. We discuss the nearest λ_q-multiple continued fractions and their duals for $\lambda_q = 2\cos(\pi/q)$ which are closely related to the Hecke triangle groups G_q, $q = 3, 4, \dots$. They have been introduced in the case $q = 3$ by Hurwitz and for even q by Nakada. These continued fractions are generated by interval maps f_q respectively $f_q^\star$ which are conjugate to subshifts over infinite alphabets. We generalize to arbitrary q a result of Hurwitz concerning the G_q- and f_q-equivalence of points on the real line. The natural extension of the maps f_q and $f_q^\star$ can be used as a Poincaré map for the geodesic flow on the Hecke surfaces $G_q \setminus \mathbb{H}$ and allows to construct the transfer operator for this flow.

1. Introduction

In the transfer operator approach to Selberg's zeta-function for Fuchsian groups G [12], [13] this function is expressed through the Fredholm-determinant of the generalized Perron – Frobenius operator $\mathcal{L}_\beta$ for the geodesic flow on the corresponding surface $G \setminus \mathbb{H}$ of constant negative curvature. This operator is constructed through an expanding interval map $f\colon I \to I$ closely related to a Poincaré map of the flow. In all the cases treated up to now this interval map generates some kind of continued fraction expansion like the Gauss expansion or its extensions such that the length spectrum of the flow can be completely characterized by the periodic orbits of f respectively the purely periodic continued fraction expansions. This program has been carried out in full detail for the modular surfaces $\Gamma \setminus \mathbb{H}$ defined by subgroups $\Gamma \subset \mathrm{PSL}(2, \mathbb{Z})$ of the full modular group.

For these groups the transfer operator has another rather important property: its eigenfunctions with eigenvalue 1 can be directly related to their automorphic forms, that is real analytic Eisenstein series and Maass wave forms respectively the holomorphic modular forms. This relation gave rise to the theory of period functions [2, 10] which generalize the Eichler – Manin – Shimura cohomology theory for holomorphic modular forms.

2010 *Mathematics Subject Classification.* Primary 11A55; Secondary 11J70, 30B70.

Key words and phrases. Continued fractions, Hecke triangle groups, discrete dynamical systems, natural extension, subshifts of infinite type, sofic systems.

The second named author was supported in part by the Deutsche Forschungsgemeinschaft through the DFG Research Project "Transfer operators and non arithmetic quantum chaos" (Ma 633/16-1).

This is the final form of the paper.

The physical interpretation of these relations between the transfer operator and the spectral properties of the Laplacian for these groups G is within the theory of quantum chaos [17,20]: the transfer operator encodes the classical length spectrum of the geodesic flow and relates these data to the quantum data, namely eigenvalues and eigenfunctions respectively resonances of its quantized system. In this sense this transfer operator approach extends the more common approach to quantum chaos via the Selberg–Gutzwiller trace formula [7, Theorem 13.8, p. 209; 11].

Obviously it is necessary to work out the transfer operator for more general Fuchsian groups, especially nonarithmetic ones, for which the Hecke triangle groups G_q are good examples, since up to the cases $q = 3, 4, 6$ all of them are indeed nonarithmetic. In [14] the authors constructed a symbolic dynamics for the geodesic flow on the Hecke surfaces $G_q \setminus \mathbb{H}$ for arbitrary q, the case $q = 3$ was treated earlier in [9]. In both cases the authors used the nearest λ_q-multiple continued fraction expansion, denoted for short by λ_q-CF, and its dual expansion. Another approach was discussed also in [21]. Some of the ergodic properties of these λ_q-CFs for q even have been worked out in [15] by H. Nakada. In the present paper we discuss the λ_q-CFs and their duals for arbitrary q via their generating interval maps f_q and $f_q^\star$, which allow us to derive also a transfer operator for the Hecke triangle groups G_q, whose Fredholm determinant is closely related to the Selberg function for the groups G_q as we will discuss in a forthcoming paper.

In [8] Hurwitz introduced nearest integer continued fraction expansions of the form

$$a_0 + \cfrac{-1}{a_1 + \cfrac{-1}{a_2 + \cfrac{-1}{a_3 + \cfrac{-1}{\cdots}}}} \tag{1.1}$$

where a_0 is an arbitrary integer and the a_i, $i \geq 1$, are integers satisfying $|a_i| \geq 2$ and $a_i a_{i+1} < 0$ if $|a_i| = 2$. They are generated by the interval map

$$f_3 \colon I_3 \to I_3; \quad x \mapsto \frac{-1}{x} - \left\langle \frac{-1}{x} \right\rangle,$$

where $I_3 = [-\frac{1}{2}, \frac{1}{2}]$ and $\langle x \rangle$ denotes the nearest integer to x, by the usual algorithm:

(0) $a_0 = \langle x \rangle$ and $x_1 := x - a_0$,
(1) $a_1 = \langle -1/x_1 \rangle$ and $x_2 := -1/x_1 - a_1 = f_3(x_1)$,
(i) $a_i = \langle -1/x_i \rangle$ and $x_{i+1} := -1/x_i - a_i = f_3(x_i)$,
($\star$) the algorithm terminates if $x_{i+1} = 0$.

Let $\mathrm{PSL}(2,\mathbb{Z}) = \mathrm{SL}(2,\mathbb{Z}) \bmod \{\pm\mathbf{1}\}$ denote the full modular group. Elements of the group can be identified with 2×2-matrices with integer entries and determinant 1, up to a common sign. The group acts on the projective real line $\mathbb{R} \cup \{\infty\}$ by Möbius transformations $\left(\begin{smallmatrix} a & b \\ c & d \end{smallmatrix}\right) z = (az+b)/(cz+d)$. The group $\mathrm{PSL}(2,\mathbb{Z})$ is generated by the elements S and T corresponding to the actions $z \mapsto -1/z$ and $z \mapsto z+1$. The generators satisfy the relations $S^2 = (ST)^3 = \mathbf{1}$. In particular, the elements T^a and ST^a correspond to the actions $z \mapsto z + a$ and $z \mapsto -1/(a+z)$. Hence we can write the continued fraction expansion in (1.1) in terms of a (formal) Möbius transformation as $T^{a_0} S T^{a_1} S T^{a_2} S T^{a_3} \cdots 0$.

Hurwitz found in [8], that equivalence of two points $x, y \in \mathbb{R}$ under the generating map f_3 and its extension to $\mathbb{R}$ is not the same as equivalence under the group action of $\mathrm{PSL}(2,\mathbb{Z})$. This is obviously in contrast with the case of the Gauss map $f_G\colon [0,1] \to [0,1]$ with $f_G(x) = 1/x \bmod 1$ and the modular group $\mathrm{PSL}(2,\mathbb{Z})$.

In [15] Nakada introduced for even integers $q \geq 4$ the nearest λ_q-multiple continued fractions with $\lambda_q = 2\cos(\pi/q)$, which we will denote by λ_q-CFs. They are similar to the Rosen continued fractions introduced in [16] and discussed in detail in [1]. The extension to the case $q \geq 3$ odd is straightforward, where $q = 3$ corresponds to the nearest integer continued fractions of Hurwitz. These λ_q-CFs and their dual expansions, introduced for $q = 3$ also by Hurwitz, can be generated by interval maps f_q and $f_q^\star$ closely related to the Hecke triangle groups G_q. Both maps are conjugate to subshifts over infinite alphabets, which when reduced to certain sofic systems, determine completely the properties of the corresponding λ_q-CF and its dual expansion. It turns out, that Hurwitz's result on equivalence of points on the real axis under f_3 and the group action of $\mathrm{PSL}(2,\mathbb{Z})$ is true for general $q \geq 3$: there exists for every $q \geq 3$ exactly one pair of points $(r_q, -r_q)$ which are equivalent under G_q but not under the map f_q. The natural extension F_q of the interval map $f_q\colon I_q \to I_q$ can be easily constructed from the symbolic dynamics of the maps f_q and $f_q^\star$ as sofic systems. It can be used to construct a Poincare section for the geodesic flow on the Hecke surface $G_q\backslash\mathbb{H}$, and hence also a transfer operator for the group G_q and its Selberg zeta function. The properties of this operator will be discussed elsewhere.

The structure of this article is as follows: In Section 2 we introduce the Hecke triangle groups and the λ_q-CFs respectively the dual λ_q-CFs. In Section 3 we discuss the interval maps f_q and $f_q^\star$ generating the nearest λ_q-multiple continued fractions and construct Markov partitions for these maps. In Section 4 we show that the maps f_q and $f_q^\star$ are conjugate to subshifts over infinite alphabets and introduce sofic systems closely related to the λ_q-CF and its dual. This allows a simple construction of the natural extension F_q of the map f_q. In Section 5 we relate the natural extension F_q to the geodesic flow on the Hecke surfaces $G_q\backslash\mathbb{H}$ and derive the transfer operator for this flow. The final Section 6 contains a discussion of the convergence properties of the λ_q-CFs by relating them to reduced Rosen λ-fractions as discussed in [16].

2. Nearest λ_q-multiple continued fractions

2.1. Hecke triangle groups. Hecke triangle groups are Fuchsian groups of the first kind, all except three are nonarithmetic. The reader may wish to consult [6, pp. 591, 627] for a discussion of Hecke triangle groups and their relation to Dirichlet series.

Denote by $\mathrm{PSL}(2,\mathbb{R})$ the projective special linear group given by

$$\mathrm{PSL}(2,\mathbb{R}) = \mathrm{SL}(2,\mathbb{R})/\{\pm\mathbf{1}\} \tag{2.1}$$

where $\pm\mathbf{1} = \left(\begin{smallmatrix} \pm 1 & 0 \\ 0 & \pm 1 \end{smallmatrix}\right)$. We denote by $\left[\begin{smallmatrix} a & b \\ c & d \end{smallmatrix}\right] = \left\{\left(\begin{smallmatrix} a & b \\ c & d \end{smallmatrix}\right), \left(\begin{smallmatrix} -a & -b \\ -c & -d \end{smallmatrix}\right)\right\}$ the elements of $\mathrm{PSL}(2,\mathbb{R})$, but identify often elements of $\mathrm{PSL}(2,\mathbb{R})$ and $\mathrm{SL}(2,\mathbb{R})$.

For a given integer $q \geq 3$ the qth *Hecke triangle group* G_q is generated by

$$S := \begin{bmatrix} 0 & -1 \\ 1 & 0 \end{bmatrix} \quad \text{and} \quad T_q := \begin{bmatrix} 1 & \lambda_q \\ 0 & 1 \end{bmatrix} \tag{2.2}$$

with relations

$$S^2 = (ST_q)^q = \mathbf{1}, \tag{2.3}$$

where λ_q is given by

$$\lambda_q := 2\cos\left(\frac{\pi}{q}\right) \tag{2.4}$$

Later on we also need the element

$$T'_q := \begin{bmatrix} 1 & 0 \\ \lambda_q & 1 \end{bmatrix} = ST_q^{-1}S \in G_q. \tag{2.5}$$

We may suppress the q-dependence in our notation when we work with a fixed value of q.

The Hecke triangle group G_q is a discrete subgroup of $\mathrm{PSL}(2,\mathbb{R})$ and its limit set is the projective line $\mathbb{P}^1_{\mathbb{R}} = \mathbb{R} \cup \{\infty\}$. It acts on the upper half-plane, the lower half-plane and on $\mathbb{P}^1_{\mathbb{R}}$ by *Möbius transformations*

$$G_q \times \mathbb{P}^1_{\mathbb{R}} \to \mathbb{P}^1_{\mathbb{R}}; \quad \left(\begin{bmatrix} a & b \\ c & d \end{bmatrix}, x\right) \mapsto \begin{bmatrix} a & b \\ c & d \end{bmatrix} x := \begin{cases} \dfrac{ax+b}{cx+d} & \text{if } x \in \mathbb{R}, \\ \dfrac{a}{c} & \text{if } x = \infty. \end{cases} \tag{2.6}$$

The points $x, y \in \mathbb{P}^1_{\mathbb{R}}$ are *G_q-equivalent* denoted by $x \sim_{G_q} y$, if there exists an element $g \in G_q$ such that $y = g\,x$. Obviously, this is an equivalence relation.

2.2. Nearest λ_q-multiple continued fractions and their duals. Consider finite or infinite sequences $(a_i)_i$. We denote periodic parts of the sequences by overlining the period part and finitely repeated patterns are denoted by a power where a 0th power vanishes:

$$(a_1, \overline{a_2, a_3}) = (a_1, a_2, a_3, a_2, a_3,\ a_2, a_3, \dots),$$
$$(a_1, (a_2, a_3)^i, a_4, \dots) = (a_1, \underbrace{a_2, a_3, a_2, a_3, \dots, a_2, a_3}_{i \text{ times } a_2,\, a_3}, a_4, \dots),$$
$$(a_1, (a_2)^0, a_3, \dots) = (a_1, a_3, \dots).$$

We use also $-(a_1, \dots) = (-a_1, \dots)$.

Put

$$h_q := \begin{cases} \dfrac{q-2}{2} & \text{for } q \text{ even}, \\ \dfrac{q-3}{2} & \text{for } q \text{ odd}. \end{cases} \tag{2.7}$$

We define the set $\mathcal{B}_q$ of *forbidden blocks* as

(2.8)
$$\mathcal{B}_q := \begin{cases} \{(\pm 1)\} \cup \bigcup_{m=1}^{\infty} \{(\pm 2, \pm m)\} & \text{for } q = 3, \\ \{((\pm 1)^{h_q+1})\} \cup \bigcup_{m=1}^{\infty} \{((\pm 1)^{h}_{q}, \pm m)\} & \text{for } q \text{ even}, \\ \{((\pm 1)^{h_q+1})\} \cup \bigcup_{m=1}^{\infty} \{((\pm 1)^{h_q}, \pm 2, (\pm 1)^{h_q}, \pm m)\} & \text{for } q \text{ odd}, q \geq 5. \end{cases}$$

The choice of the sign is the same within each block. For example $(2,3), (-2,-3) \in \mathcal{B}$ and $(2,-3) \notin \mathcal{B}$ for $q = 3$.

We call a sequence $(a_1, a_2, a_3, \dots)$ *q-regular* if $(a_l, a_{l+1}, \dots, a_L) \notin \mathcal{B}_q$ for all $1 \leq l < L$ and *dual q-regular* if $(a_L, a_{L-1}, \dots, a_l) \notin \mathcal{B}_q$ for all $1 \leq l < L$. Denote by

$\mathcal{A}_q^{\text{reg}}$ respectively by $\mathcal{A}_q^{\text{dreg}}$ the set of infinite q-regular respectively dual q-regular sequences $(a_i)_{i\in\mathbb{N}}$.

A *nearest λ_q-multiple continued fraction*, or λ_q-CF, is a formal expansion of the type

$$(2.9)\qquad [a_0; a_1, a_2, a_3, \dots] := a_0\lambda_q + \cfrac{-1}{a_1\lambda_q + \cfrac{-1}{a_2\lambda_q + \cfrac{-1}{a_3\lambda_q + \dots}}}$$

with $a_i \in \mathbb{Z}_{\neq 0}$, $i \geq 1$ and $a_0 \in \mathbb{Z}$.

We say that $[a_0; a_1, a_2, a_3, \dots]$ *converges* if either $[a_0; a_1, a_2, a_3, \dots]$ has finite length or $\lim_{L\to\infty}[a_0; a_1, a_2, a_3, \dots, a_L]$ exists in $\mathbb{R}$.

We adopt the same notations as introduced for sequences earlier. For example we write $[a_0; a_1, \overline{a_2, a_3}]$ for a periodic tail of the expansion and $-[a_0; a_1, \dots]$ for $[-a_0; -a_1, \dots]$.

A λ_q-CF is *regular* respectively *dual regular* if the sequence $(a_1, a_2, a_3, \dots)$ is q-regular respectively dual q-regular. Regular and dual regular λ_q-CFs are denoted by $[\![a_0; a_1, \dots]\!]$ respectively $[\![a_0; a_1, \dots]\!]^\star$.

Proposition 2.1. *Regular and dual regular λ_q-CFs converge.*

PROOF. This follows immediately from Lemmas 4 and 34 in [14]. □

An alternative proof of Proposition 2.1 for infinite regular and dual regular expansions with leading 0 follows also from the lemmas in the Sections 4.2 and 4.4.

Converging λ_q-CFs can be rewritten in terms of elements of the Hecke triangle group G_q: if the expansion (2.9) is finite it can be written as follows

$$(2.10)\qquad \begin{aligned}[a_0; a_1, a_2, a_3, \dots, a_L] &= a_0\lambda_q + \cfrac{-1}{a_1\lambda_q + \cfrac{-1}{a_2\lambda_q + \cfrac{-1}{a_3\lambda_q + \dots \cfrac{-1}{a_L\lambda_q}}}} \\ &= T^{a_0}ST^{a_1}ST^{a_2}ST^{a_3}\cdots ST^{a_L}\,0,\end{aligned}$$

since $-1/(a\lambda_q + x) = ST^a\,x$ For infinite converging λ_q-CF the expansion has to be interpreted as

$$\begin{aligned}[a_0; a_1, a_2, a_3, \dots] &= \lim_{L\to\infty}[a_0; a_1, a_2, a_3, \dots, a_L] \\ &= \lim_{L\to\infty} T^{a_0}ST^{a_1}ST^{a_2}ST^{a_3}\cdots ST^{a_L}\,0 \\ &= T^{a_0}ST^{a_1}ST^{a_2}ST^{a_3}\cdots\,0.\end{aligned}$$

An immediate consequence of this is

Lemma 2.2. *For a finite regular λ_q-CF one finds for q even*

$$[\![a_0; a_1, \dots, a_n, (1)^h]\!] = [\![a_0; a_1, \dots, a_n - 1, (-1)^h]\!]$$

respectively for q odd

$$[\![a_0; \dots, a_n, (1)^h, 2, (1)^h]\!] = [\![a_0; \dots, a_n - 1, (-1)^h, -2, (-1)^h]\!].$$

PROOF. Assume the left-hand side to be regular. This implies $a_n \neq 1$ and hence the right-hand side is regular, too. Conversely, assume the right-hand side to be regular and hence $a_n - 1 \neq -1$. Therefore the expansions on the left-hand side are regular.

The identity now follows by writing λ_q-CFs in terms of Möbius transformations and using the identity $(ST)^{h_q}\,0 = T^{-1}(ST^{-1})^{h_q}ST^{-1}S\,0 = T^{-1}(ST^{-1})^{h_q}\,0$ since 0 is a fixed point of $ST^{-1}S = T'$. □

Remark 2.3. For $q = 3$ the nearest λ_q-multiple continued fractions are in fact the well-known nearest integer fractions extensively discussed by Hurwitz in [8]. In particular, Theorem 2.10 for $q = 3$ was proved by him there. We include his results for the sake of completeness and to show how this special case $q = 3$ fits well into the discussion of the case of odd $q \geq 5$. See also Remark 2.6.

Remark 2.4. For $q \geq 4$ the regular λ_q-CFs correspond to Rosen's λ_q-fractions introduced in [16] and discussed in [1]. We will discuss this relation in more detail in Section 5.1.

2.3. Special values and their expansions. The following results are shown in [14]:

The point $x = \mp\lambda_q/2$ has the regular λ_q-CF

$$\mp\frac{\lambda_q}{2} = \begin{cases} [\![0; (\pm 1)^{h_q}]\!] & \text{for even } q, \\ [\![0; (\pm 1)^{h_q}, \pm 2, (\pm 1)^{h_q}]\!] & \text{for odd } q. \end{cases} \tag{2.11}$$

Put

$$R_q := \lambda_q + r_q \quad \text{with } r_q := \begin{cases} [\![0; \overline{3}]\!] & \text{for } q = 3, \\ [\![0; \overline{(1)^{h_q-1}, 2}]\!] & \text{for } q \text{ even}, \\ [\![0; \overline{(1)^{h_q}, 2, (1)^{h_q-1}, 2}]\!] & \text{for } q \text{ odd}, q \geq 5. \end{cases} \tag{2.12}$$

whose expansion is periodic of length κ_q with

$$\kappa_q := \begin{cases} h_q = \dfrac{q-2}{2} & \text{for even } q, \\ 2h_q + 1 = q - 2 & \text{for odd } q. \end{cases} \tag{2.13}$$

The regular respectively dual regular λ_q-CF of the point $x = R_q$ has the form

$$R_q = \begin{cases} [\![1; \overline{(1)^{h_q-1}, 2}]\!] & \text{for even } q, \\ [\![1; \overline{(1)^{h_q}, 2, (1)^{h_q-1}, 2}]\!] & \text{for odd } q \geq 5, \\ [\![1; \overline{3}]\!] & \text{for } q = 3. \end{cases} \tag{2.14}$$

$$= \begin{cases} [\![0; (-1)^h, \overline{-2, (-1)^{h_q-1}}]\!]^\star & \text{for even } q, \\ [\![0; (-1)^h, \overline{-2, (-1)^{h_q}, -2, (-1)^{h_q-1}}]\!]^\star & \text{for odd } q \geq 5, \\ [\![0; -2, \overline{-3}]\!]^\star & \text{for } q = 3. \end{cases} \tag{2.15}$$

Moreover,

$$R_q = 1 \quad \text{and} \quad -R_q = SR_q \qquad \text{for even } q, \tag{2.16}$$

$$R_q^2 + (2 - \lambda_q)R_q = 1 \quad \text{and} \quad -R_q = (TS)^{h_q+1}R_q \qquad \text{for odd } q, \tag{2.17}$$

and R_q satisfies the inequality

$$\frac{\lambda_q}{2} < R_q \leq 1. \tag{2.18}$$

Remark 2.5. For R_3 one finds

$$1 + R_3 = \frac{1+\sqrt{5}}{2}.$$

Remark 2.6. The form of the λ_q-CF of r_3 in (2.12) can be obtained from the expansions for q odd, $q \geq 5$ by interpreting it as a Möbius transformation with $(1)^{-1}$ as ST^{-1}:

$$\begin{aligned} r_3 &= [\![0; \overline{1^{h_3}, 2, (1)^{h_3-1}, 2}]\!] = [\![0; \overline{2, (1)^{-1}, 2}]\!] \\ &= ST^2ST^{-1}ST^2 \cdot ST^2ST^{-1}ST^2 \cdots 0 \\ &= ST^2TSTSST^2 \cdot ST^2TSTSST^2 \cdots 0 \\ &= ST^3ST^3 \cdot ST^3ST^3 \cdots 0 = [\![0; \overline{3}]\!]. \end{aligned}$$

2.4. A lexiographic order. Let $x, y \in I_{R_q} := [-R_q, R_q]$ have the regular λ_q-CFs $x = [\![a_0; a_1, \ldots]\!]$ and $y = [\![b_0; b_1, \ldots]\!]$. Denote by $l(x) \leq \infty$ respectively $l(y) \leq \infty$ the number of entries in the above λ_q-CFs. We introduce a *lexiographic order* "$\prec$" for λ_q-CFs in the following way: For $n \in \mathbb{Z}_{\geq 0}$ being the number of identical digits at the head of the λ_q-CFs, i.e., $a_i = b_i$ for all $0 \leq i \leq n$ and $l(x), l(y) \geq n$, we define
(2.19)

$$x \prec y :\Longleftrightarrow \begin{cases} a_0 < b_0 & \text{if } n = 0, \\ a_n > 0 > b_n & \text{if } n > 0, \text{ both } l(x), l(y) \geq n+1 \text{ and } a_n b_n < 0, \\ a_n < b_n & \text{if } n > 0, \text{ both } l(x), l(y) \geq n+1 \text{ and } a_n b_n > 0, \\ b_n < 0 & \text{if } n > 0 \text{ and } l(x) = n \text{ or} \\ a_n > 0 & \text{if } n > 0 \text{ and } l(y) = n. \end{cases}$$

We also write $x \preceq y$ for $x \prec y$ or $x = y$.

This is indeed an order on regular λ_q-CFs, since Lemmas 22 and 23 in [14] imply:

Lemma 2.7. *Let x and y have regular λ_q-CFs. Then $x \prec y$ if and only if $x < y$.*

The authors of [14] introduce a process called "rewriting" of λ_q-CFs where forbidden blocks in the λ_q-CF are replaced by allowed ones without changing its value. The rules for "rewriting" are based on the interpretation of a λ_q-CF in terms of Möbius transformations given by group elements of the Hecke group, see (2.10), and applying the group relations (2.3). We refer in particular to Lemma 11 and Lemma 13 in [14] for the details. A simple example for this rewriting is used in the proof of Lemma 2.2.

It follows from the proof of Lemma 34 in [14] that every dual regular λ_q-CF can be rewritten into a regular λ_q-CF.

Lemma 2.8. *The lexiographic order $\prec$ in (2.19) can be extended to dual regular λ_q-CFs with leading digit 0. Rewriting two dual regular λ_q-CFs satisfying $[\![0; a_1, \ldots]\!]^\star \prec [\![0; b_1, \ldots]\!]^\star$ into regular λ_q-CFs does not change their order.*

Remark 2.9. The lexiographic order "$\prec$" however cannot be defined for all dual regular λ-CFs with arbitrary leading coefficient as the following example shows: consider the dual regular λ_3-CFs of R_3 in (2.14) and (2.15). Obviously $R_3 =$

$[\![0;-2,\overline{-3}]\!]^\star = [\![1;\overline{3}]\!]^\star$. Extending naively "$\prec$" in (2.19) to this case would lead to $[\![0;-2,\overline{-3}]\!]^\star \prec [\![1;\overline{3}]\!]^\star$ and hence $[\![0;-2,\overline{-3}]\!]^\star \prec [\![0;-2,\overline{-3}]\!]^\star$.

Proof of Lemma 2.8. The λ_q-CFs $[\![0;a_1,\dots]\!]^\star$ and $[\![0;b_1,\dots]\!]^\star$ are dual regular. No rewriting is necessary if both are also regular.

Assume $[\![0;b_1,b_2,\dots]\!]^\star$ starts with a forbidden block. If it is of the form

$$[\![0;b_1,b_2,\dots]\!]^\star = \begin{cases} [\![0;(1)^{h_q},m]\!]^\star & \text{for even } q, \\ [\![0;(1)^{h_q},2,(1)^{h_q},m]\!]^\star & \text{for odd } q \ge 5, \\ [\![0;2,m]\!]^\star & \text{for } q=3, \end{cases}$$

with $m \ge 2$ for $q \ge 4$ respectively $m \ge 3$ for $q=3$, then $[\![0;a_1,a_2,\dots]\!]^\star$ must be of the form

$$[\![0;a_1,a_2,\dots]\!]^\star = \begin{cases} [\![0;(1)^{h_q},n]\!]^\star & \text{for even } q, \\ [\![0;(1)^{h_q},2,(1)^{h_q},n]\!]^\star & \text{for odd } q \ge 5, \\ [\![0;2,n]\!]^\star & \text{for } q=3, \end{cases}$$

with $n < m$ and $n \ge 2$ for $q \ge 4$ respectively $n \ge 3$ for $q=3$. Using the rewriting rules in Lemmas 11 and 13 in [14] we find

$$\begin{aligned} [\![0;a_1,a_2,\dots]\!]^\star &\to [\![\tilde{a}_0;\tilde{a}_1,\tilde{a}_2,\dots]\!] \\ &:= \begin{cases} [\![-1;(-1)^{h_q},n-1,\dots]\!] & \text{for even } q, \\ [\![-1;(-1)^{h_q},-2,(-1)^{h_q},n-1,\dots]\!] & \text{for odd } q \ge 5, \\ [\![-1;-2,n-1,\dots]\!] & \text{for } q=3, \end{cases} \end{aligned}$$

respectively

$$\begin{aligned} [\![0;b_1,b_2,\dots]\!]^\star &\to [\![\tilde{b}_0;\tilde{b}_1,\tilde{b}_2,\dots]\!] \\ &:= \begin{cases} [\![-1;(-1)^{h_q},m-1,\dots]\!] & \text{for even } q, \\ [\![-1;(-1)^{h_q},-2,(-1)^{h_q},m-1,\dots]\!] & \text{for odd } q \ge 5, \\ [\![-1;-2,m-1,\dots]\!] & \text{for } q=3, \end{cases} \end{aligned}$$

hence by (2.19) $[\![-1;\tilde{a}_1,\tilde{a}_2,\dots]\!] \prec [\![-1;\tilde{b}_1,\tilde{b}_2,\dots]\!]$.

If $[\![0;b_1,b_2,b_3,\dots]\!]^\star$ is of the form

$$[\![0;b_1,b_2,\dots]\!]^\star = \begin{cases} [\![0;(-1)^{h_q},-m,\dots]\!]^\star & \text{for even } q, \\ [\![0;(-1)^{h_q},-2,(-1)^{h_q},-m,\dots]\!]^\star & \text{for odd } q \ge 5, \\ [\![0;(-2,-m),\dots]\!]^\star & \text{for } q=3, \end{cases} \tag{2.20}$$

with $m \ge 2$ for $q \ge 4$ respectively $m \ge 3$ for $q=3$, and $[\![0;a_1,a_2,\dots]\!]^\star$ does not contain a forbidden block starting with a_1, the rewriting rules in [14] give

$$\begin{aligned} [\![0;b_1,b_2,\dots]\!]^\star &\to [\![\tilde{b}_0;\tilde{b}_1,\tilde{b}_2,\dots]\!] \\ &= \begin{cases} [\![1;(1)^{h_q},1-m,\dots]\!] & \text{for even } q, \\ [\![1;(1)^{h_q},2,(1)^{h_q},1-m,\dots]\!] & \text{for odd } q \ge 5, \\ [\![1;2,1-m,\dots]\!] & \text{for } q=3. \end{cases} \end{aligned}$$

Therefore (2.19) implies $[\![0;a_1,a_2,\dots]\!]^\star \prec [\![1;\tilde{b}_1,\tilde{b}_2,\dots]\!]$.

If $[\![0; b_1, b_2, b_3, \dots]\!]^\star$ is of the form (2.20) and $[\![0; a_1, a_2, \dots]\!]^\star$ is of the form

$$[\![0; a_1, a_2, \dots]\!]^\star = \begin{cases} [\![0; (-1)^{h_q}, -n]\!]^\star & \text{for even } q, \\ [\![0; (-1)^{h_q}, 2, (1)^{h_q}, -n]\!]^\star & \text{for odd } q \geq 5, \\ [\![0; 2, -n]\!]^\star & \text{for } q = 3, \end{cases}$$

with $n > m$ then the rewriting rules in [14] lead to

$$[\![1; \tilde{a}_1, \tilde{a}_2, \dots]\!] = \begin{cases} [\![1; (1)^{h_q}, 1-n, \dots]\!] & \text{for even } q, \\ [\![1; (1)^{h_q}, 2, (1)^{h_q}, 1-n, \dots]\!] & \text{for odd } q \geq 5, \\ [\![1; 2, 1-n, \dots]\!] & \text{for } q = 3, \end{cases}$$

$$[\![1; \tilde{b}_1, \tilde{b}_2, \dots]\!] = \begin{cases} [\![1; (1)^{h_q}, 1-m, \dots]\!] & \text{for even } q, \\ [\![1; (1)^{h_q}, 2, (1)^{h_q}, 1-m, \dots]\!] & \text{for odd } q \geq 5, \\ [\![1; 2, 1-m, \dots]\!] & \text{for } q = 3, \end{cases}$$

and hence $[\![1; \tilde{a}_1, \tilde{a}_2, \dots]\!] \prec [\![1; \tilde{b}_1, \tilde{b}_2, \dots]\!]$.

Completely analogous are the cases when $[\![0; a_1, \dots]\!]^\star$ starts with a forbidden block or the first forbidden block starts at a_n, $n > 1$ and $[\![0; b_1, b_2, \dots]\!]^\star$ is a regular λ_q-CF. If both $[\![0; a_1, \dots]\!]^\star$ and $[\![0; b_1, \dots]\!]^\star$ have the same forbidden block starting at a_1 respectively b_1 then both dual regular λ_q-CFs are rewritten in the same way and the forbidden block does not influence the order "$\prec$." □

2.5. Equivalence relations and continued fractions. Let $x, y \in \mathbb{R}$ have infinite regular λ_q-CFs $x = [\![a_0; a_1, \dots]\!]$ and $y = [\![b_0; b_1, \dots]\!]$. We say that x and y are *regular λ_q-CF-equivalent*, denoted by $x \sim_{\text{reg}} y$, if the regular λ_q-CFs of x and y have the same tail, i.e., there exists $m, n \in \mathbb{N}$ such that the sequences $(a_m, a_{m+1}, \dots)$ and $(b_n, b_{n+1}, \dots)$ coincide. Obviously, this is an equivalence relation. We can extend this equivalence relation to all regular λ_q-CFs by declaring all finite regular λ_q-CFs to be regular λ_q-CF-equivalent.

Theorem 2.10 (Equivalence relations). *For $x, y \in \mathbb{R}$ the following properties are equivalent*:

(1) $x \sim_{G_q} y$.
(2) *x and y satisfy*:
- $x \sim_{\text{reg}} y$ *or*
- $x \sim_{\text{reg}} \pm r$ *and* $y \sim_{\text{reg}} \mp r$.

To prove the proposition, we need the following lemmas:

Lemma 2.11. *If x has an infinite regular λ_q-CF and $g \in G_q$ satisfies $gx \in \mathbb{R}$, then gx has an infinite λ_q-CF with at most h_q consecutive digits ± 1. Its tail coincides with the tail of the regular λ_q-CF of x.*

Proof. Let x have the regular λ_q-CF $x = [\![a_0; a_1, \dots]\!]$. We can write g as a word in the generators S and T_q: $g = T_q^{b_0} S T_q^{b_1} S T_q^{b_2} \cdots S T_q^{b_m} S^\delta$ with $b_0 \in \mathbb{Z}$, $b_i \in \mathbb{Z}_{\neq 0}$, $i = 1, \dots, m$, and $\delta \in \{0, 1\}$. Then gx can formally be written as $gx = T^{b_0} S T^{b_1} S T^{b_2} \cdots S T^{b_m} S^\delta T^{a_0} S T^{a_1} S T^{a_2} \cdots 0$.

Consider for $n > m$ sufficiently large the element $g_n \in G_q$ given by

$$g_n := T_q^{b_0} S T_q^{b_1} S T_q^{b_2} \cdots S T_q^{b_m} S^\delta T_q^{a_0} S T_q^{a_1} S T_q^{a_2} \cdots S T_q^{a_n} \in G_q. \tag{2.21}$$

TABLE 1. Under assumptions of Lemma 2.12 we list all possibilities where the rewriting of a forbidden block generates a new forbidden block. The forbidden blocks are underlined.

even q: $(m \geq 1)$	
$\pm[a_{i-1}, \underline{1^{h_q+1}}, 2, 1^{h_q-1}, m]$	$\to \pm[a_{i-1}-1, (-1)^{h_q-1}, \underline{(1)^{h_q}, m}]$
$\pm[a_{i-1}, \underline{1^{h_q}, 2}, 1^{h_q-1}, m]$	$\to \pm[a_{i-1}-1, (-1)^{h_q}, \underline{1^{h_q}, m}]$
odd $q \geq 5$: $(m \geq 2)$	
$\pm[a_{i-1}, \underline{1^{h_q+1}}, 2, 1^{h_q}]$	$\to \pm[a_{i-1}-1, (-1)^{h_q}, \underline{1^{h_q+1}}]$
$\pm[a_{i-1}, \underline{1^{h_q+1}}, 2, 1^{h_q-1}, 2, 1^{h_q}, m]$	$\to \pm[a_{i-1}-1, (-1)^{h_q}, \underline{1^{h_q}, 2, 1^{h_q}, m}]$
$\pm[a_{i-1}, \underline{1^{h_q}, 2, 1^{h_q}, 2}, 1^{h_q}]$	$\to \pm[a_{i-1}-1, (-1)^{h_q}, -2, (-1)^{h_q}, \underline{1^{h_q+1}}]$
$\pm[a_{i-1}, \underline{1^{h_q}, 2, 1^{h_q}, 2}, 1^{h_q-1}, 2, 1^{h_q}, m]$	$\to \pm[a_{i-1}-1, (-1)^{h_q}, -2, (-1)^{h_q}, \underline{1^{h_q}, 2, 1^{h_q}, m}]$
$q = 3$: $(m \geq 3$ and $n \geq 0)$	
$\pm[a_{i-1}, \underline{2, 3}, m]$	$\to \pm[a_{i-1}-1, -2, \underline{2, m}]$
$\pm[a_{i-1}, \underline{2, 2}, 2^n, 3, m]$	$\to \pm[a_{i-1}-1, -(3+n), \underline{2, m}]$

The identities $S^2 = \mathbf{1}$, $(ST_q^{\pm 1})^q = \mathbf{1}$, $T_q^a (ST_q^{\pm 1})^{q-1} ST_q^b = T_q^{a+b\mp 1}$ and

$$T_q^a (ST_q^{\pm 1} ST_q^b)^l = T_q^{a\mp 1} (ST_q^{\mp 1})^{q-l-2} ST_q^{b\mp 1} \qquad \text{for } h_q + 1 \leq l \leq q-2$$

and a, b arbitrary follow from (2.3). But $q - l - 2 \leq q - (h_q + 1) - 2 = h_q - 1$ for q even and $q - l - 2 \leq h_q$ for q odd. We apply these identities recursively on g_n in (2.21). After a finite number of steps one arrives at a word representing g_n which contains blocks of at most h_q consecutive digits ± 1. Indeed, since each application of one of these identities reduces the length of the word, the process of applying the identities has to stop after a finite number of steps. And, since the λ_q-CF of x is reduced, there are no blocks of more than h_q consecutive ± 1 to the right of the right of part "$ST^{b_m} S^\delta T^{a_0} ST^{a_1}$" of the word g_n in (2.21).

Hence $g\,x$ can be written as a λ_q-CF of the form (2.10) without blocks of more than h_q consecutive digits ± 1 and with a tail identical to the regular tail in the λ_q-CF of x. □

Lemma 2.12. *Let $[a_0; a_1, a_2, \ldots]$ be an infinite λ_q-CF containing no blocks of more than $h_q + 1$ consecutive digits ± 1 and at most one block of $h_q + 1$ consecutive digits ± 1 for $q \geq 4$ respectively no forbidden digits ± 1 for $q = 3$. If the block $[(\pm 1)^{h_q+1}]$ exists the block has to be the first forbidden block of the λ_q-CF and has to be preceded by a digit of alternate sign.*

If the first forbidden block starts at a_i, $i \geq 1$ and its rewriting leads to a new forbidden block then this forbidden block and its rewritten version must have the form given in Table 1. The new forbidden block will appear to the right of a_i. If the new forbidden block is of the form $[(\pm 1)^{h_q+1}]$ then its preceding digit is negative.

PROOF. Without loss of generality assume that the forbidden block starting at a_i has positive digits and hence $a_{i-1} \neq 1$. For q even the forbidden block must have the form $[1^{h_q+1}]$ or $[1^{h_q}, m]$ with $m \geq 2$. The rewriting rules in Lemma 11 of [14] lead to

$$[\ldots, a_{i-1}, 1^{h_q+1}, a_{i+h_q+1}, \ldots] \to [\ldots, a_{i-1} - 1, (-1)^{h_q-1}, a_{i+h_q+1} - 1, \ldots],$$

$$[\dots, a_{i-1}, 1^{h_q}, m, a_{i+h_q+1}, \dots] \to [\dots, a_{i-1} - 1, (-1)^{h_q}, m-1, a_{i+h_q+1}, \dots]$$

for $m \geq 2$. Changing a_{i-1} to $a_{i-1} - 1$ cannot introduce a forbidden block since $a_{i-1} \neq 1$ and a digit $a_{i-1} \geq 2$ cannot follow a block of the from $[1^{h_q}]$. Hence any new forbidden block has to start with the digit $a_{i+h_q+1} - 1$ respectively $m-1$. Two cases are possible: $[a_{i+h_q+1}, \dots] = [2, 1^{h_q-1}, l, \dots]$ respectively $[m, \dots] = [2, 1^{h_q-1}, l, \dots]$ with $l \geq 1$. This shows that the block $[a_i, \dots]$ must have the form $[1^{h_q+1}, 2, 1^{h_q-1}, l]$ respectively $[1^{h_q}, 2, 1^{h_q-1}, l]$ with its rewriting leading to the form as stated in the lemma.

For q odd, $q \geq 5$, the forbidden block $[a_i, \dots]$ has either the form $[1^{h_q}, 2, 1^{h_q}, m]$ with $m \geq 2$ or the form $[1^{h_q+1}]$. Rewriting rules in Lemma 13 of [14] then give

$$[\dots, a_{i-1}, 1^{h_q+1}, a_{i+h_q+1}, \dots] \to [\dots, a_{i-1} - 1, (-1)^{h_q}, a_{i+h_q+1} - 1, \dots]$$
$$[\dots, a_{i-1}, 1^{h_q}, 2, 1^{h_q}, m, \dots] \to [\dots, a_{i-1} - 1, (-1)^{h_q}, -2, (-1)^{h_q}, m-1, \dots]$$

for $m \geq 2$, and similar arguments as for q even show that the forbidden block and the digits following it are either of the form $[1^{h_q+1}]$ followed by $[2, 1^{h_q}]$ or $[2, 1^{h_q-1}, 2, 1^{h_q}, l]$, $l \geq 2$, or $[1^{h_q}, 2, 1^{h_2}, 2]$ followed by $[1^{h_q}]$ or $[1^{h_q-1}, 2, 1^{h_q}, l]$, $l \geq 2$ respectively. The rewritten form is the as given in the lemma. For $a_{i-1} = 2$ rewriting cannot lead to a new forbidden block to the left of a_i contradicting otherwise the first forbidden block to start with a_i.

The case $q = 3$ with forbidden block $[2, 2^n, m]$, $m \geq 2$ and $n \in \mathbb{Z}_{\geq 0}$ can be handled in complete analogy by using the rewriting rule

$$[a, 2, 2^n, b] \to [a-1, -2-n, b-1]$$

with $a, b \neq 2$. □

PROOF OF PROPOSITION 2.10. We show first the implication (2) $\Rightarrow$ (1). If $x \sim_{\mathrm{reg}} y$ then x and y have regular λ_q-CFs with the same tail:

$$x = [\![a_0; a_1, \dots, a_m, a_{m+1}, \dots]\!] \quad \text{and} \quad y = [\![b_0; b_1, \dots, b_n, a_{m+1}, \dots]\!].$$

Put $g := T^{a_0} S T^{a_1} \cdots S T^{a_m} (T^{b_0} S T^{b_1} \cdots S T^{b_n})^{-1} \in G_q$. Writing x and y in terms of Möbius transformations as explained in (2.10) we find,

$$\begin{aligned} g\, y &= g\, T^{b_0} S T^{b_1} \cdots S T^{b_n} S T^{a_{m+1}} \cdots 0 \\ &= T^{a_0} S T^{a_1} \cdots S T^{a_m} S T^{a_{m+1}} \cdots 0 = x \end{aligned}$$

and hence x and y are G_q-equivalent.

Assume next $x \sim_{\mathrm{reg}} r$ and $y \sim_{\mathrm{reg}} -r$ and hence $x \sim_{G_q} r$ and $y \sim_{G_q} -r$. Since $R_q = T_q r_q$ according to (2.12) and $-R_q = S R_q$ according to (2.16) for even q respectively $-R_q = (T_q)^{h_q+1} R_q$ according to (2.17) for odd q obviously $r \sim_{G_q} -r$ and hence $x \sim_{G_q} y$.

To show implication (1) $\Rightarrow$ (2), assume there exists $g \in G_q$ with $g\, x = y$ with x and y having infinite regular λ_q-CFs. Lemma 2.11 shows that $g\, x$ can be written as an infinite λ_q-CF with regular tail satisfying the assumptions of Lemma 2.12. Using the rewriting rules in Lemmas 11 and 13 of [14] we can recursively rewrite the λ_q-CF of $g\, x$ into a regular λ_q-CF from the left to the right. We proceed to the next forbidden block if rewriting does not lead to a new forbidden block. Lemma 2.12 implies that a new forbidden block can only appear to the right of the original one which we process next. If this rewriting process stops after finitely many steps then $y = g\, x$ and x have the same tail in their λ_q-CFs and $x \sim_{\mathrm{reg}} y$.

Hence assume, the rewriting process has to be repeated again and again. Then after a sufficiently large but finite number of rewriting steps one arrives at the situation where the λ_q-CF of $g\,x$ is regular up to one forbidden block. Denote this λ_q-CF by $[a_0; a_1, \dots]$ with the remaining forbidden block starting at digit a_i, $i \geq 1$ and assume without loss of generality the forbidden block has positive digits.

Consider first the case q even: By Lemma 2.12 the forbidden block and the following digits have the form $[a_i, \dots] = [B_0, B_1, B_2, \dots]$ with the block $B_0 \in \{[1^{h_q+1}, 2], [1^{h_q}, 2]\}$ and the blocks $B_j \in \{[1^{h_q}, 2], [1^{h_q-1}, 2]\}$, for all $j \geq 1$. Since by assumption B_0 was the last forbidden block in the λ_q-CF, necessarily $B_j \neq [1^{h_q}, 2]$. Hence the λ_q-CF of $g\,x$ has the form

$$g\,x = [a_0; a_1, \dots, a_{i-1}, \underline{1^l, 2}, \overline{1^{h_q-1}, 2}] \qquad \text{with } l = h_q, h_q + 1$$

where the forbidden block at digit a_i is underlined, and whose tail, determining also the tail of x, is regular λ_q-CF-equivalent to r_q. After infinitely many further rewritings one arrives at the regular λ_q-CF of y whose tail is regular λ_q-CF-equivalent to $-r_q$.

Consider next the case $q \geq 5$ odd: Lemma 2.12 again determines the form of the forbidden block and the following digits as

$$[a_i, \dots] = [B_0, B_1, B_2, \dots]$$

with the block $B_0 \in \{[1^{h_q+1}, 2], [1^{h_q}, 2, 1^{h_q}, 2]\}$ and the blocks

$$B_j \in \{A_1 := [1^{h_q}, 2], A_2 := [1^{h_q-1}, 2, 1^{h_q}, 2]\}, \qquad j \geq 1.$$

Since the blocks $[A_1, A_1]$ and $[A_2, A_1]$ are forbidden blocks, necessarily $B_j = A_2$ for all $j \geq 2$, since otherwise B_0 would not be the last forbidden block in the λ_q-CF of $g\,x$. Hence the λ_q-CF of $g\,x$ has the form

$$g\,x = [a_0; a_1, \dots, a_{i-1}, \underline{B_0}, B_1, \overline{1^{h_q-1}, 2, 1^{h_q}, 2}]$$

where the forbidden block at digit a_i is again underlined. As in the previous case, we find x is regular λ_q-CF-equivalent to r_q and y is regular λ_q-CF-equivalent to $-r_q$.

Consider finally the case $q = 3$: Lemma 2.12 gives again the form of the forbidden block and the following digits as $[a_i, \dots] = [B_0, B_1, \dots]$ with $B_0 = [2, 2^n, 3]$, $n \geq 0$, and the blocks $B_j \in \{[2], [3]\}$, $j \geq 1$. Since the blocks $[2, 2]$ and $[2, 3]$ are forbidden, necessarily $B_j = [3]$ for all $j \geq 1$. The λ_q-CF of $g\,x$ hence has the form

$$g\,x = [a_0; a_1, \dots, a_{i-1}, \underline{2, 2^n, 3}, \overline{3}]$$

where the forbidden block at digit a_i is underlined. Again x is regular λ_q-CF-equivalent to r_q and y is regular λ_q-CF-equivalent to $-r_q$. □

3. Generating maps for the λ_q-continued fractions and their duals

Similar to the Gauss continued fractions also the λ_q-continued fractions and their duals, which for $q = 3$ have been introduced by Hurwitz in [8], can be generated by interval maps with strong ergodic properties like in the case of the Gauss maps.

3.1. The interval maps f_q and $f_q^\star$. Denote by I_q respectively I_{R_q} the intervals

$$I_q = \left[-\frac{\lambda_q}{2}, \frac{\lambda_q}{2}\right] \qquad \text{respectively } I_{R_q} = [-R_q, R_q] \tag{3.1}$$

with λ_q as in (2.4) and $R_q = \lambda_q + r_q$ as in (2.12). The *nearest λ_q-multiple map* $\langle\cdot\rangle_q$ is given by

$$\langle\cdot\rangle_q \colon \mathbb{R} \to \mathbb{Z}; \quad x \mapsto \langle x\rangle_q := \left\lfloor \frac{x}{\lambda_q} + \frac{1}{2} \right\rfloor \tag{3.2}$$

where $\lfloor\cdot\rfloor$ is the (modified) floor function

$$\lfloor x\rfloor = n \iff \begin{cases} n < x \le n+1 & \text{if } x > 0, \\ n \le x < n+1 & \text{if } x \le 0. \end{cases} \tag{3.3}$$

We also need the map $\langle\cdot\rangle_q^\star$ given by

$$\langle\cdot\rangle_q^\star \colon \mathbb{R} \to \mathbb{Z}; \quad x \mapsto \langle x\rangle_q^\star := \begin{cases} \left\lfloor \dfrac{x}{\lambda_q} + 1 - \dfrac{R_q}{\lambda_q} \right\rfloor & \text{if } x \ge 0, \\ \left\lfloor \dfrac{x}{\lambda_q} + \dfrac{R_q}{\lambda_q} \right\rfloor & \text{if } x < 0. \end{cases} \tag{3.4}$$

The interval maps $f_q \colon I_q \to I_q$ and $f_q^\star \colon I_{R_q} \to I_{R_q}$ are defined as follows:

$$f_q(x) = \begin{cases} -1/x - \langle -1/x\rangle_q \lambda_q & \text{if } x \in I_q \setminus \{0\}, \\ 0 & \text{if } x = 0 \text{ and} \end{cases} \tag{3.5}$$

$$f_q^\star(y) = \begin{cases} -1/y - \langle -1/y\rangle_q^\star \lambda_q & \text{if } y \in I_{R_q} \setminus \{0\}, \\ 0 & \text{if } y = 0. \end{cases} \tag{3.6}$$

3.2. λ_q-CFs and their generating interval maps. The maps f_q and $f_q^\star$ generate the regular respectively dual regular λ_q-CFs in the following sense:

For given $x, y \in \mathbb{R}$ the coefficients a_i and b_i, $i \in \mathbb{Z}_{\ge 0}$ are determined by the following algorithms:

(0) $a_0 = \langle x\rangle_q$ and $x_1 := x - a_0\lambda_q \in I_q$,
(1) $a_1 = \langle -1/x_1\rangle_q$ and $x_2 := -1/x_1 - a_1\lambda_q = f_q(x_1) \in I_q$,
(i) $a_i = \langle -1/x_i\rangle_q$ and $x_{i+1} := -1/x_i - a_i\lambda_q = f_q(x_i) \in I_q$,
($\star$) the algorithm terminates if $x_{i+1} = 0$

and

(0) $b_0 = \langle y\rangle_q^\star$ and $y_1 := y - b_0\lambda_q \in I_{R_q}$,
(1) $b_1 = \langle -1/y_1\rangle_q^\star$ and $y_2 := -1/y_1 - b_1\lambda_q = f_q^\star(y_1) \in I_{R_q}$,
(i) $b_i = \langle -1/y_i\rangle_q^\star$ and $y_{i+1} := -1/y_i - b_i\lambda_q = f_q^\star(y_i) \in I_{R_q}$,
($\star$) the algorithm terminates if $y_{i+1} = 0$.

By construction the coefficients form λ_q-CFs in the sense of (2.9):

$$x = [a_0; a_1, a_2, \ldots] \quad \text{and} \quad y = [b_0; b_1, b_2, \ldots]. \tag{3.7}$$

Proposition 3.1. *The λ_q-CF of x in* (3.7) *above is unique for all x not in $\bigcup_{n=0}^\infty f_q^{-n}(\pm\lambda_q/2)$ and regular whereas the one of y is unique for all $y \notin \bigcup_{n=0}^\infty (f_q^\star)^{-n}(\pm r_q)$ and dual regular.*

PROOF. A simple calculation shows that the regular λ_q-CF of all points $x = \pm 2/\big((2m-1)\lambda_q\big)$, $m = 2, 3, \dots$ and their preimages is not unique. But these points belong to the preimages of the points $\pm\lambda_q/2$ On the other hand the dual λ_q-CF of the points $y = \pm 1/(r_q + m\lambda_q)$, $m = 1, 2, \dots$ and their preimages is not unique. But these points are all the preimages of the points $\pm r_q$. □

Remark 3.2. The non-uniqueness of certain finite regular λ_q-CF in Lemma 2.2 can also be derived from Proposition 3.1.

3.3. Markov partitions for f_q and $f_q^\star$. Obviously f_q is locally expanding, that means $|f_q'(x)| > 1$ for all $x \in I_q$, if one takes the one-sided derivatives at the points of discontinuity. The same holds true for the map $f_q^\star$ for q odd. For q even $f_q^{\star\prime}(\pm R_q) = 1$ but $|(f_q^{\star 2})'(y)| > 1$ for all $y \in I_{R_q}$, and hence both maps f_q and $f_q^\star$ are locally smooth, expanding maps. Indeed both maps have the Markov property, that means that they allow for Markov partitions. To construct these partitions we use the orbits of the boundary points of the two intervals I_q and I_{R_q} respectively the monotonicity intervals of the maps f_q and $f_q^\star$.

Define the orbit of x under f_q respectively $f_q^\star$ as

$$\begin{aligned} \text{orbit}(x) &= \big\{x, f_q(x), f_q^2(x) := f_q\big(f_q(x)\big), f_q^3(x), \dots\big\} \qquad (3.8)\\ &= \{f_q^n(x); n = 0, 1, 2, \dots\} \quad \text{respectively,} \\ \text{orbit}^\star(x) &= \{(f_q^\star)^n(x); n = 0, 1, 2, \dots\}. \qquad (3.9)\end{aligned}$$

The orbits orbit$(-\lambda_q/2)$ and orbit$^\star(-R_q)$ are both finite. If $\sharp\{S\}$ denotes the cardinality of the set S, we have

$$\sharp\,\text{orbit}\left(-\frac{\lambda_q}{2}\right) = \sharp\,\text{orbit}^\star(-R_q) = \kappa_q + 1,$$

as can be seen from the regular λ_q-CF of $-\lambda_q/2$ in (2.11) and the dual regular λ_q-CF of $-R_q$ in (2.15). We denote the elements of orbit$(-\lambda_q/2)$ by ϕ_i respectively of orbit$^\star(-R_q)$ by ψ_i, $i = 0, \dots, \kappa_q$, such that

$$\begin{aligned} -R_q = -\psi_0 < -\frac{\lambda_q}{2} = \phi_0 < \psi_1 < \phi_1 < \psi_2 < \phi_2 < \cdots \qquad (3.10)\\ < \psi_{\kappa_q-2} < \phi_{\kappa_q-2} < \psi_{\kappa_q-1} < \phi_{\kappa_q-1} = -\frac{1}{\lambda_q} < \psi_{\kappa_q} < \phi_{\kappa_q} = 0 \end{aligned}$$

holds. By using the regular λ_q-CF of $-\lambda_q/2$ and the dual regular $-\lambda_q$-CF of R_q respectively the order "$\prec$" in Section 2.4 one easily verifies

Lemma 3.3. *The order in* (3.10) *is achieved for q even by defining*

$$\phi_i = f_q^i\left(-\frac{\lambda_q}{2}\right) \quad \textit{and} \quad \psi_i = (f_q^\star)^i(-R_q), \qquad 0 \le i \le h = \kappa_q, \qquad (3.11)$$

respectively for q odd by defining

$$\begin{aligned} \phi_{2i} &= f_q^i\left(-\frac{\lambda_q}{2}\right), & \phi_{2i+1} &= f_q^{h+i+1}\left(-\frac{\lambda_q}{2}\right), \\ \psi_{2i} &= (f_q^\star)^i(-R_q), & \psi_{2i+1} &= (f_q^\star)^{h+i+1}(-R_q), \qquad 0 \le i \le h_q = \frac{\kappa_q - 1}{2}. \end{aligned} \qquad (3.12)$$

In the case $q=3$ one has $\kappa_3 = 1$ and $h_3 = 0$. Therefore

$$\phi_0 = -\frac{1}{2}, \quad \phi_1 = 0, \quad \psi_0 = -R_3 = \frac{1-\sqrt{5}}{2} \quad \text{and} \quad \psi_1 = R_3 - 1 = \frac{\sqrt{5}-3}{2}.$$

Define next $\phi_{-i} = -\phi_i$, $0 \le i \le \kappa_q$, respectively $\psi_{-i} = -\psi_i$ for $0 \le i \le \kappa_q + 1$ with $\psi_{\kappa_q+1} = 0$.

Obviously the intervals

$$\Phi_i := [\phi_{i-1}, \phi_i] \quad \text{and} \quad \Phi_{-i} := [\phi_{-i}, \phi_{-(i-1)}] \qquad 1 \le i \le \kappa_q \tag{3.13}$$

respectively

$$\Psi_i := [\psi_{i-1}, \psi_i] \quad \text{and} \quad \Psi_{-i} := [\psi_{-i}, \psi_{-(i-1)}], \tag{3.14}$$

$1 \le i \le \kappa_q + 1$, define Markov partitions of the intervals I_q and I_{R_q}: this means that

$$\bigcup_{\varepsilon=+,-} \bigcup_{i=1}^{\kappa_q} \Phi_{\varepsilon i} = I_q, \qquad \Phi_{\varepsilon i}^\circ \cap \Phi_{\delta j}^\circ = \varnothing \qquad \text{for } \varepsilon i \ne \delta j,$$

$$\bigcup_{\varepsilon=+,-} \bigcup_{i=1}^{\kappa_q+1} \Psi_{\varepsilon i} = I_{R_q}, \qquad \Psi_{\varepsilon i}^\circ \cap \Psi_{\delta j}^\circ = \varnothing \qquad \text{for } \varepsilon i \ne \delta j.$$

where S° denotes the interior of the set S. To get a reasonable symbolic dynamics for the two maps f_q and $f_q^\star$ we have to construct finer partitions using the monotonicity intervals of the two maps.

Consider first the case $q=3$ such that $\lambda_3 = 1$. Define for $m = 2,3,4,\dots$ the intervals J_m as

$$J_2 = \left[-\frac{1}{2}, -\frac{2}{5}\right], \qquad J_m = \left[-\frac{2}{2m-1}, -\frac{2}{2m+1}\right], \qquad m = 3,4,\dots, \tag{3.15}$$

and set $J_{-m} := -J_m$ for $m = 2,3,4,\dots$. Since $f_3(J_{\pm 2}) = \mp[0, \frac{1}{2}]$ and $f_3(J_{\pm m}) = I_3$ for $m = 3,4,\dots$ the partition satisfies

$$\bigcup_{\varepsilon=+,-} \bigcup_{m=2}^{\infty} J_{\varepsilon m} = I_3 \quad \text{and} \quad J_{\varepsilon m}^\circ \cap J_{\delta k}^\circ = \varnothing \qquad \text{for } \varepsilon m \ne \delta k.$$

Hence this partition, which we denote by $\mathcal{M}(f_3)$, is Markovian. The maps $f_3\big|_{J_m}$ are monotone with $f_3\big|_{J_m}(x) = -1/x - m$ and locally invertible with $\left(f_3\big|_{J_m}\right)^{-1}(y) = -1/(y+m)$ for $y \in f_3(J_m)$.

For $q \ge 4$ define intervals J_m, $m = 1,2,\dots$, as

$$\begin{aligned} J_1 &= \left[-\frac{\lambda_q}{2}, -\frac{2}{3\lambda_q}\right], \\ J_m &= \left[-\frac{2}{(2m-1)\lambda_q}, -\frac{2}{(2m+1)\lambda_q}\right], \qquad m = 2,3,\dots, \end{aligned} \tag{3.16}$$

and set $J_{-m} := -J_m$ for $m \in \mathbb{N}$. For even q, $q \ge 4$, the points in orbit$(-\lambda_q/2)$ do not fall onto a boundary point of any of the intervals J_m, $m \in \mathbb{N}$. Indeed from the regular λ_q-CF of $-\lambda_q/2$ in (2.11) and the order "$\prec$" in Section 2.4 one sees easily that

$$-\frac{\lambda_q}{2} = \phi_0 < \phi_1 < \cdots < \phi_{\kappa_q - 1} < -\frac{2}{3\lambda_q} < \phi_{\kappa_q} = 0$$

with $\phi_i = f_q^i(-\lambda_q/2)$. If we hence define the intervals $J_{\pm 1_i}$ as

$$J_{\varepsilon 1_i} := J_{\varepsilon 1} \cap \Phi_{\varepsilon i} \qquad \text{for } \varepsilon = +,-,\ 1 \le i \le \kappa_q \tag{3.17}$$

and therefore $J_{\varepsilon 1_i} = \Phi_{\varepsilon i}$ for $1 \le i \le \kappa_q - 1$ we get the partition $\mathcal{M}(f_q)$, defined as

$$I_q = \bigcup_{\varepsilon=+,-} \left(\bigcup_{i=1}^{\kappa_q} J_{\varepsilon 1_i} \cup \bigcup_{m=2}^{\infty} J_{\varepsilon m} \right), \tag{3.18}$$

which is obviously again Markovian, since

$$\begin{aligned} f_q(J_{\varepsilon 1_i}) &= J_{\varepsilon 1_{i+1}}, && \varepsilon = +,-,\ i = 1, \dots, \kappa_q - 2, \\ f_q(J_{\varepsilon 1_{\kappa_q-1}}) &= J_{\varepsilon 1_{\kappa_q}} \cup \bigcup_{m=2}^{\infty} J_{\varepsilon m}, && \varepsilon = +,- \quad \text{and} \\ f_q(J_{\varepsilon \kappa_q}) &= \varepsilon \left[0, \frac{\lambda_q}{2}\right], && \varepsilon = +,- \quad \text{respectively} \\ f_q(J_{\varepsilon m}) &= I_q, && \varepsilon = +,-,\ m = 2, 3, \dots . \end{aligned}$$

The restricted maps $f_q\big|_{J_m}$ are monotone increasing with $f_q\big|_{J_m}(x) = -1/x - m\lambda_q$ and $(f_q\big|_{J_m})^{-1}(y) = -1/(y + m\lambda_q)$ for $m = \pm 1, \pm 2, \pm 3, \dots$.

Consider next the case q odd, $q \ge 5$. In this case one has, using again the regular λ_q-CF of $-\lambda_q/2$ in (2.11) and the order "$\prec$" in Section 2.4,

$$-\frac{\lambda_q}{2} = \phi_0 < \phi_1 < \dots < \phi_{\kappa_q-2} < -\frac{2}{3\lambda_q} < \phi_{\kappa_q-1} < -\frac{2}{5\lambda_q} < \phi_{\kappa_q} = 0,$$

with $\kappa_q = 2h_q + 1$ and the ϕ_i's given in (3.12), (3.13). Hence for $\varepsilon = +,-$ one finds $\phi_{\varepsilon i} \in J_{\varepsilon 1}$ for $1 \le i \le \kappa_q - 2$ and $\phi_{\varepsilon(\kappa_q-1)} \in J_{\varepsilon 2}$. If we then define for $\varepsilon = +,-$ the intervals

$$\begin{aligned} &J_{\varepsilon 1_i} := J_{\varepsilon 1} \cap \Phi_{\varepsilon i},\ 1 \le i \le \kappa_q - 1 \text{ and hence } J_{\varepsilon 1_i} = \Phi_{\varepsilon i},\ 1 \le i \le \kappa_q - 2, \\ &J_{\varepsilon 2,i} := J_{\varepsilon 2} \cap \Phi_{\varepsilon i},\ i = \kappa_q - 1, \kappa_q, \end{aligned} \tag{3.19}$$

we find that the partition $\mathcal{M}(f_q)$ defined by

$$I_q = \bigcup_{\varepsilon=+,-} \left(\bigcup_{i=1}^{\kappa_q-1} J_{\varepsilon 1_i} \cup \bigcup_{i=\kappa_q-1}^{\kappa_q} J_{\varepsilon 2_i} \cup \bigcup_{m=3}^{\infty} J_{\varepsilon m} \right) \tag{3.20}$$

is Markovian. Indeed for $\varepsilon = +,-$ one finds $f_q(\Phi_{\varepsilon 2i}) = \Phi_{\varepsilon(2i+2)}$, $1 \le i \le h_q - 2$, $f_q(J_{\varepsilon 1_{\kappa_q-1}}) = \varepsilon[0, \lambda_q/2]$, $f_q(J_{\varepsilon 2_{\kappa_q-1}}) = \varepsilon[-\lambda_q/2, \phi_1]$, $f_q(\Phi_{\varepsilon(2i-1)}) = \Phi_{\varepsilon(2i+1)}$ for $1 \le i \le h$, $f_q(J_{\varepsilon 2_{\kappa_q}}) = \varepsilon[\phi_1, \lambda_q/2]$ and $f_q(J_{\varepsilon m}) = I_q$ for $m = 3, 4, \dots$. The maps $f_q\big|_{J_m}$ are monotone with $f_q\big|_{J_m}(x) = -1/x - m\lambda_q$ and $\left(f_q\big|_{J_m}\right)^{-1}(y) = -1/(y - m\lambda_q)$ for $m = \pm 1, \pm 2, \pm 3, \dots$.

Consider next the map $f_q^\star$ in (3.6). In the case $q = 3$ and $\lambda_3 = 1$ define the intervals $J_{\pm m}^\star$, $m = 2, 3, \dots$ as

$$\begin{aligned} J_m^\star &:= \left[\frac{-1}{r_3 + m}, \frac{-1}{r_3 + m + 1}\right] \quad \text{respectively,} \\ J_{-m}^\star &:= -J_m^\star = \left[\frac{1}{r_3 + m + 1}, \frac{1}{r_3 + m}\right]. \end{aligned} \tag{3.21}$$

Since $-R_3 = -1/(2+r_3) = -1 + r_3$ and $r_3 = -1/(3+r_3)$, see Section 2.3, we find

$$\text{(3.22)} \quad I_{R_3} = [-R_3, R_3] = \bigcup_{\varepsilon=+,-} \bigcup_{m=2}^{\infty} J^\star_{\varepsilon m} \qquad \text{with } J^\star_m \cap J^\star_n = \varnothing \text{ for all } m \neq n.$$

An easy calculation shows that

$$f_3^\star(J^\star_{\varepsilon m}) = \varepsilon[r_3, R_3] \qquad \text{for all } m \geq 2,$$

where one uses $f_3^\star(-R_3) = r_3$, $\lim_{\varepsilon \searrow 0} f_3^\star(r_3 + \varepsilon) = r_3$ and $\lim_{\varepsilon \searrow 0} f_3^\star(r_3 - \varepsilon) = R_3$. Hence the intervals $\{J^\star_m\}$ define a Markov partition $\mathcal{M}(f_3^\star)$ and $f_3^\star$ is a locally expanding, smooth Markov map.

For even q define the intervals $J^\star_{\pm m}$ for $m \in \mathbb{N}$ as

$$\text{(3.23)} \quad \begin{aligned} J^\star_m &:= \left[\frac{-1}{r_q + m\lambda_q}, \frac{-1}{r_q + (m+1)\lambda_q}\right] \qquad \text{respectively} \\ J^\star_{-m} &:= -\, J^\star_m = \left[\frac{1}{r_q + (m+1)\lambda_q}, \frac{1}{r_q + m\lambda_q}\right]. \end{aligned}$$

Since according to (2.16) $R_q = 1$ for q even, a simple calculation shows that $f_q^\star(-R_q) = r_q = 1 - \lambda_q$. But according to (2.12) $r_q = [\![0; \overline{(1)^{h-1}, 2}]\!]$ and hence $(f_q^\star)^{h_q - 1}(r_q) = [\![0; \overline{2, (1)^{h_q-1}}]\!]$. This with (2.15) shows that

$$(f_q^\star)^{h_q}(R_q) = (f_q^\star)^{h_q-1}(r_q) = -\frac{1}{2\lambda_q + r_q}.$$

The order of the points $\psi_i \in \text{orbit}^\star(-R_q)$ in (3.10) is given by

$$-R_q = \psi_0 < \psi_1 < \cdots < \psi_{\kappa_q} = \frac{-1}{2\lambda_q + r_q}$$

where $\kappa_q = h_q$ by (2.13). Hence $\psi_i \in J_1^\star$ for $0 \leq i \leq \kappa_q - 1$ whereas ψ_{κ_q} is just the common boundary of $J_1^\star$ and $J_2^\star$. Define therefore the intervals $J^\star_{\varepsilon 1_i}$ as

$$\text{(3.24)} \qquad J^\star_{\varepsilon 1_i} := J^\star_{\varepsilon 1} \cap \Psi_{\varepsilon i} \qquad \text{for all } 1 \leq i \leq \kappa_q \text{ and } \varepsilon = +,-,$$

such that $J^\star_{\varepsilon 1} = \bigcup_{i=1}^{\kappa_q} J^\star_{\varepsilon 1_i}$.

Then the partition $\mathcal{M}(f_q^\star)$, defined by

$$\text{(3.25)} \qquad I_{R_q} = \bigcup_{\varepsilon=+,-} \left(\bigcup_{i=1}^{\kappa_q} J^\star_{\varepsilon 1_i} \cup \bigcup_{m=2}^{\infty} J^\star_{\varepsilon m} \right),$$

is a Markov partition, since

$$\begin{aligned} f_q^\star(J^\star_{\varepsilon 1_i}) &= J^\star_{\varepsilon 1_{i+1}} && \text{for } 1 \leq i \leq \kappa_q - 1, \\ f_q^\star(J^\star_{\varepsilon 1_{\kappa_q}}) &= \varepsilon\left[\frac{-1}{2\lambda_q + r_q}, R_q\right], \\ f_q^\star(J^\star_{\varepsilon m}) &= \varepsilon[r_q, R_q] && \text{for } m \geq 2. \end{aligned}$$

The restriction $f_q^\star\big|_{J^\star_m}$ of $f_q^\star$ to the interval $J^\star_m$ is given by

$$f_q^\star\big|_{J_m}(x) = \frac{-1}{x} - m\lambda_q \qquad \text{for } m \in \mathbb{Z}_{\neq 0}$$

and its inverse by

$$\left(f_q^\star\big|_{J_m}\right)^{-1}(y) = \frac{-1}{y + m\lambda_q} \qquad \text{for } y \in f_q^\star(J_m).$$

Remains the case q odd, $q \geq 5$. The intervals $J^\star_{\varepsilon m}$ for $\varepsilon = +,-$, $m \geq 2$ are defined as for q even in (3.23):

$$J^\star_m := \left[\frac{-1}{r_q + m\lambda_q}, \frac{-1}{r_q + (m+1)\lambda_q}\right] \quad \text{and} \quad J^\star_{-m} := -J^\star_m. \tag{3.26}$$

The intervals $J^\star_{\varepsilon 1}$ are defined as

$$J^\star_1 := \left[-R_q, \frac{-1}{r_q + 2\lambda_q}\right] \quad \text{respectively,} \tag{3.27}$$

$$J^\star_{-1} := -J^\star_1 = \left[\frac{1}{r_q + 2\lambda_q}, R_q\right].$$

According to (3.12)

$$\psi_{2h_q} = (f^\star_q)^{h_q}(-R_q) = \frac{-1}{r_q + 2\lambda_q} = [\![0; \overline{2, 1^{h_q}, 2, 1^{h_q-1}}]\!]^\star$$

and

$$\psi_{2h_q+1} = f^\star_q(-R_q) = [\![0; 2, \overline{1^{h_q-1}, 2, 1^{h_q}, 2}]\!]^\star$$
$$\leq [\![0; 3, \overline{1^{h_q}, 2, 1^{h_q-1}, 2}]\!]^\star = \frac{-1}{3\lambda_q + r_q}.$$

Hence $\psi_i \in J^\star_1 = [-R_q, -1/(r_q + 2\lambda_q)]$ for $1 \leq i \leq 2h_q = \kappa_q - 1$ whereas $\psi_{2h_q+1} = \psi_{\kappa_q} \in J^\star_2$. Define for $\varepsilon = +,-$ the intervals $J^\star_{\varepsilon 1_i}$ as

$$J^\star_{\varepsilon 1_i} := J^\star_{\varepsilon 1} \cap \Psi_{\varepsilon i} = \Psi_{\varepsilon i} \qquad \text{for } 1 \leq i \leq \kappa_q - 1 \tag{3.28}$$

and the intervals $J^\star_{\varepsilon 2_i}$ for $i = \kappa_q, \kappa_q + 1$ as

$$J^\star_{\varepsilon 2_{\kappa_q}} := J^\star_{\varepsilon 2} \cap \Psi_{\varepsilon \kappa_q} = \Psi_{\varepsilon \kappa_q} \qquad \text{respectively,}$$

$$J^\star_{\varepsilon 2_{\kappa_q+1}} := J^\star_{\varepsilon 2} \setminus \Psi_{\varepsilon \kappa_q} = \varepsilon\left[\psi_{\kappa_q}, \frac{-1}{3\lambda_q + r_q}\right].$$

Then the partition

$$I_{R_q} = \bigcup_{\varepsilon=+,-} \left(\bigcup_{i=1}^{\kappa_q-1} J^\star_{\varepsilon 1_i} \cup \bigcup_{i=\kappa_q}^{\kappa_q+1} J^\star_{\varepsilon 2_i} \cup \bigcup_{m=3}^{\infty} J^\star_{\varepsilon m} \right) \tag{3.29}$$

is a Markov partition. This follows from the following identities, which can be easily verified:

$$f^\star_q(J^\star_{\varepsilon m}) = \varepsilon[r_q, R_q] \qquad \text{for all } m = 3, 4, \dots,$$
$$f^\star_q(J^\star_{\varepsilon 1_{2i}}) = J^\star_{\varepsilon 1_{2i+2}} \qquad \text{for all } 1 \leq i \leq h_q - 1,$$
$$f^\star_q(J^\star_{\varepsilon 1_{2i-1}}) = J^\star_{\varepsilon 1_{2i+1}} \qquad \text{for all } 1 \leq i \leq h_q,$$

$$f^\star_q(J^\star_{\varepsilon 1_{2h}}) = \varepsilon[\psi_{\kappa_q}, R_q] = J^\star_{\varepsilon 2_{\kappa_q+1}} \cup \bigcup_{\delta=+,-} \bigcup_{m=3}^{\infty} J^\star_{\delta m} \cup J^\star_{-\varepsilon 2m},$$

$$f^\star_q(J^\star_{\varepsilon 2_{\kappa_q}}) = J_{\varepsilon 1_2},$$

$$f^\star_q(J^\star_{\varepsilon 2_{\kappa_q+1}}) = \varepsilon[\psi_2, R_q] = \bigcup_{i=3}^{\kappa_q-1} J_{\varepsilon 1_i} \cup \bigcup_{i=\kappa_q}^{\kappa_q+1} J_{\varepsilon 2_i} \cup \bigcup_{\delta=+,-} \bigcup_{m=3}^{\infty} J^\star_{\delta m} \cup J_{-\varepsilon 2} \cup J_{-\varepsilon 1}.$$

4. The maps f_q and $f_q^\star$ and regular respectively dual regular λ_q-CFs

We are going to use the Markov partitions $\mathcal{M}(f_q)$ respectively $\mathcal{M}(f_q^\star)$ constructed in the forgoing section for the maps $f_q\colon I_q \to I_q$ and $f_q^\star\colon I_{R_q} \to I_{R_q}$ to show that these maps can be conjugated to subshifts over infinite alphabets. By introducing sofic systems closely related to these subshifts the symbolic dynamics of the above two maps are directly related to the regular respectively dual regular λ_q-CFs.

4.1. Symbolic dynamics for f_q and a subshift of infinite type. For $q = 3$ and $f_3\colon I_3 \to I_3$ let F be the alphabet $F = \mathbb{Z} \setminus \{0, \pm 1\}$. Define the transition matrix $\mathbb{A} = (\mathbb{A}_{i,j})_{i,j\in F}$ with $\mathbb{A}_{i,j} \in \{0, 1\}$ for $\varepsilon = +, -$ as follows:

$$\begin{aligned} \mathbb{A}_{\varepsilon 2, \varepsilon m} &= 0, && m \geq 2, \\ \mathbb{A}_{\varepsilon 2, -\varepsilon m} &= 1, && m \geq 2, \\ \mathbb{A}_{\varepsilon k, m} &= 1, && k \geq 3 \text{ and all } m \in F. \end{aligned} \tag{4.1}$$

Denote by $(F_{\mathbb{A}}^{\mathbb{N}}, \tau)$ the subshift over the alphabet F with

$$F_{\mathbb{A}}^{\mathbb{N}} = \{\underline{\xi} = (\xi_i)_{i\in\mathbb{N}}, \xi_i \in F, \mathbb{A}_{\xi_i,\xi_{i+1}} = 1, i \in \mathbb{N}\}$$

and $\left(\tau(\underline{\xi})\right)_i = \xi_{i+1}$ the shift map.

Let $\partial\mathcal{M}(f_3) := \{x \in I_3 : \exists n \in \mathbb{Z}_{\geq 0} : f_3^n(x) = 0\}$. The projection map $\pi\colon I_3 \setminus \partial\mathcal{M}(f_3) \to F_{\mathbb{A}}^{\mathbb{N}}$ defined by

$$\pi(x) = \underline{\xi} = (\xi_i)_{i\in\mathbb{N}} \qquad \text{if } f_3^{i-1}(x) \in J_{\xi_i} \text{ for } i \in \mathbb{N}$$

is bijective with inverse $\pi^{-1}(\underline{\xi}) = x$, where x is the unique point with $x \in J_{\xi_1} \cap \bigcap_{l=1}^{\infty}(f_{\xi_l} \circ \cdots \circ f_{\xi_1})^{-1} J_{\xi_{l+1}}$. That the point x is uniquely defined follows from the expanding property of the local branches $f_{\xi_i} = f_3\big|_{J_{\xi_i}}$ of the map f_3, given on the interval J_m by $f_{J_m} = -1/x - m$, $m \in \mathbb{Z} \setminus \{0, \pm 1\}$. Obviously one has $\pi \circ f_3 = \tau \circ \pi$ on $I_3 \setminus \partial\mathcal{M}(f_3)$.

For q even with $q = 2h_q + 2$ define the alphabet F as $F = \{\varepsilon 1_i, \varepsilon = +, -, 1 \leq i \leq \kappa_q\} \cup \mathbb{Z} \setminus \{0, \pm 1\}$. The transition matrix $\mathbb{A} = (\mathbb{A}_{i,j})_{i,j\in F}$ is defined in this case as follows:

$$\begin{aligned} \mathbb{A}_{\varepsilon 1_l, \varepsilon 1_{l+1}} &= 1, && 1 \leq l \leq \kappa_q - 1, \\ \mathbb{A}_{\varepsilon 1_{\kappa_q - 1}, \varepsilon m} &= 1, && m = 2, 3, \ldots, \\ \mathbb{A}_{\varepsilon 1_{\kappa_q}, -\varepsilon 1_l} &= 1, && 1 \leq l \leq \kappa_q, \\ \mathbb{A}_{\varepsilon 1_{\kappa_q}, -m} &= 1, && m = 2, 3, \ldots, \\ \mathbb{A}_{m,n} &= 1, && m \in \mathbb{Z} \setminus \{0, \pm 1\},\ n \in F, \end{aligned} \tag{4.2}$$

and all the other matrix elements vanishing.

Define the set $\partial\mathcal{M}(f_q)$ and the map $\pi\colon I_q \setminus \partial\mathcal{M}(f_q) \to F_{\mathbb{A}}^{\mathbb{N}}$ in analogy to the case $q = 3$. The same arguments as there show that this map is bijective and conjugates f_q to the shift map τ with $\pi \circ f_q = \tau \circ \pi$ on $I_q \setminus \partial\mathcal{M}(f_q)$.

For $q = 2h_q + 3 = \kappa_q + 1$ finally define the alphabet F as $F = \{\varepsilon 1_i, \varepsilon = +, -, 1 \leq i \leq \kappa_q - 1\} \cup \{\varepsilon 2_i, \varepsilon = +, -, \kappa_q - 1 \leq i \leq \kappa_q\} \cup \mathbb{Z} \setminus \{0, \pm 1, \pm 2\}$. The transition matrix $\mathbb{A} = (\mathbb{A}_{i,j})_{i,j\in F}$ is given in Table 2.

The set $\partial\mathcal{M}(f_q)$ and the map $\pi\colon I_q \setminus \partial\mathcal{M}(f_q) \to F_{\mathbb{A}}^{\mathbb{N}}$ are defined similarly as in the foregoing cases $q = 3$ and q even and have the same properties. The inverse $(\pi)^{-1}\colon F_{\mathbb{A}}^{\mathbb{N}} \to I_q \setminus \partial\mathcal{M}(f_q)$ is given by $\pi^{-1}(\underline{\xi}) = x$ with $x \in J_{\xi_1} \cap \bigcap_{l=1}^{\infty}(f_{\xi_l} \circ \cdots \circ$

TABLE 2. The transition matrix $\mathbb{A} = (\mathbb{A}_{i,j})_{i,j\in F}$ with $\varepsilon = +,-$ for q-regular sequences and q odd, $q \geq 5$. All other matrix elements vanish.

$$
\begin{aligned}
\mathbb{A}_{\varepsilon 1_{2i},\varepsilon 1_{2i+1}} &= 1 && 1 \leq i \leq h_q - 2,\\
\mathbb{A}_{\varepsilon 1_{2h_q-2},\varepsilon 1_{2h_q}} &= 1, &&\\
\mathbb{A}_{\varepsilon 1_{2h_q-2},\varepsilon 2_{\kappa_q}} &= 1, &&\\
\mathbb{A}_{\varepsilon 1_{2h_q},-\varepsilon 1_i} &= 1 && 1 \leq i \leq \kappa_q - 1,\\
\mathbb{A}_{\varepsilon 1_{2h_q},-\varepsilon 2_i} &= 1 && \kappa_q \leq i \leq \kappa_q + 1,\\
\mathbb{A}_{\varepsilon 1_{2h_q},-\varepsilon m} &= 1 && m \geq 3,\\
\mathbb{A}_{\varepsilon 1_{2i-1},\varepsilon 1_{2i+1}} &= 1, && 1 \leq i \leq h_q - 1,\\
\mathbb{A}_{\varepsilon 1_{2h_q-1},\varepsilon 2_{\kappa_q+1}} &= 1, &&\\
\mathbb{A}_{\varepsilon 1_{2h_q-1},-\varepsilon m} &= 1, && m \geq 3,\\
\mathbb{A}_{\varepsilon 2_{\kappa_q},-\varepsilon 1_1} &= 1, &&\\
\mathbb{A}_{\varepsilon 2_{\kappa_q+1},\delta 1_i} &= 1, && \delta = +,-,\ 2 \leq i \leq \kappa_q - 1,\\
\mathbb{A}_{\varepsilon 2_{\kappa_q+1},\delta 2_i} &= 1, && \delta = +,-,\ 2 \leq i \leq \kappa_q - 1,\\
\mathbb{A}_{\varepsilon 2_{\kappa_q+1},\delta n} &= 1, && \delta = +,-,\ n \geq 3,\\
\mathbb{A}_{\varepsilon 2_{\kappa_q+1},-\varepsilon 1_1} &= 1, &&\\
\mathbb{A}_{m,n} &= 1, && m \in \mathbb{Z} \setminus \{0, \pm 1, \pm 2\},\ n \in F,
\end{aligned}
$$

$f_{\xi_1})^{-1} J_{\xi_{l+1}}$ where $f_{\xi_i} = f\big|_{J_{\varepsilon 1}}$ for $\xi_i = \varepsilon 1_l$, $1 \leq l \leq \kappa_q - 1$ respectively $f_{\xi_i} = f\big|_{J_{\varepsilon 2}}$ for $\xi_i = \varepsilon 2_l$, $\kappa_q \leq l \leq \kappa_q + 1$. Hence also in this case the map f_q gets conjugated by π to the shift map τ on the space $F_{\mathbb{A}}^{\mathbb{N}}$ of symbol sequences and therefore is itself a subshift of infinite type.

4.2. A sofic system related to the map f_q and the regular λ_q-CF. The transition matrix $\mathbb{A}$ in (4.1) for the subshift $f_3\colon I_3 \to I_3$ shows that a symbol sequence $\underline{a} = (a_i)_{i\in\mathbb{N}} \in F_{\mathbb{A}}^{\mathbb{N}} = \pi\big(I_3 \setminus \partial\mathcal{M}(f_3)\big)$ if and only if $(a_i, a_{i+1}) \neq (\varepsilon 2, \varepsilon m)$, $m \geq 2$ for all $i \in \mathbb{N}$. Hence this sequence is q-regular for $q = 3$ and $F_{\mathbb{A}}^{\mathbb{N}} = \mathcal{A}_3^{\mathrm{reg}}$. The inverse map $\pi^{-1}\colon F_{\mathbb{A}}^{\mathbb{N}} \to I_3 \setminus \partial\mathcal{M}(f_3)$ therefore has the form $\pi^{-1}(\underline{a}) = [\![0; a_1, a_2, \ldots]\!]$. This follows from

Lemma 4.1. *For $\underline{a} = (a_i)_{i\in\mathbb{N}} \in \mathcal{A}_3^{\mathrm{reg}}$ a 3-regular sequence the limit*

$$\lim_{n\to\infty} [\![0; a_1, a_2, \ldots, a_n]\!]$$

exists and defines a point $x \in \mathbb{R}$.

PROOF. Set $x_n := [\![0; a_1, a_2, \ldots, a_n]\!]$ and denote by $J_{(a_1,\ldots,a_n)}$ the closed interval $J_{(a_1,\ldots,a_n)} := J_{a_1} \cap \bigcap_{l=1}^{n-1} (f_{a_l} \circ \ldots f_{a_1})^{-1} J_{a_{l+1}}$ with $f_{a_i} := f_q\big|_{J_{a_i}}$. Obviously $x_n \in J_{(a_1,\ldots,a_n)}$. All these intervals are nonempty and $J_{(a_1,\ldots,a_{n+1})} \subset J_{(a_1,\ldots,a_n)}$ for all n. Hence $\bigcap_{n=1}^{\infty} J_{(a_1,\ldots,a_n)}$ is not empty. Because the map f_3 is strictly expanding this set contains exactly one point x. But this shows that $\lim_{n\to\infty} [\![0; a_1, a_2, \ldots, a_n]\!] = x$. □

In the case $q \neq 3$ the relation between the symbolic dynamics with respect to the Markov partitions $\mathcal{M}(f_q)$ and the λ_q-CF is more complicated. Indeed one has to introduce a corresponding sofic system, namely in the alphabet F the letters $\varepsilon 1_i$ respectively $\varepsilon 2_i$ have to be replaced by the letters $\varepsilon 1$ respectively $\varepsilon 2$ for all i. This corresponds to replacing the Markov partition $\mathcal{M}(f_q)$ defined in Section 3.3 by the partition $\mathcal{J}(f_q)$ defined as $I_q = \bigcup_{\varepsilon=+,-} \bigcup_{m=1}^{\infty} J_{\varepsilon m}$ with J_m given in (3.16). It is not difficult to see that this partition is generating that means $\bigcap_{i=1}^{\infty} f_q^{-(i-1)} J_{m_i}$ is either empty or consists of exactly one point. This follows again from the fact that all branches of f_q are expanding. Denote by $\partial\mathcal{J}(f_q)$ the boundary points of the intervals J_m including the point $x = 0$ together with all their preimages under the map f_q. Since $x = 0$ belongs to the orbit of $-\frac{\lambda_q}{2}$ the boundaries $\partial\mathcal{M}(f_q)$ and $\partial\mathcal{J}(f_q)$ coincide. Denote by $\widehat{\pi}\colon I_q \setminus \partial\mathcal{J}(f_q) \to \widehat{F}^{\mathbb{N}}$ with $\widehat{F} = \mathbb{Z} \setminus \{0\}$ the map $\widehat{\pi}(x) = \underline{a} = (a_i)_{i\in\mathbb{N}}$ when $f_q^{i-1}(x) \in J_{a_i}$. If $\pi(x) = \underline{\xi}$, then obviously $a_i = m$ if $\xi_i = m \in \mathbb{Z}$ and $a_i = \varepsilon 1$ if $\xi_i = \varepsilon 1_l$ for some l respectively $a_i = \varepsilon 2$ if $\xi_i = \varepsilon 2_l$ for some l. The following lemma then holds

Lemma 4.2. *The map $\widehat{\pi}\colon I_q \setminus \partial\mathcal{J}(f_q) \to \mathcal{A}_q^{\mathrm{reg}} \subset \widehat{F}^{\mathbb{N}}$ is bijective. The inverse map $\widehat{\pi}^{-1}\colon \mathcal{A}_q^{\mathrm{reg}} \to I_q \setminus \partial\mathcal{J}(f_q)$ is given by $\widehat{\pi}^{-1}(\underline{a}) = [\![0; a_1, a_2, \ldots]\!]$.*

Proof. Consider first q even: From the definition of the transition matrix $\mathbb{A}$ in (4.2) for the case $q = 2h_q+2$ it follows that there cannot be more than h_q consecutive symbols $\varepsilon 1$ in $\underline{a} = (a_i)_{i\in\mathbb{N}} = \widehat{\pi}(x)$ since $(\mathbb{A})_{\varepsilon 1_{\kappa_q}, \varepsilon m} = 0$ for all $m = 2, 3, \ldots$: indeed h_q consecutive symbols $\varepsilon 1$ are only possible for points x with $f_q^{i-1}(x) \in J_{\varepsilon 1_i}$ for $i = 1, \ldots, \kappa_q$ and $f_q^{\kappa_q} \in J_{-\varepsilon 1_l}$ for some $1 \le l \le \kappa_q$ or $f_q^{\kappa_q} \in J_{-\varepsilon m}$ for some $m \ge 2$. This shows that $\underline{a} = (a_i)_{i\in\mathbb{N}} = \widehat{\pi}(x)$ defines a q-regular sequence in $\mathcal{A}_q^{\mathrm{reg}}$.

Given on the other hand such a q-regular sequence $\underline{a} = (a_i)_{i\in\mathbb{N}}$ there exists a unique point $x \in I_q \setminus \partial\mathcal{M}(f_q)$ with $\widehat{\pi}(x) = \underline{a}$: indeed if for some $l \ge 1$ and $k \ge 0$ one has $a_l = a_{l+1} = \cdots = a_{l+k} = \varepsilon 1$ and $a_{l+k+1} = m \ne \varepsilon 1$ consider the sequence $\underline{\xi} \in F_{\mathbb{A}}^{\mathbb{N}}$ with $\xi_{l+k} = \varepsilon 1_{\kappa_q}$, $\xi_{l+k-1} = \varepsilon 1_{\kappa_q - 1}, \ldots, \xi_l = \varepsilon 1_{\kappa_q - l}$ if $\operatorname{sign}(m) \ne \varepsilon$ respectively $\xi_{l+k} = \varepsilon 1_{\kappa_q-1}$, $\xi_{l+k-1} = \varepsilon 1_{\kappa_q-2}, \ldots, \xi_l = \varepsilon 1_{\kappa_q-l-1}$ if $\operatorname{sign}(m) = \varepsilon$, whereas $\xi_i = a_i$ for all $a_i \ne \varepsilon 1$. Since $k \le \kappa_q - 1$ respectively $k \le \kappa_q - 2$ in the second case, the sequence $\underline{\xi}$ belongs indeed to $F_{\mathbb{A}}^{\mathbb{N}}$ and hence there exists a point $x \in I_q \setminus \partial\mathcal{M}(f_q)$ with $\pi(x) = \underline{\xi}$ and hence also $\widehat{\pi}(x) = \underline{a}$. The inverse map $\widehat{\pi}^{-1}$ is again given by $\widehat{\pi}^{-1}(\underline{a}) = [\![0; a_1, a_2, \ldots]\!]$. Since for any $\underline{a} \in \mathcal{A}_q^{\mathrm{reg}}$ there exists an unique $\underline{\xi} \in F_{\mathbb{A}}^{\mathbb{N}}$ which is related to $\underline{a}$ when replacing the symbols $\pm 1_i$ by the symbol ± 1, there exists therefore $x \in I_q \setminus \partial\mathcal{J}(f_q)$ with $\pi(x) = \underline{\xi}$. But $x_n := [\![0; a_1, a_2, \ldots a_n]\!] \in J_{(\xi_1, \ldots, \xi_n)}$ and hence $\lim_{n\to\infty} x_n = x \in \bigcap_{n=1}^{\infty} J_{(\xi_1, \ldots, \xi_n)}$ and hence $\widehat{\pi}^{-1}(\underline{a}) = [\![0; a_1, a_2, \ldots]\!]$.

The same reasoning can be applied in the case $q = 2h_q + 3$ odd to show that the map $\widehat{\pi}\colon I_q \setminus \partial\mathcal{J}(f_q) \to \mathcal{A}_q^{\mathrm{reg}} \in \widehat{F}^{\mathbb{N}}$ is bijective with inverse $\widehat{\pi}^{-1}(\underline{a}) = [\![0; a_1, a_2, \ldots]\!]$. □

4.3. Symbolic dynamics for $f_q^\star$ and a subshift of infinite type. Let us start again with the case $q = 3$ and recall the Markov partition $\mathcal{M}(f_3^\star)$ defined in (3.22) by $I_{R_3} = \bigcup_{\varepsilon=+,-} \bigcup_{m=2}^{\infty} J_{\varepsilon m}^\star$ with $J_{\varepsilon m}^\star = \varepsilon[-1/(r_3+m), -1/(r_3+m+1)]$. Denote by F the alphabet $F = \mathbb{Z} \setminus \{0, \pm 1\}$ and by $\mathbb{A} = (\mathbb{A}_{i,j})_{i,j\in F}$ the transition matrix with

$$(\mathbb{A})_{m,n} = 1 \qquad \text{for all } m, n \in F \text{ with } n \ne 2\operatorname{sign}(m), \tag{4.3}$$

and all the other matrix elements vanishing. Denote by $\partial\mathcal{M}(f_3^\star)$ the set

$$\partial\mathcal{M}(f_3^\star) = \{y \in I_{R_3} : \exists n \in \mathbb{Z}_{\geq 0} \text{ with } (f_3^\star)^n(y) = \pm r_3 \text{ or } (f_3^\star)^n(y) = 0\}$$

and by $\pm \underline{r}_3$ the sequence $\pm\underline{r}_3 = (\pm\overline{3})$. Then one has for $f_{\xi_i}^\star := f_3^\star|_{J_{\xi_i}^\star}$

Lemma 4.3. *The map*

$$\pi \colon I_{R_3} \setminus \partial\mathcal{M}(f_3^\star) \to F_{\mathbb{A}}^{\mathbb{N}} \setminus \{\underline{\xi} \in F_{\mathbb{A}}^{\mathbb{N}} : \exists n \in \mathbb{Z}_{\geq 0} : \tau^n(\underline{\xi}) = \pm \underline{r}_3\},$$

given by $\pi(x) = \underline{\xi} = (\xi_i)_{i\in\mathbb{N}}$ *if* $(f_3^\star)^{i-1} \in J_{\xi_i}^\star$ *for* $i \in \mathbb{N}$, *is bijective, and* $\pi \circ f_3^\star = \tau \circ \pi$. *Its inverse, the map* $\pi^{-1}\colon F_{\mathbb{A}}^{\mathbb{N}} \to I_{R_3}$ *can be defined on the entire set* $F_{\mathbb{A}}^{\mathbb{N}}$ *and is given by* $\pi^{-1}(\underline{\xi}) = x$ *with* x *the unique point in* I_{R_3} *with* $x \in J_{\xi_1}^\star \bigcap_{l=1}^{\infty} (f_{\xi_l}^\star \circ \cdots \circ f_{\xi_1}^\star)^{-1} J_{\xi_{l+1}}^\star$.

PROOF. Obviously all the preimages of the point $x = 0$ have a finite symbol sequence $(\xi_1, \dots, \xi_N)$, whereas the points $\pm r_3 \in J_{\pm 2}^\star \cup J_{\pm 3}^\star$ have the two different symbol sequences $\pi(\pm r_3) = \underline{\xi} = (\pm\overline{3})$ respectively $\pi(\pm r_3) = (\pm 2, \mp\overline{3})$. The same holds then true for all the preimages of these points. The point $x \in I_{R_3}$ is again uniquely determined because of the expansive nature of the local branches of the map $f_3^\star$. □

The map $f_3^\star \colon I_{R_3} \to I_{R_3}$ is hence a subshift of infinite type.

Consider next the case q even with $q = 2h_q + 2$ and $\kappa_q = h_q$. Recall the Markov partition $\mathcal{M}(f_q^\star)$ in (3.25). We define the alphabet F as

$$F = \{\varepsilon 1_i, \varepsilon = +,-, 1 \leq i \leq \kappa_q\} \cup \mathbb{Z} \setminus \{0, \pm 1\}$$

and by $\mathbb{A} = (\mathbb{A}_{i,j})_{i,j\in F}$ the transition matrix with matrix elements

$$\begin{aligned} (\mathbb{A})_{\varepsilon 1_i, \varepsilon 1_{i+1}} &= 1 && \text{for } \varepsilon = +,-;\ 1 \leq i \leq \kappa_q - 1, \\ (\mathbb{A})_{\varepsilon 1_{\kappa_q}, m} &= 1 && \text{for } m \in F,\ m \neq \varepsilon 1_i\ 1 \leq i \leq \kappa_q, \\ (\mathbb{A})_{m,n} &= 1 && \text{for } |m| \geq 2 \text{ and all } n \neq \operatorname{sign}(m) 1_1, \end{aligned} \tag{4.4}$$

whereas all other matrix elements vanish. If we define again $\partial\mathcal{M}(f_q^\star)$ by

$$\partial\mathcal{M}(f_q^\star) = \{y \in I_{R_q} : \exists n \in \mathbb{Z}_{\geq 0} \text{ with } (f_q^\star)^n(y) = \pm r_q \text{ or } (f_q^\star)^n(y) = 0\}$$

one shows in complete analogy with Lemma 4.3.

Lemma 4.4. *The map*

$$\pi \colon I_{R_q} \setminus \mathcal{M}(f_q^\star) \to F_{\mathbb{A}}^{\mathbb{N}} \setminus \{\underline{\xi} \in F_{\mathbb{A}}^{\mathbb{N}} : \exists n \in \mathbb{Z}_{\geq 0} : \tau^n(\underline{\xi}) = \pm \underline{r}_q\},$$

given by $\pi(x) = \underline{\xi} = (\xi_i)_{i\in\mathbb{N}}$ *if* $(f_q^\star)^{i-1} \in J_{\xi_i}^\star$ *for* $i \in \mathbb{N}$, *is bijective, and* $\pi \circ f_q^\star = \tau \circ \pi$. *Its inverse, the map* $\pi^{-1}\colon F_{\mathbb{A}}^{\mathbb{N}} \to I_{R_q}$ *can be defined on the entire set* $F_{\mathbb{A}}^{\mathbb{N}}$ *and is given by* $\pi^{-1}(\underline{\xi}) = x$ *with* x *the unique point in* I_{R_q} *with* $x \in J_{\xi_1}^\star \bigcap_{l=1}^{\infty} (f_{\xi_l}^\star \circ \cdots \circ f_{\xi_1}^\star)^{-1} J_{\xi_{l+1}}^\star$.

This shows that the map $f_q^\star \colon I_{R_q} \to I_{R_q}$ is a subshift of infinite type also for even q.

Consider finally the case $q = 2h_q + 3$ and $\kappa_q = 2h_q + 1$. The Markov partition $\mathcal{M}(f_q^\star)$ was given in this case in (3.29). Define the alphabet F as

$$\begin{aligned} F = \{\varepsilon 1_i, \varepsilon = +,-, 1 \leq i \leq \kappa_q - 1\} \\ \cup \{\varepsilon 2_i, \varepsilon = +,-, \kappa_q \leq i \leq \kappa_q + 1\} \cup \mathbb{Z} \setminus \{0, \pm 1, \pm 2\}. \end{aligned}$$

The transition matrix $\mathbb{A} = (\mathbb{A}_{i,j})_{i,j\in F}$ has now the form given in Table 3.

TABLE 3. The transition matrix $\mathbb{A} = (\mathbb{A}_{i,j})_{i,j\in F}$ with $\varepsilon = +,-$ for q-dual regular sequences and q odd, $q \geq 5$. All other matrix elements vanish.

$$
\begin{aligned}
(\mathbb{A})_{\varepsilon 1_{2i-1},\varepsilon 1_{2i+1}} &= 1 && \text{for } 1 \le i \le h_q - 1,\\
(\mathbb{A})_{\varepsilon 1_{2h_q-1},\varepsilon 2_{\kappa_q}} &= 1 &&\\
(\mathbb{A})_{\varepsilon 1_{2i},\varepsilon 1_{2i+2}} &= 1 && \text{for } 1 \le i \le h_q - 1,\\
(\mathbb{A})_{\varepsilon 1_{2h_q},\varepsilon 2_{\kappa_q+1}} &= 1 &&\\
(\mathbb{A})_{\varepsilon 1_{2h_q},m} &= 1 && m \in \mathbb{Z}\setminus\{0,\pm 1,\pm 2\},\\
(\mathbb{A})_{\varepsilon 1_{2h_q},-2} &= 1, &&\\
(\mathbb{A})_{\varepsilon 2_{\kappa_q},\varepsilon 1_2} &= 1, &&\\
(\mathbb{A})_{\varepsilon 2_{\kappa_q+1},\varepsilon 1_i} &= 1 && \text{for } 2 \le i \le \kappa_q - 1,\\
(\mathbb{A})_{\varepsilon 2_{\kappa_q+1},\varepsilon 2_i} &= 1 && \text{for } i = \kappa_q, \kappa_q + 1,\\
(\mathbb{A})_{\varepsilon 2_{\kappa_q+1},m} &= 1 && \text{for } m \in \mathbb{Z}\setminus\{0,\pm 1,\pm 2\},\\
(\mathbb{A})_{\varepsilon 2_{\kappa_q+1},-\varepsilon 1_i} &= 1 && \text{for } 1 \le i \le \kappa_q - 1,\\
(\mathbb{A})_{\varepsilon 2_{\kappa_q+1},-\varepsilon 2_i} &= 1 && \text{for } i = \kappa_q, \kappa_q + 1,\\
(\mathbb{A})_{\varepsilon m,\varepsilon 1_i} &= 1 && \text{for } 2 \le i \le \kappa_q - 1,\ \varepsilon = +.-,\ m \in \mathbb{N}\setminus\{0,1,2\},\\
(\mathbb{A})_{\varepsilon m,-\varepsilon 1_1} &= 1, && \varepsilon = +,-,\\
(\mathbb{A})_{\varepsilon m,-\varepsilon 2_i} &= 1, && \varepsilon = +,-,\ I = \kappa_q, \kappa_q + 1.
\end{aligned}
$$

If $\partial\mathcal{M}(f_q^\star)$ denotes again the set of preimages of the points $x = \pm r_q$ and the point $x = 0$ one shows as in the former cases that the map

$$\pi\colon I_{R_q} \setminus \partial\mathcal{M}(f_q^\star) \to F_{\mathbb{A}}^{\mathbb{N}} \setminus \{\underline{\xi} \in F_{\mathbb{A}}^{\mathbb{N}} : \exists n \in \mathbb{Z}_{\ge 0} : \tau^n(\underline{\xi}) = \pm\underline{r}_q\}$$

is bijective and the map $f_q^\star$ is conjugated therefore on $I_{R_q} \setminus \partial\mathcal{M}(f_q^\star)$ to the shift τ on $F_{\mathbb{A}}^{\mathbb{N}} \setminus \{\underline{\xi} \in F_{\mathbb{A}}^{\mathbb{N}} : \exists n \in \mathbb{Z}_{\ge 0} : \tau^n(\underline{\xi}) = \pm\underline{r}_q\}$.

Hence also in the case q is odd the map $f_q^\star$ is conjugate to a subshift of infinite type.

4.4. A sofic system related to $f_q^\star$ and the dual regular λ_q-CF. In the case of the map $f_3^\star$ the subshift $\tau\colon F_{\mathbb{A}}^{\mathbb{N}} \to F_{\mathbb{A}}^{\mathbb{N}}$ can be easily related to the dual λ_3-CF: from the form of the transition matrix $(\mathbb{A})$ in (4.3) it follows that the sequence $\underline{b} \in F_{\mathbb{A}}^{\mathbb{N}}$ with $\underline{b} = \pi(x)$ can be characterized by the fact that $(b_i, b_{i+1}) \neq (m, 2\,\mathrm{sign}(m))$ for all $i \in \mathbb{N}$ and hence $\underline{b} \in \mathcal{A}_3^{\mathrm{dreg}}$. On the other hand any such sequence $\underline{b} \in \mathcal{A}_3^{\mathrm{dreg}}$ belongs to $F_{\mathbb{A}}^{\mathbb{N}}$ and defines a unique point $x \in I_{R_q}$ through $J_{b_1}^\star \bigcap_{l=1}^\infty (f_{b_l}^\star \circ \cdots \circ f_{b_1}^\star)^{-1} J_{b_{l+1}}^\star$. Since $x_n := [\![0; b_1, \ldots, b_n]\!]^\star \in J_{b_1}^\star \bigcap_{l=1}^n (f_{b_l}^\star \circ \cdots \circ f_{b_1}^\star)^{-1} J_{b_{l+1}}^\star$ for all n we find $\lim_{n\to\infty} x_n = x$ and hence $\pi^{-1}(\underline{b}) = [\![0; b_1, b_2, \ldots,]\!]^\star$.

To connect the subshift for the map $f_q^\star$ in the case $q = 2h_q + 2$ with the dual λ_q-CF one has to introduce the sofic systems by replacing in $\underline{\xi}$ all the symbols $\pm 1_i$, $1 \le i \le \kappa_q$ by the symbol ± 1. This corresponds to replacing the Markov partition $\mathcal{M}(f_q^\star)$ by the generating partition $\mathcal{J}(f_q^\star)$ determined by $I_{R_3} = \bigcup_{\varepsilon=+,-} \bigcup_{m=1}^\infty J_{\varepsilon m}^\star$ with $J_{\varepsilon m}^\star$ defined in (3.23). Denote by $\partial\mathcal{J}(f_q^\star)$ the set

$$\partial\mathcal{J}(f_q^\star) = \{y \in I_{R_q} : \exists n \in \mathbb{Z}_{\ge 0} \text{ with } (f_q^\star)^n(y) = \pm r_q \text{ or } (f_q^\star)^n(y) = 0\}$$

which obviously coincides with the set $\partial\mathcal{M}(f_q^\star)$. Then for the alphabet $\widehat{F} = \mathbb{Z}\setminus\{0\}$ one shows again

Lemma 4.5. *The map*

$$\widehat{\pi}\colon I_{R_q}\setminus\partial\mathcal{J}(f_q^\star)\to\mathcal{A}_q^{\mathrm{dreg}}\setminus\{\underline{b}\in\mathcal{A}_q^{\mathrm{dreg}}:\exists n\in\mathbb{Z}_{\geq 0}:\tau^n(\underline{b})=\pm\underline{r}_q\}\subset\widehat{F}^{\mathbb{N}}$$

defined by $\widehat{\pi}(x)=\underline{b}=(b_i)_{i\in\mathbb{N}}$ if $(f_q^\star)^{i-1}\in J_{b_i}^\star$ for $i\in\mathbb{N}$, is bijective, and $\widehat{\pi}\circ f_q^\star = \tau\circ\widehat{\pi}$. Its inverse, the map $\widehat{\pi}^{-1}\colon\mathcal{A}_q^{\mathrm{dreg}}\to I_{R_q}$ can be defined on the entire set $\mathcal{A}_q^{\mathrm{dreg}}$ and is given by $\widehat{\pi}^{-1}(\underline{b})=[\![0;b_1,b_2,\dots,]\!]^\star$.

Proof. Since $\pm\psi_{\kappa_q} = (f_q^\star)^{h_q-1}(\pm r_q)\in J_{\pm 1}^\star\cap J_{\pm 2}^\star$ this point has two different dual regular sequences $\underline{b}=\pm(1,\overline{(-1)^{h_q},-2,(-1)^{h_q}-1})$ respectively $\underline{b}=\pm(\overline{1^{h_q-1},2})$. Hence also all preimages of this point have two different dual regular sequences, but these points all have the same tail as the point $\pm r_q$. If $\underline{b}=(b_i)_{i\in\mathbb{N}}=\widehat{\pi}(x)$ assume that $\underline{b}$ contains for some $k\geq 0$ and some $l\geq 0$ a subsequence $b_{k+1}=\dots=b_{k+l}=\varepsilon 1$ with either $k=0$ or $b_k\neq\varepsilon 1$ and $b_{k+l+1}\neq\varepsilon 1$. Then the sequence $\underline{\xi}=(\xi_i)_{i\in\mathbb{N}}$ related to $\underline{b}$ by replacing the symbols $\pm 1_i$ by the symbol ± 1 must be of the form $\xi_{k+l}=\varepsilon 1_{\kappa_q}$ and hence $\xi_{k+i}=\varepsilon 1_{\kappa_q-(l-i)}$ for $1\leq i\leq l$. This shows that $l\leq\kappa_q$. The case $l=\kappa_q$ however is only possible if either $k=0$ or $b_k=-\varepsilon m$, $m\geq 1$. This shows that $\underline{b}=\widehat{\pi}(x)\in\mathcal{A}_q^{\mathrm{dreg}}$.

Given on the other hand $\underline{b}\in\mathcal{A}_q^{\mathrm{dreg}}$ with a subsequence $b_{k+1}=\dots=b_{k+l}=\varepsilon 1$ and $b_{k+l+1}\neq\varepsilon 1$ for some k and some l then define the sequence $\underline{\xi}$ such that $\xi_{k+i}=\varepsilon 1_{\kappa_q-(l-i)}$, $1\leq i\leq l$. Since $l\leq\kappa_q$ respectively $l\leq\kappa_q-1$ the sequence $\underline{\xi}$ belongs to $F_{\mathbb{A}}^{\mathbb{N}}$ and hence there exists a point $x\in I_{R_q}$ with $\pi(x)=\underline{\xi}$ and therefore by construction also $\widehat{\pi}(x)=\underline{b}$. The inverse map $\widehat{\pi}^{-1}$ is obviously defined for all $\underline{b}\in\mathcal{A}_q^{\mathrm{dreg}}$. An argument completely analogous to the one in the case $q=3$ then shows that $\widehat{\pi}^{-1}(\underline{b})=[\![0;b_1,b_2,\dots]\!]^\star$. □

Introduce finally in the case $q=2h_q+3$ for the map $f_q^\star$ the sofic system defined by replacing in $\underline{\xi}\in F_{\mathbb{A}}^{\mathbb{N}}$ all the symbols $\pm 1_i$ by the symbol ± 1 and the symbols $\pm 2_i$ by the symbol ± 2. Denote by $\mathcal{J}(f_q^\star)$ the corresponding generating partition $I_{R_q}=\bigcup_{\varepsilon=+,-}\bigcup_{m=1}^{\infty}J_{\varepsilon m}^\star$ and by $\partial\mathcal{J}(f_q^\star)$ the set of preimages of the points $x=\pm r_q$ and $x=0$ which obviously coincides with the set $\partial\mathcal{M}(f_q^\star)$. As in the previous cases one shows also for $q=2h_q+3$

Lemma 4.6. *The map*

$$\widehat{\pi}\colon I_{R_q}\setminus\partial\mathcal{J}(f_q^\star)\to\mathcal{A}_q^{\mathrm{dreg}}\setminus\{\underline{b}\in\mathcal{A}_q^{\mathrm{dreg}}:\underline{b}\text{ has the tail }\pm\underline{r}_q\}\subset\widehat{F}^{\mathbb{N}}$$

defined by $\widehat{\pi}(x)=\underline{b}=(b_i)_{i\in\mathbb{N}}$ if $(f_q^\star)^{i-1}\in J_{b_i}^\star$ for $i\in\mathbb{N}$, is bijective and $\widehat{\pi}\circ f_q^\star=\tau\circ\widehat{\pi}$. Its inverse, the map $\widehat{\pi}^{-1}\colon\mathcal{A}_q^{\mathrm{dreg}}\to I_{R_q}$ can be defined on the entire set $\mathcal{A}_q^{\mathrm{dreg}}$ and is given by $\widehat{\pi}^{-1}(\underline{b})=[\![0;b_1,b_2,\dots]\!]^\star$.

Proof. From the form of the transition matrix $\mathbb{A}$ in Table 3 it is clear that there are only restrictions on the symbol sequence $\underline{b}$ for $\underline{b}=(b_i)_{i\in\mathbb{N}}=\widehat{\pi}(x)$ if it contains subsequences of consecutive symbols ± 1 since $(\mathbb{A})_{i,j}=1$ for all $j\in F$ if $|i|\geq 3$. Assume $b_k=m\neq 1$ and $b_{k+1}=\dots=b_{k+l}=\pm 1$, $b_{k+l+1}\neq\pm 1$ for some $k\geq 0$ and some $l\geq 1$, where $k=0$ means that $b_1=\pm 1$. Then either $(f_q^\star)^{k+l-1}(x)\in J_{\pm 1_{2h_q-1}}^\star$ or $(f_q^\star)^{k+l-1}(x)\in J_{\pm 1_{2h_q}}^\star$ and hence $\xi_{k+l}=\pm 1_{2h_q-1}$ or $\xi_{k+l}=\pm 1_{2h_q}$. In the first case $\xi_{k+i}=\pm 1_{2h_q-2(l-i)-1}$, $1\leq i\leq l$ and hence $\xi_{k+1}=\pm 1_{2h_q-(2l-1)}$. If $m=\pm n$ for some $n\geq 3$ then necessarily $l\leq h_q-1$ since

$(\mathbb{A})_{m,\pm 1_1} = 0$ for all $m \in F$. If on the other hand $m = \mp n$ for some $n \geq 3$ or $k = 0$ then $l < h_q$ with $l = h_q$ iff $\xi_{k+1} = \pm 1_1$. In the case $(f_q^\star)(x) \in J^\star_{\pm 1_{2h_q}}$ we find $\xi_{k+i} = \pm 1_{2h_q - 2(l-i)}$, $1 \leq i \leq l$ and hence $\xi_{k+1} = \pm 1_{2+2(h_q - l)}$. This shows that also in this case $l \leq h_q$. In the symbol sequence $\underline{b}$ there can appear therefore no subsequence of more than h_q consecutive symbols ± 1.

Assume next that there exists in $\underline{b}$ a subsequence of h_q consecutive symbols ± 1 such that $b_{k+1} = \cdots = b_{k+h_q} = \pm 1$ and $b_{k+h_q+1} \neq= \pm 1$. Then either $k = 0$, that means $b_{k+1} = b_1$, or $b_k = \mp n$ for some $n \neq 1$. Then $(f_q^\star)^{k+h_q-1}(x) \in J^\star_{\pm 1_{2h_q-1}}$ or $(f_q^\star)^{k+h_q-1}(x) \in J^\star_{\pm 1_{2h_q}}$ and hence $\xi_{k+h_q} = \pm 1_{2h_q-1}$ respectively $\xi_{k+h_q} = \pm 1_{2h_q}$. The transition matrix $\mathbb{A}$ in Table 3 then shows that in the first case $\xi_{k+h_q+1} = \pm 2_{\kappa_q}$ and in the second case

$$\xi_{k+h_q+1} \in \{n \in F, n \neq \pm 1_i, 1 \leq i \leq \kappa_q - 1, n \neq \pm 2_{\kappa_q}\}.$$

If $\xi_{k+h_q+1} = \pm 2_{\kappa_q}$ then $\xi_{k+h_q+1+i} = \pm 1_{2i}$ for $1 \leq i \leq h_q$ and hence $\xi_{k+2h_q+1} = \pm 1_{2h_q}$.

If in the second case $\xi_{k+h_q+1} = \pm 2_{\kappa_q+1}$ then the maximal number of consecutive symbols ± 1 in $\underline{b}$ is $h_q - 1$ since in this case $\xi_{k+h_q+2} = \pm 1_i$ for some $i \geq 3$ and only for $i = 3$ one has $\xi_{k+2h_q} = \pm 1_{2h_q-1}$. In all other cases when $b_{k+h_q+1} \neq \pm 2$ the number of consecutive symbols ± 1 is certainly bounded by h_q. This shows that in the sequence $\underline{b} = \widehat{\pi}(x)$ the subsequence $(\pm m, (\pm 1)^h_q, \pm 2, (\pm 1)^h_q, \pm 2)$ and the subsequence $(\pm 1)^{h_q+1}$ cannot appear. Hence $\underline{b} = \widehat{\pi}(x) \in \mathcal{A}_q^{\mathrm{dreg}}$. Since also in this case to every $\underline{b} \in \mathcal{A}_q^{\mathrm{dreg}}$ there exist a unique $\underline{\xi} \in F_{\mathbb{A}}^{\mathbb{N}}$ which is related to $\underline{b}$ by replacing the symbols $\pm 1_i$ respectively the symbols $\pm 2_i$ by the symbols ± 1 respectively ± 2, the same arguments as in the previous cases apply to show, that the inverse map $\widehat{\pi}^{-1}$ is given by $\widehat{\pi}^{-1}(\underline{b}) = [\![0; b_1, b_2, \ldots]\!]^\star$. □

4.5. Symbolic dynamics and the natural extension F_q of the map f_q. Consider the maps f_q and $f_q^\star$. Since

$$f_q([\![0; a_1, a_2, a_3, a_4, \ldots]\!]) = [\![0; a_2, a_3, a_4, \ldots]\!],$$
$$f_q^\star([\![0; a_0, a_{-1}, a_{-2}, a_{-3}, \ldots]\!]^\star) = [\![0; a_{-1}, a_{-2}, a_{-3}, \ldots]\!]^\star,$$

f_q and $f_q^\star$ are equivalent to the shift map τ on one sided infinite sequences $\underline{a}_> := (a_i)_{i \in \mathbb{N}} \in \mathcal{A}_q^{\mathrm{reg}}$ respectively $\underline{a}_< := (a_i)_{i \in \mathbb{Z}_{\leq 0}} \in \mathcal{A}_q^{\mathrm{dreg}}$. Denote by $\mathcal{A}_q$ the set of two-sided infinite sequences

$$\mathcal{A}_q = \{\underline{a} = (a_i)_{i \in \mathbb{Z}} : \forall l \in \mathbb{Z}, \forall k > 0 : (a_l, a_{l+1}, \ldots, a_{l+k}) \notin \mathcal{B}_q\},$$

where $\mathcal{B}_q$ was defined in (2.8). The natural extension of the one-sided shift map τ is the two-sided shift $\tau \colon \mathcal{A}_q \to \mathcal{A}_q$ with

$$(4.5) \qquad \begin{aligned} \left(\tau(\underline{a})\right)_i &= a_{i+1}, \; i \in \mathbb{Z} \quad \text{respectively its inverse,} \\ \left(\tau^{-1}(\underline{a})\right)_i &= a_{i-1}, \; i \in \mathbb{Z} \quad \text{if } \underline{a} = (a_i)_{i \in \mathbb{Z}}. \end{aligned}$$

The natural extension F_q of the map f_q respectively its inverse F_q^{-1} can then be identified simply with the corresponding induced maps on pairs of points (x, y) with regular respectively dual regular λ_q-CF

$$x = [\![0; a_1, a_2, a_3, a_4, \ldots]\!] \quad \text{and} \quad y = [\![0; a_0, a_{-1}, a_{-2}, a_{-3}, a_{-4}, \ldots]\!]^\star,$$

as long as the two-sided sequence $\underline{a} = (a_i)_{i\in\mathbb{Z}}$ belongs to $\mathcal{A}_q$. Then, F_q and F_q^{-1} satisfy

$$(4.6)\qquad \begin{aligned} F_q(\llbracket 0; a_1, a_2, \ldots \rrbracket, \llbracket 0; a_0, \ldots \rrbracket^\star) &= (\llbracket 0; a_2, \ldots \rrbracket, \llbracket 0; a_1, a_0, \ldots \rrbracket^\star), \\ F_q^{-1}(\llbracket 0; a_1, \ldots \rrbracket, \llbracket 0; a_0, a_{-1}, \ldots \rrbracket^\star) &= (\llbracket 0; a_0, a_1, \ldots \rrbracket, \llbracket 0; a_{-1}, \ldots \rrbracket^\star). \end{aligned}$$

To characterize the set Ω_q of pairs (x, y) with the above property, define in a first step $I_q^\star := I_q \setminus \{x \text{ has a finite regular } \lambda_q\text{-CF}\}$. Obviously $I_q^\star$ has full measure. Denote next by $\Pi_1 \colon \mathcal{A}_q \to I_q^\star$ the map

$$(4.7)\qquad \Pi_1(\ldots, a_{-1}, a_0; a_1, a_2, a_3 \ldots) = \llbracket 0; a_1, a_2, a_3, \ldots \rrbracket.$$

By construction the following lemma holds.

Lemma 4.7. *The map Π_1 is surjective and satisfies $\Pi_1 \circ f_q = \tau \circ \Pi_1$.*

Next, define $I_{R_q}^\star := I_{R_q} \setminus \{y \text{ has a finite dual regular } \lambda_q\text{-CF}\}$ which has full measure. Similar to (4.7) the map $\Pi_2 \colon \mathcal{A}_q \to I_{R_q}^\star$ given by

$$(4.8)\qquad \Pi_2(\ldots, a_{-2}, a_{-1}, a_0; a_1, \ldots) = \llbracket 0; a_0, a_{-1}, a_{-2}, \ldots \rrbracket^\star.$$

is well defined, surjective and satisfies $\Pi_2 \circ f_q^\star = \tau^{-1} \circ \Pi_2$. The following Lemma characterizes the domain of definition Ω_q of the natural extension F_q:

Lemma 4.8. *For $\underline{a} \in \mathcal{A}_q$ we have*

$$\Pi_2(\underline{a}) \in \pm[\psi_{\kappa_q - i+1}, R_q] \qquad \text{if } \Pi_1(\underline{a}) \in \pm\Phi_i = \pm[\phi_{i-1}, \phi_i],\ i \in \{1, \ldots, \kappa_q\}.$$

The set $\Omega_q \subset I_q \times I_{R_q}$ hence is given by

$$\Omega_q = \bigcup_{i=1}^{\kappa_q} ([\phi_{i-1}, \phi_i] \times [\psi_{\kappa_q-i+1}, R_q]) \cup ([-\phi_i, -\phi_{i-1}] \times [-R_q, -\psi_{\kappa_q-i+1}]).$$

Proof. Without loss of generality assume $\Pi_1(\underline{a}) \in \Phi_i \subset I_q$ where Φ_i is defined in (3.13).

For q even Lemma 3.3 and the λ_q-CF of $-\frac{\lambda}{2}$ in (2.11) show

$$\llbracket 0; (1)^{h_q-i+1} \rrbracket \preceq \Pi_1(\underline{a}) \preceq \llbracket 0; (1)^{h_q-i} \rrbracket.$$

Hence $\Pi_1(\underline{a})$ has a λ_q-CF of the form $\Pi_1(\underline{a}) = \llbracket 0; (1)^{h_q-i}, m, \ldots \rrbracket$ for some $m \geq 2$. Since $\underline{a}$ is q-regular we have

$$\underline{a} = (\ldots, a_{-i-1}, a_{-i}, a_{-i+1}, \ldots, a_0; (1)^{h_q-i}, m, a_{h_q-i+2}, \ldots)$$

with at most $i-1$ consecutive 1's in the sequence $(a_{-i+2}, \ldots, a_0)$. The point $\Pi_2(\underline{a})$ hence is bounded by the largest and smallest number whose dual regular λ_q-CF starts with at most $i-1$ consecutive 1's and hence

$$\llbracket 0; (1)^{i-1}, 2, \overline{(1)^{h_q-i}, 2} \rrbracket^\star \preceq \Pi_2(\underline{a}) \preceq \llbracket 0; (-1)^{h_q}, -2, \overline{(-1)^{h_q-1}, -2} \rrbracket^\star.$$

But (3.11) and (2.15) show that these bounds are just ψ_{κ_q-i+1} and R_q.

The case q odd, $q \geq 5$, is slightly more complicated. First, assume i to be even and put $j = i/2$. Then by Lemma 3.3 and the λ_q-CF of $-\lambda/2$ in (2.11)

$$\Pi_1(\underline{a}) \in \Phi_{2j} = \left[\llbracket 0; (1)^{h_q-j+1} \rrbracket, \llbracket 0; (1)^{h_q-j}, 2, (1)^{h_q} \rrbracket\right].$$

Hence $\Pi_1(\underline{a})$ has a λ_q-CF of the form $\Pi_1(\underline{a}) = \llbracket 0; (1)^{h_q-j}, 2, (1)^{h_q}, m, \ldots \rrbracket$ for some $m \geq 2$. Since $\underline{a} \in \mathcal{A}_q$ the sequence $(a_{-j+1}, \ldots, a_0)$ in

$$\underline{a} = (\ldots, a_{-j}, a_{-j+1}, \ldots, a_0; (1)^{h_q-j}, 2, (1)^{h_q}, m, a_{2h_q-j+2}, \ldots)$$

cannot contain more than $j-1$ consecutive digits 1. Hence $\Pi_2(\underline{a})$ is bounded by the points

$$[\![0;(1)^{j-1},2,\overline{(1)^h,2,(1)^{h-1},2}]\!]^\star \preceq \Pi_2(A) \preceq [\![0;(-1)^h,-2,\overline{(-1)^h,-2,(-1)^{h-1},-2}]\!]^\star.$$

which by Lemma 3.3 and (2.15) are just ψ_{κ_q-2j+1} and R_q.

Next, consider the case i odd and put $j=(i-1)/2$ for $1\le i\le h_q$. Again, by Lemma 3.3 and (2.15)

$$\Pi_1(\underline{a})\in\Phi_{2j+1}=\left[[\![0;(1)^{h_q-j},2,(1)^{h_q}]\!],[\![0;(1)^{h_q-j}]\!]\right],$$

and therefore $\Pi_1(\underline{a})$ has a λ_q-CF of the form $\Pi_1(\underline{a})=[\![0;(1)^{h_q-j},m,\dots]\!]$ for some $m\ge 2$. Hence there is again a restriction on the sequence $\underline{a}$:

$$\underline{a}=(\dots,a_{-h-j-1},\underbrace{a_{-h-j},\dots,a_0;(1)^{h-j},m}_{\text{not a forbidden block}},a_{h-j+1},\dots),$$

and therefore $\Pi_2(\underline{a})$ is bounded by

$$\begin{aligned}[\![0;(1)^j,2,\overline{(1)^{h_q-1},2,(1)^{h_q},2}]\!]^\star &\preceq \Pi_2(A)\\ &\preceq [\![0;(-1)^{h_q},-2,[0;(-1)^{h_q},-2,\overline{(-1)^{h_q},-2,(-1)^{h_q-1},-2}]\!]^\star,\end{aligned}$$

which by Lemma 3.3 and (2.11) respectively (2.15) are just ψ_{κ_q-2j} and R_q .

Finally for $q=3$ Lemma 3.3 and (2.15) show $[\![0;2]\!]\preceq\Pi_1(\underline{a})\prec[\![0;]\!]$, and therefore $\Pi_1(\underline{a})$ has a λ_3-CF of the form $\Pi_1(\underline{a})=[\![0;m,\dots]\!]$ for some $m\ge 2$. Hence $\Pi_2(\underline{a})$ must not have a leading digit 2. This implies the bounds $r_3=[0;\overline{3}]\preceq\Pi_2(\underline{a})\preceq[0;-2,\overline{-3}]=R_3$.

The case $\Pi_1(\underline{a})\in\Phi_{-i}$ for some $1\le i\le\kappa_q$ follows from $\Pi_1(-\underline{a})=-\Pi_1(\underline{a})$. □

Recall the definition of the domain Ω_q in Lemma 4.8 and define the set $\Omega_q^\star=\Omega_q\cap(I_q^\star\times I_{R_q}^\star)$, which obviously is dense in Ω_q. Then one has

Lemma 4.9. *The map* $\Pi\colon\mathcal{A}_q\to\Omega_q^\star$ *with* $\Pi(\underline{a})=\big(\Pi_1(\underline{a}),\Pi_2(\underline{a})\big)$ *is a bijection. For* $F_q\colon\Omega_q^\star\to\Omega_q^\star$ *and* $F_q^{-1}\colon\Omega_q^\star\to\Omega_q^\star$ *given by* (4.6) *the diagrams*

$$\begin{array}{ccc}\mathcal{A}_q & \xrightarrow{\ \tau\ } & \mathcal{A}_q\\ {\scriptstyle\Pi}\downarrow & & \downarrow{\scriptstyle\Pi}\\ \Omega_q^\star & \xrightarrow{\ F_q\ } & \Omega_q^\star\end{array}\qquad\textit{and}\qquad\begin{array}{ccc}\mathcal{A}_q & \xrightarrow{\ \tau^{-1}\ } & \mathcal{A}_q\\ {\scriptstyle\Pi}\downarrow & & \downarrow{\scriptstyle\Pi}\\ \Omega_q^\star & \xrightarrow{\ F_q^{-1}\ } & \Omega_q^\star\end{array}$$

commute.

PROOF. Obviously, the map Π is well defined. Commutativity of the diagrams follows from combining Lemma 4.7 and

$$\begin{aligned}\Pi_2(\tau(\underline{a}))&=\Pi_2(\tau(\dots,a_{-1},a_0;a_1,a_2,\dots))=\Pi_2(\dots,a_{-1},a_0,a_1;a_2,\dots)\\ &=[\![0;a_1,a_0,a_{-1},\dots]\!]^\star=\frac{-1}{[\![a_0,a_{-1},\dots]\!]^\star+a_1\lambda_q}=\frac{-1}{\Pi_2(\underline{a})+a_1\lambda_q}\end{aligned}$$

respectively

$$\begin{aligned}\Pi_1(\tau^{-1}(\underline{a}))&=\Pi_1(\tau^{-1}(\dots,a_{-1},a_0;a_1,\dots))=\Pi_1(\dots,a_{-1};a_0,a_1,\dots)\\ &=[\![0;a_0,a_1,a_2,\dots]\!]=\frac{-1}{[\![a_1,a_2,\dots]\!]+a_0\lambda_q}=\frac{-1}{\Pi_1(\underline{a})+a_0\lambda_q}.\end{aligned}$$

Since the map $\Pi\colon \mathcal{A}_q \to \Omega_q^\star$ is obviously injective we only need to show $\Pi(\mathcal{A}) = \Omega_q^\star$. For this take $(x,y) \in \Omega_q^\star$ with

$$x = [\![0; a_1, a_2, a_3, \dots]\!] \quad \text{and} \quad y = [\![0; a_0, a_{-1}, a_{-2}, \dots]\!]^\star.$$

If $(x,y) \in ([\phi_{i-1}, \phi_i] \cap I_q^\star) \times ([\psi_{\kappa_q - i+1}, \psi_0] \cap I_{R_q}^\star)$ for some $i \in \{1, \dots, \kappa_q\}$ the definitions of ϕ_i and ψ_i as elements of orbit$(-\lambda_q/2)$ and orbit$(-R_q)$ respectively imply that the bi-infinite sequence

$$\underline{a} := (\dots, a_{-2}, a_{-1}, a_0; a_1, a_2, a_3, \dots)$$

does not contain forbidden blocks, and hence $\underline{a} \in \mathcal{A}_q$. □

It is well known that the map $\Pi\colon \mathcal{A}_q \to \mathbb{R}^2$ is continuous (see for instance [19]) when $\mathcal{A}_q$ is equipped with the usual metric of the shift space.

5. Some applications

5.1. Reduced geodesics on $G_q \setminus \mathbb{H}$ and the natural extension of f_q. The Poincaré upper half-plane, equipped with the hyperbolic metric $\mathrm{d}s$ given by $\mathrm{d}s^2 = (\mathrm{d}x^2 + \mathrm{d}y^2)/y^2$, is denoted by $\mathbb{H} = \{z \in \mathbb{C}; \mathrm{Im}(z) > 0\}$. The group of isometries of this space is given by $\mathrm{PSL}(2,\mathbb{R})$. The boundary of $\mathbb{H}$ is the projective line $\mathbb{P}^1_{\mathbb{R}}$.

We consider oriented geodesics on $\mathbb{H}$. Geodesic lines on $\mathbb{H}$ are half-circles perpendicular to $\mathbb{R}$ or straight lines parallel to the imaginary axis $\mathrm{Re}(z) = 0$. An oriented geodesic ω on $\mathbb{H}$ will be represented by the two base points $\omega_-, \omega_+ \in \mathbb{R} \cup \{\mathrm{i}\infty\}$ with its orientation from ω_- towards ω_+. We denote such a geodesic by $\omega = (\omega_-, \omega_+)$.

We call two oriented geodesics ω and υ *G_q-equivalent* if there exists an element $g \in G_q$ with $g\,\omega_- = \upsilon_-$ and $g\,\omega_+ = \upsilon_+$.

Then one can show

Theorem 5.1. *Let $\omega = (\omega_-, \omega_+)$ be a geodesic with ω_- having an infinite regular respectively ω_+ having an infinite dual regular λ_q-CF. Then there exist a geodesic $\omega' = (\omega'_-, \omega'_+)$ such that*

- *ω and ω' are G_q-equivalent and*
- *$(S\,\omega'_+, -\omega'_-) \in \Omega_q$.*

PROOF. We prove the theorem first for $q \geq 4$. Using translations by powers of T_q we may assume that for $\omega = (\omega_-, \omega_+)$ either

- $\omega_+ > 0$ and $\omega_- \in [-R_q, -r_q] \subset I_{R_q}$ or
- $\omega_+ < 0$ and $\omega_- \in [r_q, R_q] \subset I_{R_q}$.

Assume $\omega_+ > 0$ and $\omega_- \in [-R_q, -r_q]$ with infinite λ_q-CFs

$$\omega_+ = [\![a_0; a_1, a_2, \dots]\!] \quad \text{and} \quad \omega_- = [\![0; b_1, b_2, \dots]\!]^\star. \tag{5.1}$$

For $x = S\,\omega_+$ and $y = -\omega_-$ we have

$$x = \begin{cases} [0; a_0, a_1, a_2, \dots] & \text{if } a_0 \neq 0, \\ [\![a_1; a_2 \dots]\!] & \text{if } a_0 = 0, \end{cases} \qquad y = [\![0; -b_1, -b_2, \dots]\!]^\star. \tag{5.2}$$

The following three cases have to be discussed: "$a_0 \geq 2$," "$a_0 = 1$" and "$a_0 = 0$."

If $a_0 \geq 2$ then $\omega_+ \geq 2\lambda - \frac{\lambda}{2}$ and the λ_q-CF of x in (5.2) is regular. If the two-sided sequence

$$\underline{a} := (\dots, -b_2, -b_1; a_0, a_1, \dots) \tag{5.3}$$

belongs to $\mathcal{A}_q$ then by Lemma 4.9 $(S\,\omega_+, -\omega_-) = (x, y) = \Pi(\underline{a}) \in \Omega_q^\star \subset \Omega_q$.

Assume therefore $\underline{a} \notin \mathcal{A}_q$. The forbidden block must appear around ";" in (5.3). This can only happen for q even if $(-b_{h_q}, \cdots - b_1) = (1^{h_q})$ respectively for q odd if $(-b_{2h_q+1}, \cdots - b_1) = (1^{h_q}, 2, 1^{h_q})$. Using the lexicographic order "$\prec$" in Section 2.4 we find

$$-\omega_- = [\![0; -b_1, -b_2, \dots]\!]^\star = \begin{cases} [\![0; 1^{h_q}, \dots]\!]^\star & \text{for } q \text{ even} \\ [\![0; 1^{h_q}, 2, 1^{h_q}, \dots]\!]^\star & \text{for } q \text{ odd} \end{cases}$$
$$\prec \begin{cases} [\![0; \overline{(1)^{h_q-1}, 2}]\!] & \text{for } q \text{ even} \\ [\![0; \overline{(1)^{h_q}, 2, (1)^{h_q-1}, 2}]\!] & \text{for } q \text{ odd, } q \geq 5 \end{cases} = r_q.$$

Then Lemma 2.8 implies $[\![0; -b_1, -b_2, \dots]\!]^\star < r_q$.

On the other hand, $-\omega_- \in [r_q, R_q]$ implies $-\omega_- \geq r_q$. This leads to a contradiction.

If $\boldsymbol{a_0 = 0}$ then $\omega_+ \in (0, \lambda_q/2] \subset (0, 2/\lambda_q)$. Hence $a_1 < 0$ in the λ_q-CF (5.1). For $m \in \mathbb{Z}_{\geq 2}$ the sequence $(-m, a_1, a_2, \dots)$ is q-regular and also $(-m, b_1, b_2, \dots)$ is dual q-regular, since $r_q \leq -\omega_-$ implies

$$r_q = \begin{cases} [\![0; \overline{(1)^{h_q-1}, 2}]\!]^\star & \text{if } q \text{ is even} \\ [\![0; \overline{(1)^{h_q}, 2, (1)^{h_q-1}, 2}]\!]^\star & \text{if } q \text{ is odd} \end{cases}$$
$$\preceq [\![0; -b_1, -b_2, \dots]\!]^\star = -\omega_-.$$

For $g := ST_q^{-m}$ define $(\omega'_-, \omega'_+) = \omega' := g\,\omega$ with

$$\omega'_+ = [\![0; -m, a_1, a_2, \dots]\!] > 0,$$
$$\omega'_- = [\![0; -m, b_1, b_2, \dots]\!]^\star \in (0, -r_q] \subset [r_q, R_q].$$

The corresponding bi-infinite sequence $\underline{a}' := (\dots, -b_1, m; -m, a_1, \dots)$ is then q-regular and Lemma 4.9 hence implies $(S\,\omega'_+, -\omega'_-) = \Pi(\underline{a}') \in \Omega_q^\star \subset \Omega_q$.

For $\boldsymbol{a_0 = 1}$

$$\omega_+ \geq \lambda_q - \frac{\lambda_q}{2} = \begin{cases} [\![1; (1)^{h_q}]\!] & \text{for } q \text{ even,} \\ [\![1; (1)^{h_q}, 2, (1)^{h_q}]\!] & \text{for } q \text{ odd} \end{cases}$$

$$-R_q \leq \omega_- \leq -r_q = \begin{cases} [\![0; \overline{(-1)^{h_q-1}, -2}]\!]^\star & \text{for } q \text{ even,} \\ [\![0; \overline{(-1)^{h_q}, -2, (-1)^{h_q-1}, -2}]\!]^\star & \text{for } q \text{ odd.} \end{cases}$$

Then the dual regular λ_q-CF of y cannot start for q even with a block of the form (1^{h_q}) respectively for q odd $(1^{h_q}, 2, 1^{h_q})$.

Assume $\underline{a} \notin \mathcal{A}_q$, with $\underline{a}$ defined as in (5.3). Then a block of the form $(1^l; 1, 1^t)$ with $0 \leq l \leq h$ and $0 \leq t \leq h$ must exist around the ";" in $\underline{a}$ such that $a_{t+1} \neq 1$, $-b_{l+1} \neq 1$. As in the case $a_0 = 0$, choose an $m \in \mathbb{Z} \setminus \{-2, -1, 0, 1\}$ with $(m + 1, a_1, a_2, \dots)$ being q-regular and $(m, b_1, b_2, \dots)$ being dual q-regular: indeed any m with $\operatorname{sign}(m) = -\operatorname{sign}(b_1)$ can be used. Define $g := ST_q^m$ and $(\omega'_-, \omega'_+) = \omega' := g\,\omega$ with

$$\begin{aligned} \omega'_+ &= [\![0; m+1, a_1, a_2, \dots]\!], \\ \omega'_- &= [\![0; m, b_1, b_2, \dots]\!]^\star \in [r_q, -r_q]. \end{aligned} \tag{5.4}$$

Then the bi-infinite sequence $\underline{a}' := (\dots, -b_2, -b_1, -m; m+1, a_1, a_2, \dots)$ is q-regular and by Lemma 4.9 $(S\omega'_+, -\omega'_-) = \Pi(\underline{a}') \in \Omega_q$.

The case $\omega_+ < 0$ and $\omega_- \in [r_q, R_q] \subset I_{R_q}$ can be treated in the same way.

The proof for $q = 3$ is similar to the case $q \geq 4$, however there are the four cases $a_0 \geq 3$, $a_0 = 2$, $a_0 = 1$ and $a_0 = 0$ to be considered.

If $a_0 \geq 3$ then we can argue as in the case $a_0 \geq 2$ before. Since $\omega_- < -r_3 = [\![0; \overline{-3}]\!]^\star$, the bi-infinite sequence $\underline{a}$ in (5.3) is q-regular and $(S\omega_+, -\omega_-) = (x, y) = \Pi(\underline{a}) \in \Omega_q$.

The cases $a_0 = 2$ and $a_0 = 1$ are similar to the case $a_0 = 1$ for $q \geq 4$: we just take the integer $|m| \geq 5$ with $m\, b_1 < 0$. Then ω' is defined as $(\omega'_-, \omega'_+) = \omega' := g\,\omega$ with $g := ST^m$ and hence

$$\omega'_+ = \begin{cases} [\![0; m+2, a_1, a_2, \dots]\!] & \text{if } a_0 = 2, \\ [\![0; m+1, a_1, a_2, \dots]\!] & \text{if } a_0 = 1, \end{cases}$$

$$\omega'_- = [\![0; m, b_1, b_2, \dots]\!]^\star \in [r_3, -r_3].$$

The case $a_0 = 0$ is similar to the case $a_0 = 0$ for $q \geq 4$, if we choose there the integer $m \geq 3$ and recall $r_3 = [\![0; \overline{3}]\!]$ in (2.12). □

5.2. The transfer operator for G_q. The authors of [14] have constructed a Poincaré section Σ for the geodesic flow $\Phi_t\colon S_1\, G_q \setminus \mathbb{H} \to S_1\, G_q \setminus \mathbb{H}$ on the Hecke surfaces $G_q \setminus \mathbb{H}$ for which the Poincaré map $P\colon \Sigma \to \Sigma$ is basically given by the natural extension F_q of the map $f_q\colon I_q \to I_q$. The periodic orbits of this geodesic flow can therefore be characterized by the periodic orbits of F_q and therefore also by the periodic orbits of the map f_q respectively its periodic points which determine the ones of F_q uniquely. Indeed, Theorem 2.10 implies an almost one-to-one correspondence between the periodic orbits of the geodesic flow on the Hecke surfaces $G_q \setminus \mathbb{H}$ and the periodic orbits of the map f_q, only the periodic orbits of the points r_q and $-r_q$ which are not equivalent under the map f_q lead to the same periodic orbit of the geodesic flow since these points are G_q equivalent. This shows already that the Selberg zeta function Z_{G_q} for the Hecke triangle groups defined as

$$Z_{G_q}(\beta) = \prod_{k=0}^{\infty} \prod_{\gamma \text{ prime}} \left(1 - \mathrm{e}^{-(\beta+k)l(\gamma)}\right),$$

where the product is over the prime periodic orbits γ of the geodesic flow and $l(\gamma)$ denotes its period (and hence the length of the corresponding closed geodesic), cannot be expressed in terms of the transfer operator for the map f_q alone. Indeed, to relate the above Selberg zeta function to the Poincaré map P one uses the following lemma by Ruelle [18]:

Lemma 5.2.

$$Z_{G_q}(\beta) = \prod_{k=0}^{\infty} \mathrm{e}^{-\sum_{n=1}^{\infty}(1/n) Z_n(\beta+k)}$$

for $\mathrm{Re}(\beta) > 1$ *where* $Z_n(\beta)$ *is the so-called partition function*

$$Z_n(\beta) = \sum_{x \in \mathrm{Fix}\, P^n} \mathrm{e}^{-\beta \sum_{l=0}^{\infty} r(P^l(x))}$$

and $r\colon \Sigma \to \mathbb{R}^+$ *denotes the recurrence time of the geodesic flow with respect to the Poincaré section* Σ.

In the transfer operator approach to the dynamical zeta functions the partition functions $Z_n(\beta)$ get expressed in terms of the traces of an operator constructed from the Poincaré map $P\colon \Sigma \to \Sigma$ respectively its restriction to the unstable directions. In our case the unstable direction is one-dimensional and the restriction of P to it is basically just the map $f_q\colon I_q \to I_q$. On the other hand one knows that the recurrence time $r\colon \Sigma \to \mathbb{R}^+$ in our case is given by $r(x) = \log|f_q'(x)|$. The Ruelle transfer operator $\mathcal{L}_\beta$ then has the following form

$$\mathcal{L}_\beta g(x) = \sum_{y \in f_q^{-1}(x)} \mathrm{e}^{-\beta r(y)} g(y) \tag{5.5}$$

where $g\colon I_q \to \mathbb{C}$ is some complex valued function and $\mathrm{Re}(\beta) > 1$ to ensure convergence of the series. To get an explicit form for the operator $\mathcal{L}_\beta$ one has to determine the preimages y of any point $x \in I_q$. For this recall the Markov partition $I_q = \bigcup_{i \in A_{\kappa_q}} \Phi_i$ with $A_{\kappa_q} = \{\pm 1, \ldots, \pm\kappa_q\}$ in (3.13), determined by the intervals Φ_i, and the local inverses $\vartheta_{\pm m}(x) := \big(f_q\big|_{J_{\pm m}}\big)^{-1}(x) = -1/(x \pm m\lambda_q)$ on the intervals $J_{\pm m}$, $1 \leq m \leq \infty$, respectively $2 \leq m \leq \infty$ for $q = 3$, defined in Section 3.3. For $1 \leq i \leq \kappa_q$ denote by $\mathcal{N}_i$ the set $\mathcal{N}_i := \{n \in \mathbb{Z} \setminus \{0\}$ such that there exists $j \in A_{\kappa_q}$ with $\vartheta_n(\Phi_i) \subset \Phi_j\}$. But then $\mathcal{N}_i = \bigcup_{j \in A_{\kappa_q}} \mathcal{N}_{i,j}$ with $\mathcal{N}_{i,j} := \{n \in \mathbb{Z} \setminus \{0\}$ such that $\vartheta_n(\Phi_i) \subset \Phi_j\}$. Using these sets we can rewrite the transfer operator $\mathcal{L}_\beta$ in (5.5) as

$$\mathcal{L}_\beta g(x) = \sum_{i \in A_{\kappa_q}} \chi_{\Phi_i}(x) \sum_{n \in \mathcal{N}_i} \big(\vartheta_n'(x)\big)^\beta g\big(\vartheta_n(x)\big), \tag{5.6}$$

with χ_{Φ_i} the characteristic function of the set Φ_i. With $g_i := g\big|_{\Phi_i}$ this can be written also as follows

$$(\mathcal{L}_\beta g)_i(x) = \sum_{j \in A_{\kappa_q}} \sum_{n \in \mathcal{N}_{i,j}} \big(\vartheta_n'(x)\big)^\beta g_j\big(\vartheta_n(x)\big), \qquad x \in I_i. \tag{5.7}$$

Thereby we used the Markov property of the partition $I_q = \bigcup_{i \in A_{\kappa_q}} \Phi_i$. If g_i is continuous on Φ_i for all $i \in A_{\kappa_q}$ then also $(\mathcal{L}_\beta g)_i$ is continuous on Φ_i, that means $\mathcal{L}_\beta$ maps piecewise continuous functions to piecewise continuous functions. Unfortunately on the Banach space $B = \oplus_{i \in A_{\kappa_q}} C(\Phi_i)$ of piecewise continuous functions the operator $\mathcal{L}_\beta$ is not trace class, it is even not compact. Much better spectral properties however can be achieved by defining $\mathcal{L}_\beta$ on a space of piecewise holomorphic functions. This is possible since all the maps $\vartheta_{\pm m}$, $m \geq 1$ have holomorphic extensions to a complex neighbourhood of I_q. Indeed one shows

Lemma 5.3. *There exist open discs $D_i \subset \mathbb{C}$, $i \in A_{\kappa_q}$ with $\Phi_i \subset D_i$ such that for all $n \in \mathcal{N}_{i,j}$ we have $\vartheta_n(\overline{D_i}) \subset D_j$.*

Consider therefore the Banach space $B = \oplus_{i \in A_{\kappa_q}} B(D_i)$ with $B(D_i)$ the Banach space of holomorphic functions on the disc D_i with the sup norm. On this space the transfer operator $\mathcal{L}_\beta$ has the form

$$(\mathcal{L}_\beta g)_i(z) = \sum_{j \in A_{\kappa_q}} \sum_{n \in \mathcal{N}_{i,j}} \left(\frac{1}{z + n\lambda_q}\right)^{2\beta} g_j\left(\frac{-1}{z + n\lambda_q}\right), \qquad z \in D_i \tag{5.8}$$

which is well defined for $\mathrm{Re}(\beta) > \frac{1}{2}$. In a forthcoming paper we will discuss the spectral properties of this operator and its relation to the Selberg zeta function for the Hecke triangle groups G_q. Here we give the explicit form of this operator for the case $q = 3$ and $q = 4$.

For $q = 3$ one has $\kappa_3 = 1$ and therefore $A_{\kappa_3} = \{\pm 1\}$. The index sets $\mathcal{N}_{i,j}$, $i, j \in A_{\kappa_3}$ are given by $\mathcal{N}_{1,1} = \mathbb{Z}_{\geq 3}$, $\mathcal{N}_{1,-1} = \mathbb{Z}_{\leq -2}$, $\mathcal{N}_{-1,1} = \mathbb{Z}_{\geq 2}$ and $\mathcal{N}_{-1,-1} = \mathbb{Z}_{\leq -3}$.

For $q = 4$ one has also $\kappa_4 = 1$ and hence $A_{\kappa_4} = \{\pm 1\}$. The index sets $\mathcal{N}_{i,j}$, $i, j \in A_{\kappa_4}$ are given by $\mathcal{N}_{1,1} = \mathbb{Z}_{\geq 2}$, $\mathcal{N}_{1,-1} = \mathbb{Z}_{\leq -1}$, $\mathcal{N}_{-1,1} = \mathbb{Z}_{\geq 1}$ and $\mathcal{N}_{-1,-1} = \mathbb{Z}_{\leq -2}$. This leads in these two cases to the following transfer operators

$$
\begin{aligned}
(\mathcal{L}_\beta g)_1(z) = & \sum_{n=3\,(2)}^{\infty} \left(\frac{1}{z+n\lambda_q}\right)^{2\beta} g_1\left(\frac{-1}{z+n\lambda_q}\right) \\
& + \sum_{n=2\,(1)}^{\infty} \left(\frac{1}{z-n\lambda_q}\right)^{2\beta} g_{-1}\left(\frac{-1}{z-n\lambda_q}\right), \quad z \in D_1, \\
(\mathcal{L}_\beta g)_{-1}(z) = & \sum_{n=2\,(1)}^{\infty} \left(\frac{1}{z+n\lambda_q}\right)^{2\beta} g_1\left(\frac{-1}{z+n\lambda_q}\right) \\
& + \sum_{n=3\,(2)}^{\infty} \left(\frac{1}{z-n\lambda_q}\right)^{2\beta} g_{-1}\left(\frac{-1}{z-n\lambda_q}\right), \quad z \in D_2
\end{aligned} \tag{5.9}
$$

where $\lambda_3 = 1$ and $\lambda_4 = \sqrt{2}$ and the summation index in brackets belongs to the case $q = 4$. For $q = 3, 4$ the discs D_i, $i = \pm 1$ can be taken as $D_{\pm 1} = \pm\{z \in \mathbb{C}; |z - (\lambda_q - 2)/4| < (\lambda_q + 2)/4\}$.

For $q = 3$ this operator and its eigenfunctions with eigenvalue $\rho = 1$ have been discussed in [3] where it was shown that these eigenfunctions are directly related to the eigenfunctions with eigenvalues $\rho = \pm 1$ of the transfer operator for the modular group G_3 derived from a symbolic dynamics for the geodesic flow using the Gauss continued fractions in [12].

6. λ_q-CFs and Rosen λ-fractions

6.1. Regular λ_q-CFs and reduced Rosen λ-fractions ($q \geq 4$). In [16] Rosen discussed continued fractions of the form

$$
[r_0; (\varepsilon_1 : r_1), (\varepsilon_2 : r_2), (\varepsilon_3 : r_3), \ldots] = r_0\lambda_q + \cfrac{\varepsilon_1}{r_1\lambda_q + \cfrac{\varepsilon_2}{r_2\lambda_q + \cfrac{\varepsilon_3}{r_3\lambda_q + \cdots}}} \tag{6.1}
$$

with $r_0 \in \mathbb{Z}$ and $\varepsilon_i = \pm 1$, $r_i \geq 1$ for $i \in \mathbb{N}$. We call such expansions *Rosen λ-fraction.*

Rosen λ-fractions and λ_q-CFs can easily be transformed into each other using the relations

$$(6.2)\quad [r_0; (\varepsilon_1 : r_1), (\varepsilon_2 : r_2), (\varepsilon_3 : r_3), \dots]$$
$$= r_0\lambda_q + \cfrac{\varepsilon_1}{r_1\lambda_q + \cfrac{\varepsilon_2}{r_2\lambda_q + \cfrac{\varepsilon_3}{r_3\lambda_q + \cdots}}}$$
$$= r_0\lambda_q + \cfrac{-1}{-\varepsilon_1 r_1\lambda_q + \cfrac{-1}{\varepsilon_1\varepsilon_2 r_2\lambda_q + \cfrac{-1}{-\varepsilon_1\varepsilon_2\varepsilon_3 r_3\lambda_q + \cdots}}}$$
$$= [r_0; -\varepsilon_1 r_1, \varepsilon_1\varepsilon_2 r_2, -\varepsilon_1\varepsilon_2\varepsilon_3 r_3, \dots, (-1)^i\varepsilon_1\cdots\varepsilon_i r_i, \dots]$$

and

$$(6.3)\quad [a_0; a_1, a_2, a_3, \dots]$$
$$= a_0\lambda_q + \cfrac{-1}{a_1\lambda_q + \cfrac{-1}{a_2\lambda_q + \cfrac{-1}{a_3\lambda_q + \cdots}}}$$
$$= a_0\lambda_q + \cfrac{-\operatorname{sign}(a_1)}{|a_1|\lambda_q + \cfrac{-\operatorname{sign}(a_1)\operatorname{sign}(a_2)}{a_2\lambda_q + \cfrac{-\operatorname{sign}(a_2)\operatorname{sign}(a_3)}{a_3\lambda_q + \dots}}}$$
$$= [a_0; (-\operatorname{sign}(a_1) : |a_1|), (-\operatorname{sign}(a_1)\operatorname{sign}(a_2) : |a_2|),$$
$$(-\operatorname{sign}(a_2)\operatorname{sign}(a_3) : |a_3|), \dots].$$

As claimed in [14, Remark 15] these relations imply directly

Lemma 6.1. *Given a Rosen λ-fraction $[r_0; \dots, (\varepsilon_i : r_i), (\varepsilon_{i+1} : r_{i+1}), \dots]$ and its corresponding λ_q-CF $[r_0; \dots, a_i, a_{i+1}, \dots]$ in* (6.3) *we have*

$$\varepsilon_1 = -\operatorname{sign}(a_1) \quad \textit{and} \quad \varepsilon_{i+1} = -\operatorname{sign}(a_i a_{i+1}), \quad (i \in \mathbb{N}).$$

PROOF. Equation (6.2) shows that the sign of the ith digit $(-1)^i\varepsilon_1\cdots\varepsilon_i r_i$ in the formal λ_q-CF is determined by $(-1)^i\varepsilon_1\cdots\varepsilon_i$. Hence, the ratio

$$\frac{(-1)^{i+1}\,\varepsilon_1\cdots\varepsilon_i\varepsilon_{i+1}}{(-1)^i\varepsilon_1\cdots\varepsilon_i} = -\varepsilon_{i+1}$$

determines whether a_i and a_{i+1} have the same or opposite signs. □

Equations (6.2) and (6.3) indeed relate regular λ_q-CFs and reduced Rosen λ-fractions as we show next. Set

$$(6.4)\qquad h_{\mathrm{R}} = h_{\mathrm{Rosen}} := \left\lfloor \frac{q-3}{2} \right\rfloor = \begin{cases} h_q - 1 & \text{if } q \text{ is even,} \\ h_q & \text{if } q \text{ is odd.} \end{cases}$$

The above h_{R} coincides with "h" used in [16, above Definition 1]. Recall from [16, p. 555]

Definition 6.2 (Reduced Rosen λ-fractions). The Rosen λ-fraction in (6.1) is called *reduced* if it satisfies the following conditions:

(1) Blocks of the form
$$(* : 1), \underbrace{(-1 : 1), \ldots, (-1, 1)}_{h_{\mathrm{R}} \text{ times}}, (-1 : *)$$
do not appear.

(2) For q odd, blocks of the form
$$(* : 1), \underbrace{(-1 : 1), \ldots, (-1 : 1)}_{h_{\mathrm{R}} \text{ times}}$$
do not appear.

(3) For q odd, blocks of the form
$$(* : 1), \underbrace{(-1 : 1), \ldots, (-1 : 1)}_{h_{\mathrm{R}} - 1 \text{ times}}, (-1 : 2), \underbrace{(-1 : 1), \ldots, (-1 : 1)}_{h_{\mathrm{R}} \text{ times}}, (-1, *)$$
do not appear.

(4) For q odd, a finite Rosen λ-fraction expansion does not terminate in a block of the form
$$(* : 1), \underbrace{(-1 : 1), \ldots, (-1 : 1)}_{h_{\mathrm{R}} \text{ times}}.$$

(5) The value $\pm\lambda_q/2$ of the tail $[(* : r_i), (\varepsilon_{i+1} : r_{i+1}), \ldots, (\varepsilon_{i+k} : r_{i+k})]$ of a finite Rosen λ-fractions leads because of $r_{i-1}\lambda_q \pm \lambda_q/2 = (r_{i-1} \pm 1)\lambda_q \mp \lambda_q/2$ to non-uniqueness of the expansion. We always choose the first possibility.

Then one shows

Lemma 6.3. *The following three statements hold:*

- *The λ_q-CF associated to a reduced Rosen λ-fraction expansion in* (6.2) *is regular.*
- *The Rosen λ-fraction expansion corresponding to a regular λ_q-CF in* (6.3) *satisfies properties* (1) – (4) *of Definition* 6.2.
- *The two expansions of the finite Rosen λ-fractions in* (5) *of Definition* 6.2 *correspond to the identities of the finite regular λ_q-CFs in Lemma* 2.2.

PROOF. Let $x \in \mathbb{R}$ have the regular Rosen λ-fraction expansion (6.1). We have to show that the corresponding λ_q-CF in (6.2) does not contain any forbidden block from $\mathcal{B}_q$. We consider the cases q even and q odd separately.

Let q be even. Using Lemma 6.1 and the identity $h_{\mathrm{R}} = h_q - 1$ in (6.4) we see that property (1) of Definition 6.2 corresponds to the absence of blocks of the form $[(\pm 1)^{h_q}, \pm m]$ for any $m \in \mathbb{Z}_{\geq 1}$.

Consider next q odd. Using again Lemma 6.1 and the identity $h_{\mathrm{R}} = h_q$ we see that property (2) of Definition 6.2 corresponds to the absence of blocks of the form $[(\pm 1)^{h_q + 1}]$. Similarly, property (3) corresponds to the absence of blocks of the form $[(\pm 1)^{h_q}, \pm 2, (\pm 1)^{h_q}, \pm m]$ for any $m \in \mathbb{Z}_{\geq 1}$.

This shows that no forbidden block from $\mathcal{B}_q$ appears in the λ_q-CF in (6.2).

Next, let $x \in \mathbb{R}$ have the regular λ_q-CF $x = [\![a_0; a_1, a_2, a_3, \ldots]\!]$. We have to show that the corresponding formal Rosen λ-fraction in (6.3) satisfies properties (1) - (4) of Definition 6.2. Again, we discuss the cases q even and q odd separately.

Consider first q even. Using Lemma 6.1 and the identity $h_{\mathrm{R}} = h_q - 1$ in (6.4) we find that forbidden blocks of the form $[(\pm 1)^{h_q}, \pm m]$ for any $m \in \mathbb{Z}_{\geq 1}$ imply property (1). Property 5 corresponds just to the ambiguity of finite λ_q-CFs given in Lemma 2.2 since the tails $[(\pm 1)^{h_q}]$ correspond to $\mp\lambda_q/2$.

Consider next q odd. Using Lemma 6.1 and recalling the identity $h_{\mathrm{R}} = h_q$ in (6.4) we see that forbidden blocks $[(\pm 1)^{h_q+1}]$ imply property (2) and property (4) of Definition 6.2. Property (2) also implies property (1). Property (3) follows from the forbidden blocks $[(\pm 1)^{h_q}, \pm 2, (\pm 1)^{h_q}, \pm m]$ and for $m \in \mathbb{Z}_{\geq 1}$. The ambiguity of the regular λ_q-CFs in Lemma 2.2 implies property (5).

To finish the proof of Lemma 6.3 consider q even and the finite regular λ_q-CF $[\![a_0; a_1, \ldots, a_n, (1)^{h_q}]\!]$. Using Equation (6.3) we rewrite it as the Rosen λ-fraction

$$[a_0; (-\operatorname{sign}(a_1) : |a_1|), \ldots, \\ (-\operatorname{sign}(a_{n-1})\operatorname{sign}(a_n) : |a_n|), (-\operatorname{sign}(a_n) : 1), (-1 : 1)^{h_q-1}].$$

Since by the equation following (4.2) in [16] and by [1, (4)] the identity $[0; (1:1), (-1:1)^{h-1}] = \lambda_q/2$ holds, we are in the situation of Property (5). If $a_n < 0$ we choose the "+"-sign in Property (5). If $a_n > 0$ we use Lemma 2.2 to rewrite the finite regular λ_q-CF such that its tail ends in $[a_n - 1, (-1)^{h_q}]$ with $\operatorname{sign}(a_n - 1) = \operatorname{sign}(a_n)$. Using Equation (6.3) we arrive at the Rosen λ-fraction

$$[a_0; (-\operatorname{sign}(a_1) : |a_1|), \ldots, \\ (-\operatorname{sign}(a_{n-1})\operatorname{sign}(a_n - 1) : |a_n - 1|), (\operatorname{sign}(a_n - 1) : 1), (-1 : 1)^{h_q-1}]$$

with the correct tail.

The case q odd is analogous to q even, with the only difference that the reduced Rosen λ-fraction $\lambda_q/2 = [0; (1:1), (-1:1)^{h_q-1}, (-1:2), (-1,1)^{h_q}]$ as given in [1, (4)] has the corresponding tail $[(1)^{h_q}, 2, (1)^{h_q}]$. □

Remark 6.4. Consider the λ_q-CF of $\pm r_q$ in (2.12). Their corresponding Rosen λ-fractions according to formula (6.2) are

$$r_q = \begin{cases} [0; (-1:1)^{h_q-1}, \overline{(-1:2), (-1,1)^{h_q-1}}] & \text{for } q \text{ even}, \\ [0; (-1,1), \overline{(-1:1)^{h_q-1}, (-1:2), (-1:1)^{h_q-1}, (-1:2), (-1,1)}] & \\ & \text{for } q \text{ odd} \end{cases}$$

and

$$-r_q = \begin{cases} [0; (1:2), \overline{(-1:2)}] & \text{for } q = 4, \\ [0; (1:1), (-1:1)^{h_q-2}, \overline{(-1:2), (-1,1)^{h_q-1}}] & \text{for even } q \geq 6, \\ [0; (1,1), \overline{(-1:1)^{h_q-1}, (-1:2), (-1:1)^{h_q-1}, (-1:2), (-1,1)}] & \\ & \text{for } q \text{ odd}, \end{cases}$$

where $(-1:1)^0$ means that the digit $(-1:1)$ is absent. The Rosen λ-fractions hence have the same tail.

Remark 6.5. The generating map $f_q^\star$ for the dual regular λ_q-CF and the generating map f_q^{R} for the Rosen λ-fractions in [1] satisfy

$$f_q^\star(-x) = \frac{1}{x} - \lambda_q \left\lfloor \frac{1}{x\lambda_q + R_q/2} \right\rfloor = \frac{1}{x} - \lambda_q \left\lfloor \frac{1}{x\lambda_q + 1 + r_q/2} \right\rfloor = f_q^{\mathrm{R}}(x)$$

for all $x \in (0, \lambda_q/2)$.

Formally, we find also

$$f_q^\star(-x) = \frac{1}{x} - \lambda_q \left\lfloor \frac{1}{x\lambda_q + R_q/2} \right\rfloor = \frac{1}{x} - \lambda_q \left\lfloor \frac{1}{x\lambda_q + 1 + r_q/2} \right\rfloor = T_{-r/\lambda_q}(x)$$

for all $x \in (0, \lambda_q/2)$ where T_α is the generating map of the α-Rosen fractions discussed in [5]. However, the parameter $\alpha = -r_q/\lambda_q$ lies outside the range $\alpha \in [\frac{1}{2}, 1/\lambda_q]$ discussed in [5], since by (2.12) and (2.18) $-r_q = \lambda_q - R_q \in (0, \lambda_q/2)$.

6.2. Regular λ_q-CFs and convergents. We define the nth convergent in the following way:

Definition 6.6. Given a regular λ_q-CF $[\![a_0; a_1, \ldots, a_n, \ldots]\!]$ of length at least n we define its *nth convergent* as the fraction p_n/q_n where the numerator p_n and denominator q_n are given as entries in the vector

$$\begin{pmatrix} p_n \\ q_n \end{pmatrix} = T_q^{a_0} S T_q^{a_1} S T_q^{a_2} \cdots S T_q^{a_n} \begin{pmatrix} 0 \\ 1 \end{pmatrix}. \tag{6.5}$$

The convergents then satisfy the recursion relation

$$\begin{pmatrix} p_n \\ q_n \end{pmatrix} = \begin{pmatrix} p_{n-2} & p_{n-1} \\ q_{n-2} & q_{n-1} \end{pmatrix} S T_q^{a_n} \begin{pmatrix} 0 \\ 1 \end{pmatrix} = \begin{pmatrix} a_n \lambda_q p_{n-1} - p_{n-2} \\ a_n \lambda_q q_{n-1} - q_{n-2} \end{pmatrix} \tag{6.6}$$

which holds also for $n = 0$ and $n = 1$ if we define $p_{-1} = 1$, $p_{-2} = 0$, $q_{-1} = 0$ and $q_{-2} = -1$.

Remark 6.7. In the case $q = 3$ Definition 6.6 of the nth convergent coincides with the usual definition as the ratio $p_n/q_n = [\![a_0; a_1, a_2, \ldots, a_n]\!]$, since

$$[\![a_0; a_1, a_2, \ldots, a_n]\!] = \begin{pmatrix} * & p_n \\ * & q_n \end{pmatrix} 0$$

where the last expression is to be understood as a Möbius transformation.

The following lemma will show that regular λ_q-CFs are indeed well defined and determine real numbers. This obviously is true for finite regular λ_q-CFs.

Lemma 6.8. *Let $[a_0; a_1, a_2, \ldots]$ be an infinite regular λ_q-CF and denote by p_n/q_n its nth convergent. Then for $q \geq 4$ the fraction* $\operatorname{sign}(q_n)p_n/|q_n|$ *is the nth convergent of the corresponding reduced Rosen λ-fraction as defined in* [16, Definition 3]; *for $q = 3$ the fraction p_n/q_n is a "Näherungsbruch" in the sense of Hurwitz* [8, Section 2].

PROOF. The case $q = 3$ has been shown in [8].

Hence assume $q \geq 4$. Since the regular λ_q-CF is infinite, we don't have the ambiguities in Lemma 6.3 which shows that the corresponding Rosen λ-fraction

$$[a_0; (-\operatorname{sign}(a_1) : |a_1|), (-\operatorname{sign}(a_1)\operatorname{sign}(a_2) : |a_2|), \ldots]$$

is reduced. The nth convergent $\frac{P_n}{Q_n}$ of the reduced Rosen λ-expansion is well defined and satisfies $Q_n \geq 1$ by [16, Lemma 4]. We have

$$\frac{P_n}{Q_n} = [a_0; (-\operatorname{sign}(a_1) : |a_1|), (-\operatorname{sign}(a_1)\operatorname{sign}(a_2) : |a_2|), \ldots, (-\operatorname{sign}(a_{n-1})\operatorname{sign}(a_n) : |a_n|)]$$

$$= [a_0; a_1, a_2, \ldots, a_n] = \frac{p_n}{q_n}$$

with p_n and q_n satisfying (6.5). Hence we find indeed $P_n = \operatorname{sign}(q_n)p_n$ and $Q_n = |q_n|$. □

This lemma shows that the results on convergents in [16] hold also for the regular λ_q-CFs. We collect the relevant results in [16] and [8] in the following

Lemma 6.9. *The convergents p_n/q_n of an infinite regular λ_q-CF satisfy:*

- $q_n \neq 0$ *and* $|q_n| \geq |q_{n-1}|$. *For* $q = 3$ *we have* $|q_n| > |q_{n-1}|$.
- $\lim_{n\to\infty} |q_n| \to \infty$.
- *The sequence* $(p_n/q_n)_{n\in\mathbb{N}}$ *is a Cauchy sequence.*

PROOF. This follows from Lemmas 4 and 5, Theorem 4 and the proof of Theorem 5 in [16] for $q \geq 4$ and for $q = 3$ from Sections 2 and 3 in [8]. □

Now we can (re-)define infinite regular λ_q-CFs in the following way: Let $[\![a_0; a_1, a_2, \ldots]\!]$ be a regular λ_q-CF. We assign the value x to the regular λ_q-CF expansion and write $x = [\![a_0; a_1, a_2, \ldots]\!]$ where x is the limit of the sequence of convergents (see Definition 6.4 and Lemma 6.9) of the corresponding Rosen λ-fraction.

Then the following estimate for the approximation of x by the convergents holds:

Lemma 6.10. *Let $[\![a_0; a_1, a_2, \ldots]\!]$ be an infinite regular λ_q-CF and denote its nth convergents by p_n/q_n. There exists a constant $\kappa_q > 0$, independent of x, such that*

$$\left| x - \frac{p_n}{q_n} \right| \leq \frac{1}{\kappa_q q_n^2}$$

holds for all n.

PROOF. The lemma follows for $q \geq 4$ from Theorem 4.6 in [1] and for $q = 3$ from Satz on p. 383 in [8]. □

Remark 6.11. Obviously, Lemma 6.10 implies that infinite regular λ_q-CFs converge. This gives another proof of part of Proposition 2.1.

Acknowledgement. The authors thank Fredrik Strömberg for many helpful discussions.

References

1. R. M. Burton, C. Kraaikamp, and T. A. Schmidt, *Natural extensions for the Rosen fractions*, Trans. Amer. Math. Soc. **352** (2000), no. 3, 1277–1298.
2. R. W. Bruggeman, J. Lewis, and D. Zagier, *Period functions for Maass wave forms.* II: *Cohomology*, in preparation.
3. R. W. Bruggeman and T. Mühlenbruch, *Eigenfunctions of transfer operators and cohomology*, J. Number Theory **129** (2009), no. 1, 158–181.

4. C.-H. Chang and D. H. Mayer, *The transfer operator approach to Selberg's zeta function and modular and Maass wave forms for* PSL(2, **Z**), Emerging Applications of Number Theory (Minneapolis, MN, 1996), IMA Vol. Math. Appl., vol. 109, Springer, New York, 1999, pp. 73–141.
5. K. Dajani, C. Kraaikamp, and W. Steiner, *Metrical theory for α-Rosen fractions*, J. Eur. Math. Soc. (JEMS) **11** (2009), no. 6, 1259–1283.
6. E. Hecke, *Mathematische Werke*, Herausgegeben im Auftrage der Akademie der Wissenschaften zu Göttingen, Vandenhoeck & Ruprecht, Göttingen, 1959.
7. D. A. Hejhal, *The Selberg trace formula for* PSL(2, **R**). Vol. 2, Lecture Notes in Math., vol. 1001, Springer, Berlin, 1983.
8. A. Hurwitz, *Über eine besondere Art der Kettenbruch-Entwicklung reeller Grössen*, Acta Math. **12** (1889), no. 1, 367–405.
9. S. Katok and I. Ugarcovici, *Symbolic dynamics for the modular surface and beyond*, Bull. Amer. Math. Soc. (N.S.) **44** (2007), no. 1, 87–132 (electronic).
10. J. Lewis and D. Zagier, *Period functions for Maass wave forms.* I, Ann. of Math. (2) **153** (2001), no. 1, 191–258.
11. J. Marklof, *Selberg's trace formula: An introduction*, Quantum Chaos on Hyperbolic Manifolds (Günzburg, 2003), Lecture Notes in Phys., Springer, Berlin, to appear, available at `arXiv:math/0407288`.
12. D. H. Mayer, *The thermodynamic formalism approach to Selberg's zeta function for* PSL(2, **Z**), Bull. Amer. Math. Soc. (N.S.) **25** (1991), no. 1, 55–60.
13. ______, *Transfer operators, the Selberg-zeta function and Lewis-Zagier theory of period functions*, Quantum Chaos on Hyperbolic Manifolds (Günzburg, 2003), Lecture Notes in Phys., Springer, Berlin, to appear, available at `http://www.dynamik.tu-clausthal.de/research/preprints/Mayer.001.ps`.
14. D. H. Mayer and F. Strömberg, *Symbolic dynamics for the geodesic flow on Hecke surfaces*, J. Mod. Dyn. **2** (2008), no. 4, 581–627.
15. H. Nakada, *Continued fractions, geodesic flows and Ford circles*, Algorithms, Fractals, and Dynamics (Okayama/Kyoto, 1992), Plenum, New York, 1995, pp. 179–191.
16. D. Rosen, *A class of continued fractions associated with certain properly discontinuous groups*, Duke Math. J. **21** (1954), 549–563.
17. Z. Rudnick, *What is... quantum chaos?*, Notices Amer. Math. Soc. **55** (2008), no. 1, 32–34.
18. D. Ruelle, *Dynamical zeta functions for piecewise monotone maps of the interval*, CRM Monogr. Ser., vol. 4, Amer. Math. Soc., Providence, RI, 1994.
19. ______, *Thermodynamic formalism*, 2nd ed., Cambridge Math. Lib., Cambridge Univ. Press, Cambridge, 2004.
20. P. Sarnak, *Arithmetic quantum chaos*, The Schur Lectures (Tel Aviv, 1992)), Israel Math. Conf. Proc., vol. 8, Bar-Ilan Univ., Ramat Gan, 1995, pp. 183–236.
21. T. A. Schmidt and M. Sheingorn, *Length spectra of the Hecke triangle groups*, Math. Z. **220** (1995), no. 3, 369–397.

Institute of Theoretical Physics, TU Clausthal, 38678 Clausthal-Zellerfeld, Germany

E-mail address: `dieter.mayer@tu-clausthal.de`

Department of Mathematics and Computer Science, FernUniversität in Hagen, 58084 Hagen, Germany

E-mail address: `Tobias.Muehlenbruch@fernuni-hagen.de`

Centre de Recherches Mathématiques
CRM Proceedings and Lecture Notes
Volume **52**, 2010

Comparing Length Functions on Free Groups

Richard Sharp

ABSTRACT. The aim of this note is to study a pair of length functions on a free group, associated to a point in the unprojectivized outer space. We discuss various characteristics which compare them and describe how these may be unified by a "Manhattan curve" analogous to those used to study pairs of hyperbolic surfaces. Our approach involves the thermodynamic formalism used in the ergodic theory of hyperbolic dynamical systems.

Introduction

Let F be a free group on $k \geq 2$ generators and let $\mathcal{C}(F)$ denote the set of its non-trivial conjugacy classes. A natural class of length functions on F, or, more precisely, $\mathcal{C}(F)$, may be obtained from free isometric actions on metric trees (i.e., simplicial $\mathbb{R}$-trees). More precisely, if F acts freely on the metric tree $\mathcal{T}$ then we may define a length function $l\colon \mathcal{C}(F) \to \mathbb{R}^+$ by

$$l(w) = \inf_{o \in \mathcal{T}} d_{\mathcal{T}}(o, ox),$$

where $x \in w$. It is easy to see that l is well-defined. Furthermore, the image of l is strictly positive. An equivalent formulation is to consider a metric graph Γ with fundamental group identified with F (and some additional non-degeneracy conditions). Then $w \in \mathcal{C}(F)$ corresponds to a unique closed path in Γ (provided backtracking is not allowed) and $l(w)$ is equal to the sum of edge lengths around this path. Such length functions are parametrized by a set $\mathrm{cv}(F)$, which is an unprojectivized version of the Culler–Vogtmann outer space, defined in Section 1.

A natural quantity associated to $l \in \mathrm{cv}(F)$ is the *critical exponent* $\delta = \delta(l)$. This is defined by

$$\delta = \inf\Bigl\{ s \in \mathbb{R} : \sum_{w \in \mathcal{C}(F)} \mathrm{e}^{-sl(w)} < +\infty \Bigr\}.$$

It is a standard result that $\delta > 0$. We also have that

$$\delta = \limsup_{T \to +\infty} \frac{1}{T} \log \#\{ w \in \mathcal{C}(F) : l(w) \leq T \}.$$

2010 *Mathematics Subject Classification.* Primary 20E05, 20E08, 20F67, 20F69, 37C30, 37C35, 37D35; Secondary 05C05, 37A35, 37D40.

This is the final form of the paper.

In fact, there are more precise results for counting conjugacy classes with respect to l; these are described in Section 3.

The aim of this paper is to compare two arbitrary length functions $l_1, l_2 \in \mathrm{cv}(F)$. It is convenient to describe this in terms of the so-called Manhattan curve, an object first defined by Burger to compare two convex co-compact representations of a group as isometries of the hyperbolic space $\mathbb{H}^{n+1}$ [6]. A particularly interesting case is given by two points in the Teichmüller space associated to a compact surface [6, 36].

Here we define the Manhattan curve $\mathfrak{M}(l_1, l_2)$ associated to $l_1, l_2 \in \mathrm{cv}(F)$ in the following way:

$$\mathfrak{M}(l_1, l_2) = \partial\left\{(a,b) \in \mathbb{R}^2 : \sum_{w \in \mathcal{C}(F)} \mathrm{e}^{-al_1(w) - bl_2(w)} < +\infty\right\}.$$

(This may be thought of as the natural analogue of the critical exponent for a single length function.) Many numerical characteristics which compare l_1 and l_2 may be read off from $\mathfrak{M}(l_1, l_2)$:

(i) the *intersection*

$$i(l_1, l_2) := \lim_{T \to +\infty} \frac{1}{\#\{w \in \mathcal{C}(F) : l_1(w) \le T\}} \sum_{l_1(w) \le T} \frac{l_2(w)}{l_1(w)};$$

(ii) the *distortion interval*

$$D(l_1, l_2) := \overline{\left\{\frac{l_2(w)}{l_1(w)} : w \in \mathcal{C}(F)\right\}};$$

(iii) the *correlation number*

$$\alpha_1(l_1, l_2) := \lim_{\epsilon \to 0} \limsup_{T \to +\infty} \frac{1}{T} \log \#\left\{w \in \mathcal{C}(F) : l_1(w) \le T, \frac{l_2(w)}{l_1(w)} \in (1-\epsilon, 1+\epsilon)\right\}.$$

We shall study these in Section 4 and establish the relation to $\mathfrak{M}(l_1, l_2)$ in Section 5.

The earlier parts of the paper are devoted to preliminaries. In Section 1, we give some background on length functions and the outer space. In Section 2, we introduce ergodic theoretic ideas in the shape of the shift of finite type Σ^+ formed by infinite reduced words in the generators of F. As we mentioned above, a length function $l \in \mathrm{cv}(F)$ is associated (up to equivalence) to a metric graph Γ. It is a crucial point that we always use one fixed shift of finite type associated to a preferred generating set, rather than the shift formed by paths in the oriented line graph of Γ. In Section 3, we show that a length function in $\mathrm{cv}(F)$ may be encoded in a cohomology class of locally constant functions on Σ^+. So, different points in $\mathrm{cv}(F)$ correspond to different cohomology classes of functions on the fixed shift Σ^+. As mentioned in the preceding paragraph, the main results of the paper are in Sections 4 and 5. An appendix gives an introduction to the thermodynamic formalism we use.

1. Length functions and the outer space

Let $\mathcal{A} = \{a_1, \ldots, a_k\}$ be a free generating set for F. Any $x \in F$ $(x \ne 1)$ may be written uniquely as a reduced word in these generators:

$$x = x_0 x_1 \cdots x_n,$$

with $x_i \in \mathcal{A} \cup \mathcal{A}^{-1}$ and $x_{i+1} \neq x_i^{-1}$, $i = 0, \ldots, n-1$, and we call $|x| = |x|_{\mathcal{A}} = n$ the word length of x (with respect to $\mathcal{A}$). We may then define a function $\|\cdot\| \colon \mathcal{C}(F) \to \mathbb{R}$ on conjugacy classes by

$$\|w\| = \min_{x \in w} |x|.$$

The above is only one of many length functions which may be defined and in this section we characterize the class of them that we shall study. (See [14] for a discussion of a more general class of length functions.) These correspond to isometric actions of F on simplicial trees or, equivalently, to equivalence classes of metric graphs with fundamental group isomorphic to F.

To give some background, we briefly digress to discuss automorphisms of free groups. Let $\mathrm{Aut}(F)$ denote the group of automorphisms of F. An automorphism $\alpha \colon F \to F$ is called *inner* if it takes the form $\alpha(x) = yxy^{-1}$, for some $y \in F$. Let $\mathrm{Inn}(F)$ denote the the set of all inner automorphisms of F; this is a normal subgroup of $\mathrm{Aut}(F)$. The quotient $\mathrm{Out}(F) = \mathrm{Aut}(F)/\mathrm{Inn}(F)$ is called the group of outer automorphisms of F. In [10], Culler and Vogtmann studied $\mathrm{Out}(F)$ by constructing a space $\mathrm{CV}(F)$, called the outer space, on which it acts in a natural way. (This is analogous to the study of the mapping class group of a surface via its action on Teichmüller space.)

The outer space is defined in the following way [10, 38]. Let $\mathcal{G}$ be a fixed graph with one vertex $*$ and k edges and identify F with $\pi_1(\mathcal{G})$ so that each a_i, $i = 1, \ldots, k$, corresponds to an (oriented) edge. A metric graph is a graph together with an assignment of a positive length to each edge, making it into a metric space in the obvious way. Let Γ be a metric graph with fundamental group F such that each vertex has valency at least three together with a homotopy equivalence $g \colon \mathcal{G} \to \Gamma$; we call (Γ, g) a marked metric graph. Consider the set of all marked metric graphs whose edge lengths sum to one. We say that (Γ, g) and (Γ', g') are equivalent if there is an isometry $h \colon \Gamma \to \Gamma'$ such that $g \circ h$ is homotopic to g'. Then the outer space $\mathrm{CV}(F)$ is defined to be the set of equivalence classes. (An alternative definition is the set of equivalence classes of marked metric graphs under the relation $(\Gamma, g) \sim (\Gamma', g')$ if there is a homothety $h \colon \Gamma \to \Gamma'$ such that $g \circ h$ is homotopic to g'.)

There is a natural action of $\mathrm{Out}(F)$ on $\mathrm{CV}(F)$ as follows. Any $\alpha \in \mathrm{Aut}(F)$ induces a map $A \colon \mathcal{G} \to \mathcal{G}$. Define $(\Gamma, g) \cdot \alpha = (\Gamma, g \circ A)$. Since the action of inner automorphisms is trivial, this gives a well-defined action of $\mathrm{Out}(F)$.

The universal cover of a marked metric graph Γ (with the lifted metric) is a metric tree (simplicial $\mathbb{R}$-tree) $\mathcal{T}$ on which F acts on the right by isometries. For a choice of $o \in \mathcal{T}$, define a based length function $L \colon F \to \mathbb{R}$ by

$$L(x) = d_{\mathcal{T}}(o, ox).$$

If we change base point then we obtain a different function but it is easy to see that L only depends the image of o under the covering map $\pi \colon \mathcal{T} \to \Gamma$. We may eliminate the dependence on the base point by defining a new length function l by

$$l(x) = \inf_{o \in \mathcal{T}} d_{\mathcal{T}}(o, ox).$$

We have $d_{\mathcal{T}}(o, oyxy^{-1}) = d_{\mathcal{T}}(oy, oyx)$, so it is easy to see that $l(x)$ depends only on the conjugacy class of x. Thus we think of l as a function $l \colon \mathcal{C}(F) \to \mathbb{R}$. Furthermore, l only depends on the point in $\mathrm{CV}(F)$ represented by (Γ, g). (Alternatively, identifying $x \in F$ with a homotopy class in $\pi_1(\mathcal{G})$, $l(x)$ may be defined to be the length of the shortest loop in Γ freely homotopic to $g(x)$.)

Lemma 1 ([9]). *The map from* $\mathrm{CV}(F)$ *to the set of functions from* $\mathcal{C}(F)$ *to* $\mathbb{R}^+$ *is injective.*

Thus we may identify $\mathrm{CV}(F)$ with a subset of $(\mathbb{R}^+)^{\mathcal{C}(F)}$. In this paper, we consider a larger space

$$\mathrm{cv}(F) = \{\lambda l : l \in \mathrm{CV}(F),\ \lambda > 0\}.$$

Remark. The functions $l\colon \mathcal{C}(F) \to \mathbb{R}$ are called hyperbolic length functions [1] or translation length functions [9]. The notation $\mathrm{CV}(F)$ and $\mathrm{cv}(F)$ has been used in [18] and [19], for example. Clearly, $\mathrm{CV}(F)$ is the projectivization of $\mathrm{cv}(F)$.

The based length function $L\colon F \to \mathbb{R}$ which arise in the above construction are also completely characterized by the following conditions [7, 24]. Write

$$(x,y)_L = \big(L(x) + L(y) - L(xy^{-1})\big)/2.$$

For $x, y, z \in F$,

(L1) $L(x) = 0$ if and only if $x = 1$;
(L2) $L(x^{-1}) = L(x)$;
(L3) $(x,y)_L \geq 0$;
(L4) $(x,y)_L < (x,z)_L$ implies that $(y,z)_L = (x,y)_L$; and
(L5) $(x,y)_L + (x^{-1},y^{-1})_L > L(x) = L(y)$ implies that $x = y$;
(L6) $L(x^2) > L(x)$ for every $x \neq 1$.

(Note that $L(x) = (x,x)_L$, so (L3) implies that $L(x) > 0$ for every $x \neq 1$. A function satisfying properties (L1)–(L5) is called a Lyndon length function. (L6) is called the Archimedean property.)

Remark. The functions $l \in \mathrm{cv}(F)$ are also characterized by a set of axioms [9, 27].

We shall use the following facts later. For $x, y \in F$ and $o, p, q \in \mathcal{T}$, define Gromov products

$$(x,y) = \tfrac{1}{2}(|x| + |y| - |xy^{-1}|)$$

and

$$(p,q)_o = \tfrac{1}{2}\big((d_{\mathcal{T}}(o,p) + d_{\mathcal{T}}(o,q) - d_{\mathcal{T}}(p,q)\big).$$

As above, write $L(x) = d_{\mathcal{T}}(o, ox)$. Then $(x,y)_L = (ox, oy)_o$. Furthermore, there exist constants $A_1, A_2 > 0$ such that, for all $x \in F$,

$$A_1|x| \leq d_{\mathcal{T}}(o, ox) \leq A_2|x| \tag{1}$$

and constants $B_1, B_2, K > 0$ such that, for all $x, y \in F$,

$$B_1(x,y) - K \leq (x,y)_L \leq B_2(x,y) + K. \tag{2}$$

These results follow because the map $f\colon F \to \mathcal{T}$ given by $f(x) = ox$ is a quasi-isometry between F equipped with the word metric $d_{\mathrm{word}}(x,y) = |x^{-1}y|$ and $\mathcal{T}$. Then (1) is standard and (2) follows from Proposition 15 in Chapter 5 of [12].

2. Shifts and free groups

In this section we shall consider the shift of finite type $\sigma\colon \Sigma^+ \to \Sigma^+$ associated to the free group F and free basis $\mathcal{A} = \{a_1, \ldots, a_k\}$. (Shifts of finite type and their properties are discussed in greater detail in the appendix.) We define

$$\Sigma^+ = \left\{ x = (x_n)_{n=0}^{\infty} \in \prod_{n=0}^{\infty} (\mathcal{A} \cup \mathcal{A}^{-1}) : x_{n+1} \neq x_n^{-1} \ \forall n \in \mathbb{Z}^+ \right\},$$

i.e., Σ^+ is the space of infinite reduced words in $\mathcal{A} \cup \mathcal{A}^{-1}$. Of course, Σ^+ may be identified with the Gromov boundary of F. The shift map $\sigma\colon \Sigma^+ \to \Sigma^+$ is defined by $(\sigma x)_n = x_{n+1}$.

In the notation of the appendix, $\Sigma^+ = \Sigma_A^+$, where the index set is $\mathcal{I} = \mathcal{A} \cup \mathcal{A}^{-1}$ and the zero-one $\mathcal{I} \times \mathcal{I}$ matrix A is defined by $A(i,j) = 1$ unless j is the inverse of i. It is easy to see that A is aperiodic. Furthermore, the set $W^{(n)} = W_A^{(n)}$ of allowed words of length n may be identified with the set $\{x \in F : |x| = n\}$.

A cylinder set is defined by

$$[y_0, \ldots y_{n-1}]_m = \{(x_n)_{n=0}^{\infty} \in \Sigma^+ : x_{m+j} = y_j, j = 0, \ldots, n-1\}.$$

The cylinder sets generate the topology and Borel σ-algebra on Σ^+.

A simple calculation shows that the topological entropy of $\sigma\colon \Sigma^+ \to \Sigma^+$ is $h(\sigma) = \log(2k-1)$ and the measure of maximal entropy is the measure μ_0 defined on cylinder sets by

$$\mu_0([x_0, x_1, \ldots, x_{n-1}]_m) = \frac{1}{2k(2k-1)^{(n-1)}}$$

and extended to Borel sets by the Kolmogorov extension theorem.

Recall that $\mathcal{C}(F)$ denotes the set of non-trivial conjugacy classes in F. A conjugacy class $w \in \mathcal{C}(F)$ contains a cyclically reduced word in $\mathcal{A} \cup \mathcal{A}^{-1}$, i.e., a reduced word $x_0 x_1 \cdots x_{n-1}$ such that $x_{n-1} \neq x_0^{-1}$. The only other cyclically reduced elements of w are obtained from this by cyclic permutation (and also have word length n) and non-cyclically reduced elements of w have word length greater than n. Therefore it is natural to define the length $\|w\|$ of w (with respect to $\mathcal{A}$) to be

$$\|w\| := n = \min_{x \in w} |x|.$$

It is immediate from the definition that, for $m \geq 1$,

$$\|w^m\| = m\|w\|,$$

where w^m is the conjugacy class $\{x^m : x \in w\}$. Furthermore, it is clear from the preceding discussion that there is a natural bijection between $\mathcal{C}(F)$ and the set of periodic orbits of $\sigma\colon \Sigma^+ \to \Sigma^+$, such that, if $x, \sigma x, \ldots, \sigma^{n-1} x$ $(\sigma^n x = x)$ corresponds to $w \in \mathcal{C}(F)$ then $\|w\| = n$.

In order to represent elements of F as elements of a shift space (rather than simply as finite words), it is convenient to augment Σ^+ by adding an extra "dummy" symbol 0. Introduce a square matrix A_0, with rows indexed by $\mathcal{A} \cup \mathcal{A}^{-1} \cup \{0\}$, such that

$$A_0(i,j) = \begin{cases} A(i,j) & \text{if } i, j \in \mathcal{A} \cup \mathcal{A}^{-1}, \\ 1 & \text{if } i \in \mathcal{A} \cup \mathcal{A}^{-1} \cup \{0\} \text{ and } j = 0, \\ 0 & \text{if } i = 0 \text{ and } j \in \mathcal{A} \cup \mathcal{A}^{-1}. \end{cases}$$

We then define $\Sigma_0^+ = \Sigma_{A_0}^+$. In other words, an element of Σ_0^+ is either an element of Σ^+ or an element of $W^{(n)}$, for some $n \geq 1$, followed by an infinite string of 0s, which we shall denote by $\dot{0}$. Of course, this procedure does not introduce any extra periodic points.

3. Locally constant functions

Let $l \in \mathrm{cv}(F)$. We wish to encode the lengths $\{l(w) : w \in \mathcal{C}(F)\}$ in terms of the periodic points of $\sigma : \Sigma^+ \to \Sigma^+$ and a function $r\colon \Sigma^+ \to \mathbb{R}$. In fact, it will turn out that r is locally constant.

The construction of r is somewhat technical. Let $\mathcal{T}$ be a metric tree associated to l. We need to introduce an arbitrarily chosen basepoint $o \in \mathcal{T}$ and write $L(x) = d_{\mathcal{T}}(o, ox)$. The function r that we construct will depend on o only up to the addition of a coboundary.

We begin by relating the value $L(x)$ to the description of $x \in F$ as a word in the generators $\{a_1, \ldots, a_k\}$. For $n \geq 1$ and $m \leq n$, we define the following sets of n-tuples:

$$W_n^{(0)} = \{(0, 0, \ldots, 0)\},$$
$$W_n^{(m)} = \{(x_0, x_1, \ldots, x_{m-1}, 0, \ldots, 0) : (x_0, x_1, \ldots, x_{m-1}) \in W^{(m)}\}$$

and

$$W_n = \bigcup_{m=0}^{n} W_n^{(m)}.$$

The map $\iota\colon W_n \to \{x \in F : |x| \leq n\}$ defined by $\iota((0, 0, \ldots, 0)) = 1$ and, for $x_0 \neq 0$,

$$\iota\big((x_0, x_1, \ldots, x_{m-1}, 0, \ldots, 0)\big) = x_0 x_1 \cdots x_{m-1}$$

is a bijection. We shall abuse notation by using x to denote the sequence in W_n corresponding to $x \in F$. Clearly, ι restricts to a bijection $\iota\colon W_n^{(n)} = W^{(n)} \to \{x \in F : |x| = n\}$.

We begin with the following lemma, which should be compared with Proposition 4 of [30].

Lemma 2. *There exists an integer $N \geq 1$ such that if $n \geq N$ and $x_0 x_1 \cdots x_{n-1}$ is a reduced word then*

$$L(x_0 x_1 \cdots x_{n-1}) - L(x_1 \cdots x_{n-1}) = L(x_0 x_1 \cdots x_{N-1}) - L(x_1 \cdots x_{N-1}).$$

PROOF. Note that

$$(x_0 x_1 \cdots x_{N-1}, x_0 x_1 \cdots x_{n-1}) = \tfrac{1}{2}(N + n - (n - N)) = N.$$

Write $x = x_0 x_1 \cdots x_{n-1}$ and $y = x_0 x_1 \cdots x_{N-1}$. We need to show that $L(y) - L(x_0^{-1} y) = L(x) - L(x_0^{-1} x)$ or, equivalently

$$L(x_0) + L(y) - L(x_0^{-1} y) = L(x_0) + L(x) - L(x_0^{-1} x).$$

Using property (L2), this may be rewritten as

$$L(x_0^{-1}) + L(y^{-1}) - L(y^{-1}(x_0^{-1})^{-1}) = L(x_0^{-1}) + L(x^{-1}) - L(x^{-1}(x_0^{-1})^{-1})$$

or as

$$(y^{-1}, x_0^{-1})_L = (x^{-1}, x_0^{-1})_L.$$

By property (L4), this will hold provided

$$(x^{-1}, x_0^{-1})_L < (x^{-1}, y^{-1})_L.$$

By inequality (2), this will be satisfied if $N = (x^{-1}, y^{-1}) \geq B_1^{-1}(x^{-1}, x_0^{-1})_L + B_1^{-1}K$. Since

$$B_1^{-1}(x^{-1}, x_0^{-1})_L + B_1^{-1}K \leq B_1^{-1}B_2(x^{-1}, x_0^{-1}) + 2B_1^{-1}K$$

and $(x^{-1}, x_0^{-1}) = 1$, this will be satisfied provided $N \geq B_1^{-1}(B_2 + 2K)$. □

We now introduce a function on Σ_0^+, which "encodes" the values $\{L(x) : x \in F\}$.

Definition. Define a locally constant function $r\colon \Sigma_0^+ \to \mathbb{R}$ by

$$r((x_n)_{n=0}^{\infty}) = L(x_0 \cdots x_{N-1}) - L(x_1 \cdots x_{N-1}), \tag{3}$$

where N is chosen so that the conclusion of Lemma 2 holds.

The following result is immediate from the definition.

Lemma 3. *Suppose that $|x| = n$ and $x = x_0 x_1 \cdots x_{n-1}$. Then*

$$\begin{aligned} L(x) = r\big((x_0, x_1, \ldots, x_{N-1}), (x_1, x_2, \ldots, x_N)\big) + \cdots \\ + r\big((x_{n-N}, x_{n-N+1}, \ldots, x_{n-1}), (x_{n-N+1}, \ldots, x_{n-1}, 0)\big) + \cdots \\ + r\big((x_{n-2}, x_{n-1}, 0, \ldots, 0), (x_{n-1}, 0, \ldots, 0)\big) \\ + r\big((x_{n-1}, 0, \ldots, 0), (0, 0, \ldots, 0)\big). \end{aligned}$$

PROOF. We have

$$\begin{aligned} L(x) = \big(L(x_0 x_1 \cdots x_{n-1}) - L(x_1 x_2 \cdots x_{n-1})\big) \\ + \big(L(x_1 x_2 \cdots x_{n-1}) - L(x_2 x_3 \cdots x_{n-1})\big) + \cdots \\ + \big(L(x_{n-2} x_{n-1}) - L(x_{n-1})\big) + \big(L(x_{n-1}) - L(1)\big). \end{aligned}$$

The result then follows from the definition of r. □

Remark. This construction has been used by Lalley [23], Bourdon [5] and Pollicott and Sharp [31, 32] in the context of convex co-compact group actions on hyperbolic space or, more generally, CAT(-1) spaces to obtain Hölder continuous functions.

We now describe how we may recover the l-lengths of conjugacy classes. For a point $x \in \Sigma_0^+$, $r^n(x) = r(x) + r(\sigma x) + \cdots + r(\sigma^{n-1} x)$.

Lemma 4. *Let $\sigma^n x = x$ correspond to the conjugacy class w containing the cyclically reduced word $x_0 x_1 \cdots x_{n-1}$. Then*

$$l(w) = r^n(x).$$

In particular, the cohomology class of r only depends on l.

PROOF. Let $x^{(m)}$ denote the reduced word obtained from the m-fold concatenation of $x_0 x_1 \cdots x_{n-1}$. Since r is locally constant, there exists $N \geq 1$ such that

$$|r^{mn}(x) - r^{mn}(x^{(m)}, 0, 0, \ldots)| \leq 2N \|r\|_\infty.$$

Noting that $r^{mn}(x) = m r^n(x)$, the above estimate gives us that

$$r^n(x) = \lim_{m \to +\infty} \frac{1}{m} r^{mn}(x^{(m)}, 0, 0, \ldots) = \lim_{m \to +\infty} \frac{1}{m} L(x^{(m)}).$$

Thus it remains to show that this last quantity is equal to $l(w)$. We shall do this by proving inequalities in both directions.

First observe that $x^{(m)}$ is a cyclically reduced word in w^m. Therefore

$$ml(w) = l(w^m) \leq L(x^{(m)})$$

and so $l(w) \leq \lim_{m\to+\infty} m^{-1}L(x^{(m)})$.

Now suppose that $v \in w$. It is clear that

$$\lim_{m\to+\infty} m^{-1}L(v^m) = \lim_{m\to+\infty} m^{-1}L(x^{(m)}).$$

On the other hand, since $\{L(v^m)\}_{m\geq 1}$ is a subadditive sequence,

$$\lim_{m\to+\infty} m^{-1}L(v^m) = \inf_{m\geq 1} m^{-1}L(v^m)$$

and so $\lim_{m\to+\infty} m^{-1}L(x^{(m)}) \leq L(v)$. Hence

$$\lim_{m\to+\infty} m^{-1}L(v^m) \leq \inf_{v\in w} L(v) = l(w).$$

The final statement follows from Proposition 2 in the appendix, i.e., the cohomology class of r only depends on the data $r^n(x)$, where $\sigma^n x = x$. □

Lemma 5. *$r\colon \Sigma^+ \to \mathbb{R}$ is cohomologous to a strictly positive locally constant function.*

PROOF. It is clear from Lemma 3 that that r^N is strictly positive. Furthermore, r is cohomologous to $N^{-1}r^N$. □

Let $\mathrm{LC}(\Sigma^+)$ denote the linear space of cohomology classes $[f]$ locally constant functions $f\colon \Sigma^+ \to \mathbb{R}$. We say that a class $[f] \in \mathrm{LC}(\Sigma^+)$ is positive if $f^n(x) > 0$, whenever $\sigma^n x = x$, for any (and hence all) $f \in [f]$. Let $\mathrm{LC}^+(\Sigma^+)$ denote the set of positive classes in $\mathrm{LC}(\Sigma^+)$. By Lemmas 4 and 5, the above construction gives us a well-defined map

$$\mathfrak{S}\colon \mathrm{cv}(F) \to \mathrm{LC}^+(\Sigma^+).$$

Lemma 6. *The map $\mathfrak{S}\colon \mathrm{cv}(F) \to \mathrm{LC}^+(\Sigma^+)$ is injective and, for $l \in \mathrm{cv}(F)$, $c > 0$, $\mathfrak{S}(cl) = c\mathfrak{S}(l)$.*

PROOF. If $l_1, l_2 \in \mathrm{cv}(F)$ and $[r_1] = \mathfrak{S}(l_1) = \mathfrak{S}(l_2) = [r_2]$ then $r_1^n(x) = r_2^n(x)$, whenever $\sigma^n x = x$. Hence, by Lemma 4, $l_1(w) = l_2(w)$, for all $w \in \mathcal{C}(F)$, and so, by Lemma 1, $l_1 = l_2$. □

Remark. $\mathfrak{S}\colon \mathrm{cv}(F) \to \mathrm{LC}^+(\Sigma^+)$ is not a surjection. It is shown in [9] that a function $l\colon F \to \mathbb{R}$ is a hyperbolic length function for an isometric action of F on an $\mathbb{R}$-tree only if it is well-defined as a function $l\colon \mathcal{C}(F) \to \mathbb{R}$ and

(i) for $x, y \in F$, either

$$l(xy) = l(xy^{-1}) \quad \text{or} \quad \max\{l(xy), l(xy^{-1})\} \leq l(x) + l(y);$$

(ii) for $x, y \in F$ with $l(x) > 0$ and $l(y) > 0$, either

$$l(xy) = l(xy^{-1}) < l(x) + l(y) \quad \text{or} \quad \max\{l(xy), l(xy^{-1})\} = l(x) + l(y).$$

(Here the action is not assumed to be free. The conditions were shown to be sufficient in [27].) It is easy to construct elements of $\mathrm{LC}^+(\Sigma^+)$ for which these are violated.

It follows from the above construction that there are rather precise results for counting conjugacy classes with respect to l. We say that $l \in \mathrm{cv}(F)$ is mixing if $\{l(w) \,;\, w \in \mathcal{C}(F)\}$ is not contained in a discrete subgroup of $\mathbb{R}$. Let $\pi_l(T) = \#\{w \in \mathcal{C}(F) : l(w) \leq T\}$. If l is mixing then

$$\pi_l(T) \sim \frac{\mathrm{e}^{\delta T}}{\delta T}, \qquad \text{as } T \to +\infty.$$

If l is not mixing then

$$\pi_l(T) \sim \frac{\mathrm{e}^{2\pi\delta/a}}{\mathrm{e}^{2\pi\delta/a} - 1} \frac{2\pi \mathrm{e}^{2\pi\delta[aT/2\pi]/a}}{aT}, \qquad \text{as } T \to +\infty,$$

where $a > 0$ is the smallest number such that $\{l(w) : w \in \mathcal{C}(F)\} \subset a\mathbb{Z}$. This may be derived from the results in [28, 29] and an alternative direct proof may be found in [13].

Remark. There is an alternative way of relating a length $l \in \mathrm{cv}(F)$ to a shift of finite type and locally constant function (and of thus obtaining the asymptotics for $\pi_l(T)$). To describe this, let Γ be a choice of metric graph corresponding to l. Let E denote the set of *oriented* edges in Γ and let l_e denote the length of an edge $e \in E$. (Of course, $l_e = l_{\bar{e}}$, where $\bar{e}$ is the same geometric edge as e but with the reversed orientation.) The *oriented line graph* associated to Γ is the directed graph with vertices E and a directed edge (e, e') if and only if the final vertex of e coincides with the initial vertex of e', and $e' \neq \bar{e}$. We may then define a shift of finite type $\Sigma^+(\Gamma)$ to be the subset of $E^{\mathbb{Z}^+}$ formed by all (one-sided) infinite paths in the oriented line graph. A locally constant function $\omega \colon \Sigma^+(\Gamma) \to \mathbb{R}^+$ is then defined by $\omega((e_n)_{n=0}^{\infty}) = l_{e_0}$. Despite it simplicity, this construction has disadvantages when one wishes to compare two length functions. Then we would have two different shifts and, in particular, the periodic orbits corresponding to a given conjugacy class would generally have different periods.

4. Intersection, distortion and correlation

In this section we discuss a number of quantities which may be used to compare two length functions $l_1, l_2 \in \mathrm{cv}(F)$. First we define the intersection $i(l_1, l_2)$ to be the quantity

$$i(l_1, l_2) = \lim_{T \to +\infty} \frac{1}{\pi_{l_1}(T)} \sum_{l_1(w) \leq T} \frac{l_2(w)}{l_1(w)}. \tag{4}$$

(Note that this is not necessarily symmetric.) This is a direct analogue of the intersection of two Riemannian metrics on a smooth manifold M considered in [3, 4, 6, 8, 11, 21, 41]. If M is a compact and the metrics have negative sectional curvature then the intersection is given by an expression of the form (4), involving the lengths of the closed geodesic in each free homotopy class with respect to the two metrics.

The intersection may be characterized directly in terms of functions $r_i \in \mathfrak{S}(l_i)$, $i = 1, 2$, or, more precisely their integrals. We write δ_1, δ_2 for the critical exponents of l_1, l_2, respectively, and $\mu^1 = \mu_{-\delta_1 r_1}$ for the equilibrium state of $-\delta_1 r_1$. (Of course, this measure only depends on l_i or, equivalently, the cohomology class $[r_i] = \mathfrak{S}(l_i)$.)

Theorem 1. *For $l_1, l_2 \in \mathrm{cv}(F)$, let $r_1, r_2 \colon \Sigma^+ \to \mathbb{R}$ be locally constant functions with $r_i \in \mathfrak{S}(l_i)$, $i = 1, 2$. Then the limit defining $i(l_1, l_2)$ exists and is equal to*

$$\frac{\int r_2 \, \mathrm{d}\mu^1}{\int r_1 \, \mathrm{d}\mu^1}.$$

Furthermore, if L_1 and L_2 are based length functions corresponding to l_1 and l_2, then

$$i(l_1, l_2) = \lim_{n \to +\infty} \frac{L_2(x_0 x_1 \cdots x_{n-1})}{L_1(x_0 x_1 \cdots x_{n-1})} \qquad \text{for } \mu^1\text{-a.e. } (x_n)_{n=0}^{\infty} \in \Sigma^+.$$

Proof. Consider the r_1-suspended semi-flow $\sigma_t^{r_1} : \Sigma^{r_1} \to \Sigma^{r_1}$ over Σ^+. Choose $R \colon \Sigma^{r_1} \to \mathbb{R}$ so that $\mathfrak{I}(R) = r_2$. If the periodic $\sigma_t^{r_1}$-orbit τ corresponds to the σ-orbit $x, \sigma x, \ldots, \sigma^{n-1} x$ (and the conjugacy class w) then τ has period $\lambda(\tau) = r_1^{|w|}(x) = l_1(w)$ and we have

$$\pi_{l_1}(T) = \#\{\tau : \lambda(\tau) \le T\} \quad \text{and} \quad \int_0^{l_1(w)} R(\sigma_t(x, 0)) \, \mathrm{d}t = r_2^{|w|}(x) = l_2(w).$$

It is a standard result for suspended flows that

$$\lim_{T \to +\infty} \frac{1}{\#\{\tau : \lambda(\tau) \le T\}} \sum_{\lambda(\tau) \le T} \int R \, \mathrm{d}m_\tau = \int R \, \mathrm{d}m_0 = \frac{\int r_2 \, \mathrm{d}\mu^1}{\int r_1 \, \mathrm{d}\mu^1}$$

[29], where m_0 is the measure of maximal entropy for $\sigma_t^{r_1}$, giving the first part of the result.

Now consider $x = (x_n)_{n=0}^{\infty} \in \Sigma^+$. Choose the r_i, so that they are related by (3) to the based length functions L_i, $i = 1, 2$. Then

$$L_i(x_0 x_1 \cdots x_{n-1}) = r_i^n(x_0 x_1 \cdots x_{n-1}, \dot{0}).$$

It is easy to see that

$$\lim_{n \to +\infty} \frac{r_i^n(x_0 x_1 \cdots x_{n-1}, \dot{0})}{n} = \lim_{n \to +\infty} \frac{r^n(x)}{n}$$

and, by the ergodic theorem, for μ^1-a.e. x, the limit is equal to $\int r_i \, \mathrm{d}\mu^1$. Combining these observations gives the second equality. □

Remark. Kapovich [18] has defined an intersection pairing on $\mathrm{cv}(F) \times \mathrm{Curr}(F)$, where $\mathrm{Curr}(F)$ denotes the space of geodesic currents on F. There is a natural embedding of $\mathrm{cv}(F)$ in $\mathrm{Curr}(F)$, using the Patterson–Sullivan currents described by Kapovich and Nagnibeda in [19]. However, unlike the case of surfaces, the intersection does not extend to $\mathrm{Curr}(F) \times \mathrm{Curr}(F)$. There is a natural identification between $\mathrm{Curr}(F)$ and $\mathcal{M}_\sigma$ and it is natural to conjecture that, for $l \in \mathrm{cv}(F)$ and $\mu \in \mathcal{M}_\sigma$,

$$I(l, \mu) = \int r \, \mathrm{d}\mu,$$

where $r \in \mathfrak{S}(l)$, but we have been unable to prove it. If ϕ is an automorphism of F then $i(\|\cdot\|, \|\phi(\cdot)\|)$ is equal to the generic stretch of ϕ [16, 37].

Theorem 2. *If $\{w_m\}_{m=1}^{\infty}$ is a sequence in $\mathcal{C}(F)$ such that the associated periodic orbit measures μ_{w_m} converge weak* to μ_1 then*

$$i(l_1, l_2) = \lim_{m \to +\infty} \frac{l_2(w_m)}{l_1(w_m)}.$$

PROOF. By Lemma 4, $l_i(w_m)/|w_m| = \int r_i \, d\mu_{w_m}$. Suppose that μ_{w_m} converge to μ^1. Then

$$\lim_{m\to+\infty} \int r_i \, d\mu_{w_m} = \int r_i \, d\mu^1,$$

$i = 1, 2$. The result now follows from Theorem 1. □

Theorem 3. *For $l_1, l_2 \in \mathrm{cv}(F)$, we have*

$$i(l_1, l_2) \geq \frac{\delta_1}{\delta_2}$$

with equality if and only if l_2 is a constant multiple of l_1 (i.e., l_1 and l_2 define the same point in $\mathrm{CV}(F)$.)

PROOF. The inequality may be rewritten as

$$\frac{\int r_2 \, d\mu^1}{\int r_1 \, d\mu^1} \geq \frac{h(\mu^1)}{\int r_1 \, d\mu^1} \frac{\int r_2 \, d\mu^2}{h(\mu^2)}.$$

Rearranging, this becomes

$$\frac{h(\mu^2)}{\int r_2 \, d\mu^2} \geq \frac{h(\mu^1)}{\int r_2 \, d\mu^1},$$

which is true by the variational principle. Furthermore, we have equality if and only if $\mu^1 = \mu^2$, i.e., if and only if $-\delta_2 r_2$ is cohomologous to $-\delta_1 r_1 + c$, $c \in \mathbb{R}$. If this holds then, since $P(-\delta_1 r_1) = P(-\delta_2 r_2) = 0$, we must have $c = 0$, so $-\delta_2 r_2$ is cohomologous to $-\delta_1 r_1$. By Lemma 6, this is equivalent to $l_2 = (\delta_1/\delta_2) l_1$. □

The intersection defined above quantifies the typical distortion when l_2 is compared to l_1, where "typical" is with respect to the measure $\mu_{-\delta_1 r_1}$. It is natural to generalize this notion to consider all possible distortions. As for intersection, we give a definition based to conjugacy classes and then show that is has other equivalent characterizations.

Set

$$\mathcal{D}(l_1, l_2) = \left\{ \frac{l_2(w)}{l_1(w)} : w \in \mathcal{C}(F) \right\} \quad \text{and} \quad D(l_1, l_2) = \overline{\mathcal{D}(l_1, l_2)}.$$

Proposition 1.

$$D(l_1, l_2) = I(r_1, r_2) := \left\{ \frac{\int r_2 \, d\mu}{\int r_1 \, d\mu} : \mu \in \mathcal{M}_\sigma \right\}.$$

In particular, if l_1 is not a constant multiple of l_2 then the closure $D(l_1, l_2)$ is a non-trivial interval.

PROOF. For $w \in \mathcal{C}(F)$, let $\tau = \tau_w$ be the corresponding periodic $\sigma_t^{r_1}$-orbit. The characterization of $D(l_1, l_2)$ then follows from the identity

$$\frac{l_2(w)}{l_1(w)} = \int R \, dm_\tau = \frac{\int r_2 \, d\mu_w}{\int r_1 \, d\mu_w}$$

and the fact that the periodic orbit measures m_τ are weak* dense in $\mathcal{M}_{\sigma^{r_1}}$. If l_1 is not a constant multiple of l_2 then r_1 is not cohomologous to a constant multiple of r_2. Hence, by Proposition 3, $I(r_1, r_2)$ is a non-trivial interval. □

In [37], we considered the case where $l_1(w) = \|w\|$ and $l_2(w) = \|\phi(w)\|$, where ϕ is an automorphism of F. In this setting, Kapovich [17] showed that $\|\phi(w)\|/\|w\|$ may take any rational value in $D(l_1, l_2)$ and we studied the growth rate of the number of $w \in \mathcal{C}(F)$ (counted by $\|w\|$) for which $\|\phi(w)\|/\|w\| = \rho$, for any rational $\rho \in \operatorname{int}(D(l_1, l_2))$. For arbitrary (non-discrete) l_1 and l_2, $\mathcal{D}(l_1, l_2)$ is hard to characterize. In particular, for any $\rho \in \mathcal{D}(l_1, l_2)$, it is not clear that $\#\{w \in \mathcal{C}(F) : l_1(w) \leq T, l_2(w)/l_1(w) = \rho\}$ has good asymptotic properties. Thus, in this general setting, it is more natural to study those $w \in \mathcal{C}(F)$ for which $l_2(w)/l_1(w)$ are close to some prescribed value. With this in mind, for two length functions $l_1, l_2 \in \operatorname{cv}(F)$ and $\rho \in \operatorname{int}(D(l_1, l_2))$, we define the ρ-correlation number $\alpha_\rho(l_1, l_2)$ by

$$\alpha_\rho(l_1, l_2) = \lim_{\epsilon \to 0} \limsup_{T \to +\infty} \frac{1}{T} \log \# \left\{ w \in \mathcal{C}(F) : l_1(w) \leq T, \frac{l_2(w)}{l_1(w)} \in (\rho - \epsilon, \rho + \epsilon) \right\}.$$

Note that if $\delta_1 = \delta_2$ then $1 \in \operatorname{int}(D(l_1, l_2))$. The next result, from the thesis of T. White, shows that this also holds for lengths in $\mathrm{CV}(F)$.

Lemma 7 ([40]). *Suppose that $l_1, l_2 \in \mathrm{CV}(F)$ (i.e., they are associated to graphs with total side length 1). There exist $w', w'' \in \mathcal{C}(F)$ such that $l_1(w') < l_2(w')$ and $l_1(w'') > l_2(w'')$. In particular $1 \in \operatorname{int}(D(l_1, l_2))$.*

In the following theorem, we shall characterize the ρ-correlation number in terms of a variational principle for entropy over a set of measures satisfying a constraint. A more precise asymptotic holds if we impose an extra condition on the lengths. We say that $l_1, l_2 \in \operatorname{cv}(F)$ are *independent* if, for $a, b \in \mathbb{R}$,

$$\{a l_1(w) + b l_2(w) : w \in \mathcal{C}(F)\} \subset \mathbb{Z} \implies a = b = 0.$$

(By setting $a = 0$ or $b = 0$, this implies that l_1 and l_2 are each non-discrete.)

Theorem 4. *Let $l_1, l_2 \in \operatorname{cv}(F)$.*

(i) *Then, for every $\rho \in \operatorname{int}(D(l_1, l_2))$, $\alpha_\rho(l_1, l_2) > 0$ and satisfies $\alpha_\rho(l_1, l_2) = \mathfrak{h}_{r_1, r_2}(\rho)$, where*

$$\mathfrak{h}_{r_1, r_2}(\rho) = \sup \left\{ \frac{h(\mu)}{\int r_1 \, \mathrm{d}\mu} : \mu \in \mathcal{M}_\sigma \text{ and } \frac{\int r_2 \, \mathrm{d}\mu}{\int r_1 \, \mathrm{d}\mu} = \rho \right\}.$$

(ii) *Suppose that l_1, l_2 are independent. Then, for any $\rho \in \operatorname{int}(D(l_1, l_2))$,*

$$\#\{w \in \mathcal{C}(F) : l_1(w) \in (T, T + \epsilon), l_2(w) \in (\rho T, \rho T + \epsilon)\} \sim C(\rho, \epsilon) \frac{e^{\alpha_\rho(l_1, l_2) T}}{T^{3/2}},$$

as $T \to +\infty$.

Proof. We apply results from the periodic theory of hyperbolic flows, which immediately carry over to suspended semi-flows over subshifts of finite type. For part (i), observe that

$$\# \left\{ w \in \mathcal{C}(F) : l_1(w) \leq T, \frac{l_2(w)}{l_1(w)} \in (\rho - \epsilon, \rho + \epsilon) \right\}$$
$$= \# \left\{ \tau : \lambda(\tau) \leq T, \int R \, \mathrm{d}m_\tau \in (\rho - \epsilon, \rho + \epsilon) \right\}.$$

By large deviations results for periodic orbits [20], this has growth rate

$$\begin{aligned}\lim_{T\to+\infty}\frac{1}{T}\log\#\left\{\tau:\lambda(\tau)\le T,\int R\,\mathrm{d}m_\tau\in(\rho-\epsilon,\rho+\epsilon)\right\}\\ =\sup\left\{h(m):m\in\mathcal{M}_{\sigma^{r_1}}\text{ and }\int R\,\mathrm{d}m\in(\rho-\epsilon,\rho+\epsilon)\right\}\\ =\sup\left\{\frac{h(\mu)}{\int r_1\,\mathrm{d}\mu}:\mu\in\mathcal{M}_\sigma\text{ and }\frac{\int r_2\,\mathrm{d}\mu}{\int r_1\,\mathrm{d}\mu}\in(\rho-\epsilon,\rho+\epsilon)\right\}\\ =\sup\{\mathfrak{h}_{r_1,r_2}(\rho'):\rho'\in(\rho-\epsilon,\rho+\epsilon)\}.\end{aligned}$$

Since $\mathfrak{h}_{r_1,r_2}$ is analytic, we obtain $\alpha_\rho(l_1,l_2)=\mathfrak{h}_{r_1,r_2}(\rho)$, as required. Part (ii) follows from local limit results for suspended semi-flows [2, 22, 34]. □

Corollary 1. $\alpha_1(l_1,l_2)=\alpha_1(l_2,l_1)$.

PROOF. From Theorem 4, $\alpha_1(l_1,l_2)=\mathfrak{h}_{r_1,r_2}(1)$ and $\alpha_1(l_2,l_1)=\mathfrak{h}_{r_2,r_1}(1)$. Clearly, we have

$$\begin{aligned}\mathfrak{h}_{r_1,r_2}(1)&=\sup\left\{\frac{h(\mu)}{\int r_1\,\mathrm{d}\mu}:\mu\in\mathcal{M}_\sigma\text{ and }\int r_1\,\mathrm{d}\mu=\int r_2\,\mathrm{d}\mu\right\}\\ &=\sup\left\{\frac{h(\mu)}{\int r_2\,\mathrm{d}\mu}:\mu\in\mathcal{M}_\sigma\text{ and }\int r_1\,\mathrm{d}\mu=\int r_2\,\mathrm{d}\mu\right\}=\mathfrak{h}_{r_2,r_1}(1),\end{aligned}$$

as required. □

If ϕ is an automorphism of F and $l_1=\|\cdot\|$ and $l_2=\|\phi(\cdot)\|$ then the critical exponents are both $\log(2k-1)$ and $\alpha_1(l_1,l_2)$ is related to the quantity $\mathrm{Curl}(\phi)$ [26, 37].

Remark. As noted in the appendix, $\mathfrak{h}_{r_1,r_2}$ is a continuous function on the closed interval $D(l_1,l_2)=[\rho_{\min},\rho_{\max}]$. Typically, the value of $\mathfrak{h}_{r_1,r_2}$ is zero at the endpoints but it is easy to construct examples where $\mathfrak{h}_{r_1,r_2}(\rho_{\min})>0$ or $\mathfrak{h}_{r_1,r_2}(\rho_{\max})>0$.

Consider the graph $\mathcal{G}$, with one vertex and k edges. Make this into a metric graph by assigning lengths $0<l_1^{(1)}<l_1^{(2)}<\cdots<l_1^{(k)}$ to the edges corresponding to the generators $a_1^{\pm1},a_2^{\pm1},\dots,a_k^{\pm1}$, respectively. Repeat the procedure with a different set of lengths $0<l_2^{(1)}<l_2^{(2)}<\cdots<l_2^{(k)}$ to obtain another metric graph. These give rise to two length functions $l_1,l_2\in\mathrm{cv}(F)$.

Suppose that we choose the edge lengths so that

$$\frac{l_2^{(1)}}{l_1^{(1)}}<\min_{2\le j\le k}\frac{l_2^{(j)}}{l_1^{(j)}}.$$

Let $r_1,r_2\colon\Sigma^+\to\mathbb{R}$ be functions in $\mathfrak{S}(l_1),\mathfrak{S}(l_2)$, respectively. It is easy to see that $\min D(l_1,l_2)=\min I_{r_1,r_2}$ is only attained for convex combinations of the periodic point measures supported on $(a_1,a_1,0\dots)$ and $(a_1^{-1},a_1^{-1},0\dots)$. In particular, $\mathfrak{h}_{r_1,r_2}(\rho_{\min})=0$.

Now modify the lengths so that

$$\frac{l_2^{(1)}}{l_1^{(1)}}=\frac{l_2^{(2)}}{l_1^{(2)}}<\min_{3\le j\le k}\frac{l_2^{(j)}}{l_1^{(j)}}.$$

Then $\min D(l_1, l_2) = \min I_{r_1,r_2}$ is attained, in particular, for the measure of maximal entropy for the subshift of finite type obtained by restricting to the symbols $\{a_1, a_1^{-1}, a_2, a_2^{-1}\}$. A simple calculation then shows that

$$\mathfrak{h}_{r_1,r_2}(\rho_{\min}) \geq \frac{2\log 3}{l_1^{(1)} + l_1^{(2)}} > 0.$$

Of course, a small perturbation returns us to the situation of zero entropy at the endpoints. For analogous results in the context of Hölder continuous function on subshifts of finite type, see [25].

5. The Manhattan curve

In this section, we discuss the Manhattan curve $\mathfrak{M}(l_1, l_2)$ associated to a pair of lengths $l_1, l_2 \in \mathrm{cv}(F)$. This is defined by

$$\mathfrak{M}(l_1, l_2) = \partial\Big\{(a, b) \in \mathbb{R}^2 : \sum_{w \in \mathcal{C}(F)} \mathrm{e}^{-al_1(w) - bl_2(w)} < +\infty\Big\}.$$

As noted in the introduction, this may be thought of as the natural analogue of the critical exponent for a single length function.

The following theorem describes how various quantities may be read off from $\mathfrak{M}(l_1, l_2)$.

Theorem 5. *Let $l_1, l_2 \in \mathrm{cv}(F)$.*

(i) *$\mathfrak{M}(l_1, l_2)$ is a straight line if and only if l_1 is a constant multiple of l_2.*
(ii) *$\mathfrak{M}(l_1, l_2)$ is real analytic.*
(iii) *The set of slopes of normals to $\mathfrak{M}(l_1, l_2)$ is equal to $\mathrm{int}\big(D(l_1, l_2)\big)$. Furthermore, $\mathfrak{M}(l_1, l_2)$ has asymptotes whose normals have slopes equal to $\max D(l_1, l_2)$ and $\min D(l_1, l_2)$.*
(iv) *$\mathfrak{M}(l_1, l_2)$ passes through $(\delta_1, 0)$, where its normal has slope equal to $i(l_1, l_2)$ (and through $(0, \delta_2)$, where its normal has slope equal to $1/i(l_2, l_1)$).*

If $l_1, l_2 \in \mathrm{CV}(F)$ then

(v) *there is a unique point $(a, b) \in \mathfrak{M}(l_1, l_2)$ where the normal has slope 1 and $a + b = \alpha_1(l_1, l_2)$.*

Proof. We modify the analysis of [36, 37]. Rewriting the definition of the Manhattan curve in terms of periodic points for $\sigma \colon \Sigma^+ \to \Sigma^+$, we have

$$\mathfrak{M}(l_1, l_2) = \partial\Big\{(a, b) \in \mathbb{R}^2 : \sum_{n=1}^{\infty} \frac{1}{n} \sum_{\sigma^n x = x} \mathrm{e}^{-ar_1^n(x) - br_2^n(x)} < +\infty\Big\}.$$

By (7), this may be described as the set $\{(a, b) \in \mathbb{R}^2 : P(-ar_1 - br_2) = 0\}$. Let us define $\mathfrak{q}(t)$ implicitly by

$$P(-tr_1 - \mathfrak{q}(t)r_2) = 0; \tag{5}$$

then $\mathfrak{M}(l_1, l_2)$ is the graph of $\mathfrak{q}$. Since

$$\frac{\partial}{\partial s} P(-tr_1 - sr_2) = -\int r_2 \, d\mu_{-tr_1 - sr_2} \neq 0,$$

the Implicit Function Theorem gives that $\mathfrak{q}$ is real analytic, establishing (ii).

If we define $\mathfrak{p}(s) = \mathfrak{p}_{r_1,r_2}(s)$ (i.e., $P(-\mathfrak{p}(s)r_1 + sr_2) = 0$ or, equivalently, $\mathfrak{p}(s) = P_{\sigma^{r_1}}(sR)$) then

$$\mathfrak{q}(t) = -\mathfrak{p}^{-1}(t).$$

Now $P_{\sigma^{r_1}}(sR)$ is strictly convex and $P''_{\sigma^{r_1}}(sR) > 0$ unless R is not cohomologous to a constant, c say. This latter condition gives $\int_\tau R = c\lambda(\tau)$, for all periodic σ^{r_1}-orbits τ, i.e., $l_2(w) = cl_1(w)$, for all $w \in \mathcal{C}(F)$. So $\mathfrak{M}(l_1, l_2)$ is strictly convex unless l_2 is a constant multiple of l_1, proving (i).

Let us examine the slope of the normal at a point $(a,b) = (t, \mathfrak{q}(t))$ on $\mathfrak{M}(l_1, l_2)$. At this point, the normal has slope $-1/\mathfrak{q}'(t)$. Now, differentiating (5),

$$0 = \frac{\mathrm{d}}{\mathrm{d}t} P(-tr_1 - \mathfrak{q}(t)r_2) = -\int r_1 \, \mathrm{d}\mu_{-tr_1 - \mathfrak{q}(t)r_2} - \left(\int r_2 \, \mathrm{d}\mu_{-tr_1-\mathfrak{q}(t)r_2}\right)\mathfrak{q}'(t),$$

i.e.,

$$\mathfrak{q}'(t) = -\frac{\int r_1 \, \mathrm{d}\mu_{-tr_1-\mathfrak{q}(t)r_2}}{\int r_2 \, \mathrm{d}\mu_{-tr_1-\mathfrak{q}(t)r_2}}.$$

Thus the normal to $\mathfrak{M}(l_1, l_2)$ at $\big(t, \mathfrak{q}(t)\big)$ has slope

$$\frac{-1}{\mathfrak{q}'(t)} = \frac{\int r_2 \, \mathrm{d}\mu_{-tr_1-\mathfrak{q}(t)r_2}}{\int r_1 \, \mathrm{d}\mu_{-tr_1-\mathfrak{q}(t)r_2}}.$$

Therefore, using Proposition 4, the set of slopes of normals to $\mathfrak{M}(l_1, l_2)$ is equal to

$$\left\{\frac{\int r_2 \, \mathrm{d}\mu_{-tr_1-\mathfrak{q}(t)r_2}}{\int r_1 \, \mathrm{d}\mu_{-tr_1-\mathfrak{q}(t)r_2}} : t \in \mathbb{R}\right\} = \mathrm{int}(I_{r_2,r_1}) = \mathrm{int}\big(D(l_1, l_2)\big).$$

This proves the first part of (iii) and the statement about the asymptotes follows. It is clear that $\mathfrak{M}(l_1, l_2)$ passes through $(\delta_1, 0)$ at this point the normal has slope

$$\frac{\int r_2 \, \mathrm{d}\mu^1}{\int r_1 \, \mathrm{d}\mu^1} = i(l_1, l_2).$$

The statement about $(0, \delta_2)$ follows easily. This proves (iv).

Suppose that $l_1, l_2 \in \mathrm{CV}(F)$, $l_1 \neq l_2$. Then, by Lemma 7, $1 \in \mathrm{int}\big(D(l_1, l_2)\big)$. If the normal has slope 1 at $(a, b) = \big(t, \mathfrak{q}(t)\big)$ then

$$\frac{\int r_2 \, \mathrm{d}\mu_{-tr_1-\mathfrak{q}(t)r_2}}{\int r_1 \, \mathrm{d}\mu_{-tr_1-\mathfrak{q}(t)r_2}} = 1$$

and

$$a + b = t + \mathfrak{q}(t) = \mathfrak{p}\big(-\mathfrak{q}(t)\big) + \mathfrak{q}(t) = \mathfrak{p}(\xi) - \xi,$$

where ξ is chosen so that $\mathfrak{p}'(\xi) = 1$. Then

$$\mathfrak{p}(\xi) - \xi = \frac{h(\mu_{-\mathfrak{p}(\xi)-\xi r_2})}{\int r_1 \, \mathrm{d}\mu_{-\mathfrak{p}(\xi)r_1 - \xi r_2}} = \mathfrak{h}_{r_1,r_2}(1)$$

and so

$$a + b = \mathfrak{h}_{r_1,r_2}(1) = \alpha_1(l_1, l_2),$$

as required. This proves (v). □

Appendix A. Shifts of finite type and thermodynamic formalism

We shall study length functions on free groups via a dynamical system called a subshift of finite type. In this appendix we review some background material about the ergodic theory of these systems. For basic facts about ergodic theory, see [39].

We begin with the definition of a subshift of finite type. Let A be finite matrix, indexed by a set $\mathcal{I}$, with entries zero and one. We define the shift space

$$\Sigma_A^+ = \{(x_n)_{n=0}^\infty \in \mathcal{I}^{\mathbb{Z}^+} : A(x_n, x_{n+1}) = 1 \ \forall n \in \mathbb{Z}^+\}$$

and the (one-sided) subshift of finite type $\sigma\colon \Sigma_A^+ \to \Sigma_A^+$ by $(\sigma x)_n = x_{n+1}$. We give $\mathcal{I}$ the discrete topology, $\mathcal{I}^{\mathbb{Z}^+}$ the product topology and Σ_A^+ the subspace topology. A compatible metric is given by

$$d\big((x_n)_{n=0}^\infty, (y_n)_{n=0}^\infty\big) = \sum_{n=0}^\infty \frac{1-\delta_{x_n y_n}}{2^n},$$

where δ_{ij} is the Kronecker symbol.

We say that A is irreducible if, for each $(i,j) \in \mathcal{I}^2$, there exists $n(i,j) \geq 1$ such that $A^{n(i,j)}(i,j) > 0$ and aperiodic if there exists $n \geq 1$ such that, for each $(i,j) \in \mathcal{I}^2$, $A^n(i,j) > 0$. The latter statement is equivalent to $\sigma\colon \Sigma_A^+ \to \Sigma_A^+$ being topologically mixing (i.e., that there exists $n \geq 1$ such that for any two non-empty open sets $U, V \subset \Sigma_A^+$, $\sigma^{-m}(U) \cap V \neq \varnothing$, for all $m \geq n$).

If A is aperiodic then it has a positive simple eigenvalue β which is strictly maximal in modulus (i.e., every other eigenvalue has modulus strictly less that β) and the topological entropy $h(\sigma)$ of σ is equal to $\log \beta$.

If an ordered n-tuple $(x_0, x_1, \ldots, x_{n-1}) \in \mathcal{I}^n$ is such that $A(x_m, x_{m+1}) = 1$, $m = 0, 1, \ldots, n-2$ then we say that $(x_0, x_1, \ldots, x_{n-1})$ is an allowed word of length n in Σ_A^+; the set of these is denoted $W_A^{(n)}$. If $\sigma^n x = x$ then we say that $\{x, \sigma x, \ldots, \sigma^{n-1} x\}$ is a periodic orbit for σ. Clearly any such an x is obtained by repeating a word $(x_0, x_1, \ldots, x_{n-1}) \in W_A^{(n)}$ with the additional property that $A(x_{n-1}, x_0) = 1$. Note that we regard the periodic orbits $\{x, \sigma x, \ldots, \sigma^{n-1} x\}$, $\{x, \sigma x, \ldots, \sigma^{n-1} x, x, \ldots, \sigma^{n-1} x\}$, etc., as distinct objects (even though they are identical as point sets). If $\sigma^n x = x$ but $\sigma^m x \neq x$ for $0 < m < n$ then we say that $\{x, \sigma x, \ldots, \sigma^{n-1} x\}$ is a prime periodic orbit.

A function $f\colon \Sigma_A^+ \to \mathbb{R}$ is Hölder continuous if there exists $\alpha > 0$ and $C(f,\alpha) \geq 0$ such that $|f(x) - f(y)| \leq C(f,\alpha) d(x,y)^\alpha$, for all $x, y \in \Sigma_A^+$. We say that a function $f\colon \Sigma_A^+ \to \mathbb{R}$ is locally constant if there exists $N \geq 0$ such that if $x = (x_n)_{n=0}^\infty, y = (x_n)_{n=0}^\infty$ have $x_n = y_n$ for all $n \geq N$ then $f(x) = f(y)$. Clearly, if f is locally constant then f is Hölder continuous (for any choice of exponent $\alpha > 0$).

Let $\mathcal{M}$ denote the space of all Borel probability measures on Σ_A^+, equipped with the weak* topology, and let $\mathcal{M}_\sigma$ denote the subspace consisting of σ-invariant probability measures. For $\mu \in \mathcal{M}_\sigma$, write $h(\mu)$ for the measure theoretic entropy of μ. There is a unique measure $\mu_0 \in \mathcal{M}_\sigma$, called the measure of maximal entropy, for which

$$h(\mu_0) = \sup_{\mu \in \mathcal{M}_\sigma} h(\mu)$$

and this value coincides with the topological entropy $h(\sigma)$.

For a continuous function $f\colon \Sigma_A^+ \to \mathbb{R}$, we define the pressure $P(f)$ of f by the formula

$$P(f) = \sup_{\mu \in \mathcal{M}_\sigma} \left(h(\mu) + \int f \, \mathrm{d}\mu \right) \tag{6}$$

and call any measure for which the supremum is attained an equilibrium state for f. If f is Hölder continuous then f has a unique equilibrium state which we denote by μ_f. The latter is fully supported and $h(\mu_f) > 0$. The equilibrium state of the zero function is the measure of maximal entropy, so this is consistent with our earlier notation. The pressure of f also has the following characterization in terms of periodic points:

$$P(f) = \lim_{n \to +\infty} \frac{1}{n} \log \sum_{\sigma^n x = x} \mathrm{e}^{f^n(x)}, \tag{7}$$

where $f^n = f + f \circ \sigma + \cdots + f \circ \sigma^{n-1}$.

We say that two functions $f, g\colon \Sigma_A^+ \to \mathbb{R}$ are (continuously) cohomologous if there is a continuous function $u\colon \Sigma_A^+ \to \mathbb{R}$ such that $f = g + u \circ \sigma - u$. The cohomology class of a Hölder continuous function is determined by its values around periodic orbits. More precisely, we have the following.

Proposition 2 (Livsic Theorem). *Let $f, g\colon \Sigma_A^+ \to \mathbb{R}$ be two Hölder continuous functions. Then f and g are cohomologous if and only if $f^n(x) = g^n(x)$ whenever $\sigma^n x = x$.*

If f and g are cohomologous then $P(f) = P(g)$ and if c is a real number then $P(f + c) = P(f) + c$. Now suppose that f and g are Hölder continuous and, for $t \in \mathbb{R}$, consider the function $t \mapsto P(tf)$ This function is convex and real analytic and

$$P'(tf + g) = \int f \, \mathrm{d}\mu_{tf+g}. \tag{8}$$

Furthermore, if f is not cohomologous to a constant then $P(tf+g)$ is strictly convex and $P''(tf + g) > 0$ everywhere. (If f is cohomologous to a constant c then clearly $P(tf + g) = P(g) + tc$.)

Suppose that a Hölder continuous function $r\colon \Sigma_A^+ \to \mathbb{R}$ is cohomologous to a strictly positive function. Since we are only interested in sums over periodic orbits and integrals with respect to invariant measures, we shall, in fact, assume that r itself is strictly positive. If $\delta > 0$ satisfies $P(-\delta r) = 0$ then, since $\mu_{-\delta r}$ attains the supremum in (6), we have the relation

$$\delta = \frac{h(\mu_{-\delta r})}{\int r \, \mathrm{d}\mu_{-\delta r}}. \tag{9}$$

More generally, given another Hölder continuous function $r\colon \Sigma_A^+ \to \mathbb{R}$, we define $\mathfrak{p}_{r,f}(s)$ by $P(-\mathfrak{p}_{r,f}(s)r + sf) = 0$. Note that, by (8),

$$P'(-tr + f) = \int r \, \mathrm{d}\mu_{-tr+f} > 0.$$

Therefore, by the Implicit Function Theorem, $\mathfrak{p}_{r,f}(s)$ is real analytic and

$$\mathfrak{p}'_{r,f}(s) = \frac{\int f \, \mathrm{d}\mu_{-\mathfrak{p}_{r,f}(s)r+sf}}{\int r \, \mathrm{d}\mu_{-\mathfrak{p}_{r,f}(s)r+sf}}. \tag{10}$$

We write

$$I_{f,r} := \left\{ \frac{\int f \,\mathrm{d}\mu}{\int r \,\mathrm{d}\mu} : \mu \in \mathcal{M}_\sigma \right\}.$$

Since $\int r \,\mathrm{d}\mu > 0$, for any $\mu \in \mathcal{M}_\sigma$, $\mu \mapsto \int f \,\mathrm{d}\mu / \int r \,\mathrm{d}\mu$ is continuous with respect to the weak* topology on $\mathcal{M}_\sigma$. So, as $\mathcal{M}_\sigma$ is compact and convex, $I_{f,r}$ is a closed interval.

Proposition 3. (i) *We have* $I_{r,f} = \{c\}$ *if and only if* f *is cohomologous to* cr.

(ii) *If* $\mathrm{int}(I_{r,f}) \neq \varnothing$ *and* $\int f \,\mathrm{d}\mu / \int r \,\mathrm{d}\mu$ *is an endpoint of* $I_{r,f}$ *then* μ *does not have full support.*

(iii) *If* $\mathrm{int}(I_{r,f}) \neq \varnothing$ *then*

$$\mathrm{int}(I_{r,f}) = \{\mathfrak{p}'_{r,f}(s) : s \in \mathbb{R}\} = \left\{ \frac{\int f \,\mathrm{d}\mu_{-\mathfrak{p}_{r,f}(s)r+sf}}{\int r \,\mathrm{d}\mu_{-\mathfrak{p}_{r,f}(s)r+sf}} : s \in \mathbb{R} \right\}.$$

(iv) *We have* $\mathrm{int}(I_{r,f}) \neq \varnothing$ *if and only if* $\mathfrak{p}''_{r,f}(s) > 0$ *for some (or, equivalently, all)* $s \in \mathbb{R}$.

PROOF. (i) The statement $I_{r,f} = \{c\}$ is equivalent to

$$\int f \,\mathrm{d}\mu - \int cr \,\mathrm{d}\mu = 0,$$

for all $\mu \in \mathcal{M}_\sigma$. In particular, this is equivalent to $f^n(x) = cr^n(x)$, whenever $\sigma^n x = x$. By Proposition 2, this is equivalent to f and cr being cohomologous.

(ii) Suppose that $I_{r,f} = [a,b]$ and that $\int f \,\mathrm{d}\mu / \int r \,\mathrm{d}\mu = b$. Then μ is a *maximizing measure* for $f - br$. Then $\mathrm{supp}\,\mu$ is contained in a closed subset of Σ_A^+ on which $f - br$ is cohomologous to 0 [15, Lemma 2]. By (i), this closed set cannot be Σ_A^+ itself.

(iii) Since $\mathfrak{p}_{r,f}$ is strictly convex, $\{\mathfrak{p}'_{r,f}(s) : s \in \mathbb{R}\}$ is an open interval. The following argument is taken from [15]. From the definition of $\mathfrak{p}_{r,f}(s)$, for any $\mu \in \mathcal{M}_\sigma$,

$$\mathfrak{p}_{r,f}(s) \geq \frac{h(\mu)}{\int r \,\mathrm{d}\mu} + s \frac{\int f \; d\mu}{\int r \,\mathrm{d}\mu},$$

for all $s \in \mathbb{R}$. In particular, the graph of the convex function $\mathfrak{p}_{r,f}$ lies above a line with slope $\int f \,\mathrm{d}\mu / \int r \,\mathrm{d}\mu$ in $\mathbb{R}^2$ (possibly touching it tangentially) and so $\int f \,\mathrm{d}\mu / \int r \,\mathrm{d}\mu \in \overline{\{\mathfrak{p}'_{r,f}(s) : s \in \mathbb{R}\}}$. Thus, since μ is arbitrary,

$$I_{r,f} \subset \overline{\{\mathfrak{p}'_{r,f}(s) : s \in \mathbb{R}\}}.$$

By (10), $\{\mathfrak{p}'_{r,f}(s) : s \in \mathbb{R}\} \subset I_{r,f}$. Therefore, $\mathrm{int}(I_{r,f}) = \{\mathfrak{p}'_{r,f}(s) : s \in \mathbb{R}\}$, as required.

(iv) If $\mathfrak{p}''_{r,f}(s) > 0$ for some $s \in \mathbb{R}$ then the graph of $\mathfrak{p}_{r,f}$ cannot be a straight line, so $\mathrm{int}(I_{r,f}) \neq \varnothing$. On the other hand, suppose that $\mathrm{int}(I_{r,f}) \neq \varnothing$ and choose $\xi \in \mathbb{R}$. Write $\rho = \mathfrak{p}'_{r,f}(\xi)$. Note that $\mathfrak{p}_{r,f-\rho r}(s) = \mathfrak{p}_{r,f}(s) - s\rho$, so $\mathfrak{p}'_{r,f-\rho r}(\xi) = 0$. Differentiating the identity $P(-\mathfrak{p}_{r,f-\rho r}(s)r + s(f - \rho r)) = 0$ twice at $s = \xi$ gives

$$\begin{aligned}\mathfrak{p}''_{r,f}(\xi) &= \mathfrak{p}''_{r,f-\rho r}(\xi) \\ &= \left(\int r \,\mathrm{d}\mu_{\mathfrak{p}_{r,f-\rho r}(\xi)r+\xi(f-\rho r)} \right)^{-1} \frac{\partial^2 P(-\mathfrak{p}_{r,f-\rho r}(\xi)r + s(f - \rho r))}{\partial s^2},\end{aligned}$$

which is positive unless $f - \rho r$ is cohomologous to a constant. By the choice of ρ, this constant must necessarily be zero, so $\mathfrak{p}''_{r,f}(\xi) > 0$ unless f is cohomologous to ρr, which contradicts the assumption that $\mathrm{int}(I_{r,f}) \neq \varnothing$. $\square$

We also define a weighted entropy function $\mathfrak{h}_{r,f} \colon I_{r,f} \to \mathbb{R}^+$ by

$$\mathfrak{h}_{r,f}(\rho) = \sup\left\{ \frac{h(\mu)}{\int r \, d\mu} : \mu \in \mathcal{M}_\sigma \text{ and } \int f \, d\mu = \rho \right\}. \tag{11}$$

From its definition,

$$\begin{aligned}
\mathfrak{p}_{r,f}(s) &= \sup\left\{ \frac{h(\mu)}{\int r \, d\mu} + s \frac{\int f \, d\mu}{\int r \, d\mu} : \mu \in \mathcal{M}_\sigma \right\} \\
&= \sup_{\rho \in I_{r,f}} \sup\left\{ \frac{h(\mu)}{\int r \, d\mu} + s \frac{\int f \, d\mu}{\int r \, d\mu} : \mu \in \mathcal{M}_\sigma, \frac{\int f \, d\mu}{\int r \, d\mu} = \rho \right\} \\
&= \sup_{\rho \in I_{r,f}} \left(\sup\left\{ \frac{h(\mu)}{\int r \, d\mu} : \mu \in \mathcal{M}_\sigma, \frac{\int f \, d\mu}{\int r \, d\mu} = \rho \right\} + s\rho \right) \\
&= \sup_{\rho \in I_{r,f}} (\mathfrak{h}_{r,f}(\rho) + s\rho).
\end{aligned}$$

In the language of convex analysis, $-\mathfrak{h}_{r,f}$ is the convex conjugate of $\mathfrak{p}_{r,f}$ [33, p. 104]. Assuming that f is not cohomologous to a constant multiple of r, $\mathfrak{p}_{r,f}$ is strictly convex, so $\mathfrak{h}_{r,f}$ is strictly concave. Since the entropy map $\mu \mapsto h(\mu)$ is upper semi-continuous, this implies that $\mathfrak{h}_{r,f} \colon I_{r,f} \to \mathbb{R}$ is continuous. The next result follows from the theory of convex functions [33].

Proposition 4. *Suppose that f is not cohomologous to a constant multiple of r. If, for $\rho \in \mathrm{int}(I_{r,f})$, $\xi \in \mathbb{R}$ is the unique solution to $\mathfrak{p}'(\xi) = \rho$ then $\mathfrak{p}_{r,f}(\xi) = \mathfrak{h}_{r,f}(\rho) + \xi\rho$, so that*

$$\mathfrak{h}_{r,f}(\rho) = \frac{h(\mu_{-\mathfrak{p}_{r,f}(\xi)r + \xi f})}{\int r \, d\mu_{-\mathfrak{p}_{r,f}(\xi)r + \xi f}}$$

and $\mu_{-\mathfrak{p}_{r,f}(\xi)r+\xi f}$ is the only element of $\mathcal{M}_\sigma$ realizing the supremum in (11). *Furthermore, $\mathfrak{h}_{r,f} \colon \mathrm{int}(I_{r,f}) \to \mathbb{R}$ is real analytic.*

In the remainder of this appendix, we shall reinterpret $I_{r,f}$, $\mathfrak{p}_{r,f}$ and $\mathfrak{h}_{r,f}$ in terms of a suspended semi-flow over Σ_A^+. The reason for this is that it will allow us to quote a number of results from the theory of suspended and hyperbolic flows (which also hold automatically for semi-flows.)

We define the r-suspension space

$$\Sigma_A^r = \{(x, s) : x \in \Sigma_A^+, 0 \le s \le r(x)\}/\sim,$$

where $(x, r(x)) \sim (\sigma x, 0)$. (It was for this construction to make sense that we took r to be strictly positive, rather than merely cohomologous to a strictly positive function.) This supports the suspended semi-flow $\sigma_t^r \colon \Sigma_A^r \to \Sigma_A^r$ defined (for small $t > 0$) by $\sigma_t^r(x, s) = (x, s + t)$ and respecting the identifications. A more familiar object of study is the suspended flow over the corresponding two-sided shift space $\Sigma_A = \{(x_n)_{n=-\infty}^{\infty} \in \mathcal{I}^{\mathbb{Z}} : A(x_n, x_{n+1}) = 1 \ \forall n \in \mathbb{Z}\}$. Such flows and their periodic orbits are well understood [22, 29, 35] and it is easy to see that the same results hold for semi-flows.

We let $\mathcal{M}_{\sigma^r}$ denote the space of σ_t^r-invariant probability measures on Σ_A^r. Every $m \in \mathcal{M}_{\sigma^r}$ may be written locally as

$$m = \frac{\mu \times \text{Leb}}{\int r \, \mathrm{d}\mu}, \tag{12}$$

where $\mu \in \mathcal{M}_\sigma$ and Leb is one-dimensional Lebesgue measure, and has entropy

$$h_{\sigma^r}(m) = \frac{h(\mu)}{\int r \, d\mu}. \tag{13}$$

Given a Hölder continuous function $F\colon \Sigma_A^r \to \mathbb{R}$, we may associate a Hölder continuous function $f = \mathfrak{I}(F)\colon \Sigma_A^+ \to \mathbb{R}$ by

$$\mathfrak{I}(F) := \int_0^{r(x)} F(x,t) \, \mathrm{d}t.$$

One easily sees that if m and μ are related by (12) then

$$\int F \, \mathrm{d}m = \frac{\int f \, \mathrm{d}\mu}{\int r \, \mathrm{d}\mu}.$$

Given a Hölder continuous function $f\colon \Sigma_A^+ \to \mathbb{R}$, it is easy to construct a Hölder continuous function $F\colon \Sigma_A^r \to \mathbb{R}$ such that $\mathfrak{I}(F) = f$. For example, choose a smooth function $\Delta\colon [0,1] \to \mathbb{R}^+$ with $\Delta(0) = \Delta(1) = 0$ and $\int \Delta(y) \, \mathrm{d}y = 1$. Then set

$$F(x,s) = \Delta\left(\frac{s}{r(x)}\right) \frac{f(x)}{r(x)}, \qquad 0 \le s \le r(x).$$

(The function Δ is introduced so that F respects the equivalence relation $\big(x, r(x)\big) \sim (\sigma x, 0)$.) In particular,

$$I_{r,f} = \left\{ \int F \, \mathrm{d}m : m \in \mathcal{M}_{\sigma^r} \right\}.$$

We define the pressure $P_{\sigma^r}(F)$ of a function $F\colon \Sigma_A^r \to \mathbb{R}$ by the formula

$$P_{\sigma^r}(F) = \sup_{m \in \mathcal{M}_{\sigma^r}} \left(h(m) + \int F \, \mathrm{d}m \right)$$

and call any measure for which the supremum is attained an equilibrium state for F. If F is Hölder continuous then F has a unique equilibrium state which we denote by m_F.

Proposition 5. *We have* $P(-P_{\sigma^r}(F)r + f) = 0$, *where* $f = \mathfrak{I}(F)$, *and, locally,*

$$m_F = \frac{\mu_{-P(F)r+f} \times \text{Leb}}{\int r \, \mathrm{d}\mu_{-P(F)r+f}}.$$

PROOF. The identity $P(-cr + f) = 0$ is equivalent to

$$\sup_{\mu \in \mathcal{M}_\sigma} \left(h(\mu) + \int (-cr + f) \, \mathrm{d}\mu \right) = 0,$$

which may be rewritten as

$$\sup_{\mu \in \mathcal{M}_\sigma} \left(\frac{h(\mu)}{\int r \, \mathrm{d}\mu} + \frac{\int f \, \mathrm{d}\mu}{\int r \, \mathrm{d}\mu} \right) = c.$$

By (12) and (13), this becomes

$$c = \sup_{m \in \mathcal{M}_{\sigma^r}} \left(h_{\sigma^r}(m) + \int F \, \mathrm{d}m \right) = P_{\sigma^r}(F).$$

This calculations also give the formula for m_F. □

Remark. In particular, the equilibrium state of the zero function m_0 is the measure of maximal entropy for σ_t^r and, locally,

$$m_0 = \frac{\mu_{-h(\sigma)r} \times \mathrm{Leb}}{\int r \, \mathrm{d}\mu_{-h(\sigma)r}}.$$

Applying the lemma to sF, gives $P_{\sigma^r}(sF) = \mathfrak{p}_{r,f}(s)$. Hence, by (10),

$$P'_{\sigma^r}(sF) = \mathfrak{p}'_{r,f}(s) = \frac{\int f \, \mathrm{d}\mu_{-\mathfrak{p}_{r,f}(s)r+sf}}{\int r \, \mathrm{d}\mu_{-\mathfrak{p}_{r,f}(s)r+sf}} = \int F \, \mathrm{d}m_{sF}. \tag{14}$$

There is clearly an exact correspondence between periodic orbits for σ_t^r and σ. If τ is the prime periodic σ^r-orbit passing through the prime periodic σ-orbit $x, \sigma x, \ldots, \sigma^{n-1}x$ ($\sigma^n x = x$) then the period of τ is

$$\lambda(\tau) := \inf\{t > 0 : \sigma_t^r(x,0) = (x,0)\} = r^n(x).$$

Furthermore, if we let m_τ denote the σ^r-invariant probability measure supported on τ then, for $F \colon \Sigma_A^r \to \mathbb{R}$,

$$\int F \, \mathrm{d}m_\tau := \frac{1}{\lambda(\tau)} \int_0^{\lambda(\tau)} F(\sigma_t^r(x,0)) \, \mathrm{d}t = \frac{f^n(x)}{r^n(x)}, \tag{15}$$

where $f = \mathfrak{I}(F)$. Since the periodic orbit measures m_τ are weak* dense in $\mathcal{M}_{\sigma^r}$, we have

$$I_{f,r} = \overline{\left\{ \int F \, \mathrm{d}m_\tau : \tau \text{ periodic } \sigma^r\text{-orbit} \right\}} = \overline{\left\{ \frac{f^n(x)}{r^n(x)} : \sigma^n x = x \right\}}. \tag{16}$$

References

1. R. Alperin and H. Bass, *Length functions of group actions on Λ-trees*, Combinatorial Group Theory and Topology (Alta, UT, 1984), Ann. of Math. Stud., vol. 111, Princeton Univ. Press, Princeton, NJ, 1987, pp. 265–378.
2. M. Babillot and F. Ledrappier, *Lalley's theorem on periodic orbits of hyperbolic flows*, Ergodic Theory Dynam. Systems **18** (1998), no. 1, 17–39.
3. F. Bonahon, *Bouts des variétés hyperboliques de dimension* 3, Ann. of Math. (2) **124** (1986), no. 1, 71–158.
4. ———, *The geometry of Teichmüller space via geodesic currents*, Invent. Math. **92** (1988), no. 1, 139–162.
5. M. Bourdon, *Actions quasi-convexes d'un groupe hyperbolique, flot géodésique*, Ph.D. thesis, Université Paris-Sud 11, 1993.
6. M. Burger, *Intersection, the Manhattan curve, and Patterson–Sullivan theory in rank* 2, Internat. Math. Res. Notices **7** (1993), 217–225.
7. I. M. Chiswell, *Abstract length functions in groups*, Math. Proc. Cambridge Philos. Soc. **80** (1976), no. 3, 451–463.
8. C. Croke and A. Fathi, *An inequality between energy and intersection*, Bull. London Math. Soc. **22** (1990), no. 5, 489–494.
9. M. Culler and J. W. Morgan, *Group actions on* **R**-*trees*, Proc. London Math. Soc. (3) **55** (1987), no. 3, 571–604.
10. M. Culler and K. Vogtmann, *Moduli of graphs and automorphisms of free groups*, Invent. Math. **84** (1986), no. 1, 91–119.

11. A. Fathi and L. Flaminio, *Infinitesimal conjugacies and Weil – Petersson metric*, Ann. Inst. Fourier (Grenoble) **43** (1993), no. 1, 279–299.
12. É. Ghys and P. de la Harpe (eds.), *Sur les groupes hyperboliques d'après Mikhael Gromov*, Progr. Math., vol. 83, Birkhäuser, Boston, MA, 1990.
13. L. Guillopé, *Entropies et spectres*, Osaka J. Math. **31** (1994), no. 2, 247–289.
14. M. Horsham and R. Sharp, *Lengths, quasi-morphisms and statistics for free groups*, Spectral Analysis in Geometry and Number Theory, Contemp. Math., vol. 484, Amer. Math. Soc., Providence, RI, 2009, pp. 219–237.
15. O. Jenkinson, *Rotation, entropy, and equilibrium states*, Trans. Amer. Math. Soc. **353** (2001), no. 9, 3713–3739 (electronic).
16. V. Kaimanovich, I. Kapovich, and P. Schupp, *The subadditive ergodic theorem and generic stretching factors for free group automorphisms*, Israel J. Math. **157** (2007), 1–46.
17. I. Kapovich, *The frequency space of a free group*, Internat. J. Algebra Comput. **15** (2005), no. 5-6, 939–969.
18. ———, *Currents on free groups*, Topological and Asymptotic Aspects of Group Theory, Contemp. Math., vol. 394, Amer. Math. Soc., Providence, RI, 2006, pp. 149–176.
19. I. Kapovich and T. Nagnibeda, *The Patterson – Sullivan embedding and minimal volume entropy for outer space*, Geom. Funct. Anal. **17** (2007), no. 4, 1201–1236.
20. Yu. Kifer, *Large deviations, averaging and periodic orbits of dynamical systems*, Comm. Math. Phys. **162** (1994), no. 1, 33–46.
21. G. Knieper, *Volume growth, entropy and the geodesic stretch*, Math. Res. Lett. **2** (1995), no. 1, 39–58.
22. S. P. Lalley, *Distribution of periodic orbits of symbolic and Axiom A flows*, Adv. in Appl. Math. **8** (1987), no. 2, 154–193.
23. ———, *Renewal theorems in symbolic dynamics, with applications to geodesic flows, non-Euclidean tessellations and their fractal limits*, Acta Math. **163** (1989), no. 1-2, 1–55.
24. R. C. Lyndon, *Length functions in groups*, Math. Scand. **12** (1963), 209–234.
25. I. D. Morris, *Maximizing measures of generic Hölder functions have zero entropy*, Nonlinearity **21** (2008), no. 5, 993–1000.
26. A. G. Myasnikov and V. Shpilrain, *Some metric properties of automorphisms of groups*, J. Algebra **304** (2006), no. 2, 782–792.
27. Wa. Parry, *Axioms for translation length functions*, Arboreal Group Theory (Berkeley, CA, 1988), Math. Sci. Res. Inst. Publ., vol. 19, Springer, New York, 1991, pp. 295–330.
28. Wi. Parry, *An analogue of the prime number theorem for closed orbits of shifts of finite type and their suspensions*, Israel J. Math. **45** (1983), no. 1, 41–52.
29. Wi. Parry and M. Pollicott, *Zeta functions and the periodic orbit structure of hyperbolic dynamics*, Astérisque **187-188** (1990), 268.
30. M. Pollicott and R. Sharp, *Poincaré series and zeta functions for surface group actions on* **R**-*trees*, Math. Z. **226** (1997), no. 3, 335–347.
31. ———, *Comparison theorems and orbit counting in hyperbolic geometry*, Trans. Amer. Math. Soc. **350** (1998), no. 2, 473–499.
32. ———, *Poincaré series and comparison theorems for variable negative curvature*, Topology, Ergodic Theory, Real Algebraic Geometry, Amer. Math. Soc. Transl. Ser. 2, vol. 202, Amer. Math. Soc., Providence, RI, 2001, pp. 229–240.
33. R. T. Rockafellar, *Convex analysis*, Princeton Math. Ser., vol. 28, Princeton Univ. Press, Princeton, NJ, 1970.
34. R. Schwartz and R. Sharp, *The correlation of length spectra of two hyperbolic surfaces*, Comm. Math. Phys. **153** (1993), no. 2, 423–430.
35. R. Sharp, *Prime orbit theorems with multi-dimensional constraints for Axiom A flows*, Monatsh. Math. **114** (1992), no. 3-4, 261–304.
36. ———, *The Manhattan curve and the correlation of length spectra on hyperbolic surfaces*, Math. Z. **228** (1998), no. 4, 745–750.
37. ———, *Distortion and entropy for automorphisms of free groups*, Discrete Contin. Dyn. Syst. **26** (2010), no. 1, 347–363.
38. K. Vogtmann, *Automorphisms of free groups and outer space*, Geom. Dedicata **94** (2002), 1–31.
39. P. Walters, *An introduction to ergodic theory*, Grad. Texts in Math., vol. 79, Springer, New York, 1982.

40. T. White, *The geometry of the outer space*, Ph.D. thesis, University of California, Los Angeles, 1991.
41. S. A. Wolpert, *Thurston's Riemannian metric for Teichmüller space*, J. Differential Geom. **23** (1986), no. 2, 143–174.

School of Mathematics, University of Manchester, Oxford Road, Manchester, M13 9PL, UK
E-mail address: sharp@maths.man.ac.uk

Titles in This Series

52 **Dmitry Jakobson, Stéphane Nonnenmacher, and Iosif Polterovich, Editors,** Spectrum and dynamics, 2010

51 **Javad Mashreghi, Thomas Ransford, and Kristian Seip, Editors,** Hilbert spaces of analytic functions, 2010

50 **P. Robert Kotiuga, Editor,** A celebration of the mathematical legacy of Raoul Bott, 2010

49 **Miguel Abreu, François Lalonde, and Leonid Polterovich, Editors,** New perspectives and challenges in symplectic field theory, 2009

48 **David Avis, David Bremner, and Antoine Deza, Editors,** Polyhedral computation, 2009

47 **John Harnad and Pavel Winternitz, Editors,** Groups and symmetries: From Neolithic Scots to John McKay, 2009

46 **Jean-Marie De Koninck, Andrew Granville, and Florian Luca, Editors,** Anatomy of integers, 2008

45 **Panos M. Pardalos and Pierre Hansen, Editors,** Data mining and mathematical programming, 2008

44 **Stanley Alama, Lia Bronsard, and Peter J. Sternberg, Editors,** Singularities in PDE and the calculus of variationa, 2007

43 **Andrew Granville, Melvyn B. Nathanson, and Jozsef Solymosi, Editors,** Additive combinatorics, 2007

42 **Donald A. Dawson, Vojkan Jakšić, and Boris Vainberg, Editors,** Probability and mathematical physics: a volume in honor of Stanislav Molchanov, 2007

41 **André Bandrauk, Michel C. Delfour, and Claude Le Bris, Editors,** High-dimensional partial differential equations in science and engineering, 2007

40 **V. Apostolov, A. Dancer, N. Hitchin, and M. Wang, Editors,** Perspectives in Riemannian geometry, 2006

39 **P. Winternitz, D. Gomez-Ullate, A. Iserles, D. Levi, P. J. Olver, R. Quispel, and P. Tempesta, Editors,** Group theory and numerical analysis, 2005

38 **Jacques Hurtubise and Eyal Markman, Editors,** Algebraic structures and moduli spaces, 2004

37 **P. Tempesta, P. Winternitz, J. Harnad, W. Miller, Jr., G. Pogosyan, and M. Rodriguez, Editors,** Superintegrability in classical and quantum systems, 2004

36 **Hershy Kisilevsky and Eyal Z. Goren, Editors,** Number theory, 2004

35 **H. E. A. Eddy Campbell and David L. Wehlau, Editors,** Invariant theory in all characteristics, 2004

34 **P. Winternitz, J. Harnad, C. S. Lam, and J. Patera, Editors,** Symmetry in physics, 2004

33 **André D. Bandrauk, Michel C. Delfour, and Claude Le Bris, Editors,** Quantum control: Mathematical and numerical challenges, 2003

32 **Vadim B. Kuznetsov, Editor,** The Kowalevski property, 2002

31 **John Harnad and Alexander Its, Editors,** Isomonodromic deformations and applications in physics, 2002

30 **John McKay and Abdellah Sebbar, Editors,** Proceedings on moonshine and related topics, 2001

29 **Alan Coley, Decio Levi, Robert Milson, Colin Rogers, and Pavel Winternitz, Editors,** Bäcklund and Darboux transformations. The geometry of solitons, 2001

28 **J. C. Taylor, Editor,** Topics in probability and Lie groups: Boundary theory, 2001

27 **I. M. Sigal and C. Sulem, Editors,** Nonlinear dynamics and renormalization group, 2001

26 **J. Harnad, G. Sabidussi, and P. Winternitz, Editors,** Integrable systems: From classical to quantum, 2000

25 **Decio Levi and Orlando Ragnisco, Editors,** SIDE III—Symmetries and integrability of difference equations, 2000

24 **B. Brent Gordon, James D. Lewis, Stefan Müller-Stach, Shuji Saito, and Noriko Yui, Editors,** The arithmetic and geometry of algebraic cycles, 2000

23 **Pierre Hansen and Odile Marcotte, Editors,** Graph colouring and applications, 1999

22 **Jan Felipe van Diejen and Luc Vinet, Editors,** Algebraic methods and q-special functions, 1999

21 **Michel Fortin, Editor,** Plates and shells, 1999

20 **Katie Coughlin, Editor,** Semi-analytic methods for the Navier-Stokes equations, 1999

19 **Rajiv Gupta and Kenneth S. Williams, Editors,** Number theory, 1999

18 **Serge Dubuc and Gilles Deslauriers, Editors,** Spline functions and the theory of wavelets, 1999

17 **Olga Kharlampovich, Editor,** Summer school in group theory in Banff, 1996, 1998

16 **Alain Vincent, Editor,** Numerical methods in fluid mechanics, 1998

15 **François Lalonde, Editor,** Geometry, topology, and dynamics, 1998

14 **John Harnad and Alex Kasman, Editors,** The bispectral problem, 1998

13 **Michel C. Delfour, Editor,** Boundaries, interfaces, and transitions, 1998

12 **Peter C. Greiner, Victor Ivrii, Luis A. Seco, and Catherine Sulem, Editors,** Partial differential equations and their applications, 1997

11 **Luc Vinet, Editor,** Advances in mathematical sciences: CRM's 25 years, 1997

10 **Donald E. Knuth,** Stable marriage and its relation to other combinatorial problems: An introduction to the mathematical analysis of algorithms, 1997

9 **D. Levi, L. Vinet, and P. Winternitz, Editors,** Symmetries and integrability of difference equations, 1996

8 **J. Feldman, R. Froese, and L. M. Rosen, Editors,** Mathematical quantum theory II: Schrödinger operators, 1995

7 **J. Feldman, R. Froese, and L. M. Rosen, Editors,** Mathematical quantum theory I: Field theory and many-body theory, 1994

6 **Guido Mislin, Editor,** The Hilton Symposium 1993: Topics in topology and group theory, 1994

5 **D. A. Dawson, Editor,** Measure-valued processes, stochastic partial differential equations, and interacting systems, 1994

4 **Hershy Kisilevsky and M. Ram Murty, Editors,** Elliptic curves and related topics, 1994

3 **Rémi Vaillancourt and Andrei L. Smirnov, Editors,** Asymptotic methods in mechanics, 1993

2 **Philip D. Loewen,** Optimal control via nonsmooth analysis, 1993

1 **M. Ram Murty, Editor,** Theta functions: From the classical to the modern, 1993